P9-DKE-279

 FTC Orlando Campus

000000589

Florida Technical College Library
12900 Challenger Prkwy.
Orlando, FL 32826

Florida Technical College Library
12900 Challenger Prkwy.
Orlando, FL 32826

FLORIDA TECHNICAL COLLEGE LIBRARY
12689 CHALLENGER PARKWAY
SUITE 130
ORLANDO, FL 32826

CAT

Exploring
Drafting
Fundamentals of Drafting Technology

by
John R. Walker

Publisher
THE GOODHEART-WILLCOX COMPANY, INC.
Tinley Park, Illinois

Copyright 2000

by

THE GOODHEART-WILLCOX COMPANY, INC.

Previous Editions Copyright 1996, 1991, 1987, 1982, 1978, 1975, 1972

All rights reserved. No part of this book may be reproduced, stored in a retrieval system, or transmitted in any form or by any means, electronic, mechanical, photocopying, recording, or otherwise, without the prior written permission of The Goodheart-Willcox Company, Inc. Manufactured in the United States of America.

Library of Congress Catalog Card Number 98-53651
International Standard Book Number 1-56637-565-7
2 3 4 5 6 7 8 9 10 00 03 02 01 00

Library of Congress Cataloging-in-Publication Data

Walker, John R.
   Exploring drafting: basic fundamentals / by John R. Walker.
      p.   cm.
   Includes index.
   ISBN 1-56637-565-7
   1. Mechanical drawing.   I. Title.
T353.W22 2000
604.2--dc21   95-53651
                                           CIP

The author and publisher wish to thank the following professionals who reviewed the text and provided valuable input:

Walter B. Cheever, Architect
Architectural Drafting Instructor
South Central Technical College
North Mankato, Minnesota

Bernard D. Mathis
Drafting/AutoCAD Instructor
Sterling High School
Sterling, Colorado

CAT

Introduction

A course in drafting should be a part of your education. It will help you develop the capacity to plan and problem solve in an organized fashion, to interpret the ideas of others, and to express yourself in an understandable manner.

Exploring Drafting teaches drafting fundamentals and basic geometric construction. As you proceed with the book, you will become familiar with the drafting methods and processes used by industry. You will develop and practice drafting skills and techniques.

Conventional drawings are constructed in a step-by-step fashion in this text to demonstrate drawing principles. Other more challenging drawings are also provided to help you develop your originality and creativity. The ***Test Your Knowledge*** section at the end of each unit will allow you to check your comprehension of the material covered in that unit. The ***Problems*** section at the end of the unit allows you an opportunity to practice your drafting skills and improve your techniques.

Exploring Drafting is current with the latest ANSI/ASME (American National Standards Institute/American Society of Mechanical Engineers) practices. You will learn how to use symbols to communicate in the international language–drafting.

Exploring Drafting provides you with the basic understanding necessary to allow you to progress into CADD (Computer-Aided Design and Drafting). The knowledge learned in this text will give you the background to draw and design using most CAD systems. *The principles of drafting are common to both traditional drafting and CAD.*

Drafting, the "language of industry," offers many career opportunities. Some of these exciting and rewarding futures are described in the unit on Careers in Drafting and Design (Unit 2). You will find that the ability to draw and to understand drafting will be of benefit throughout your lifetime and your career.

John R. Walker

Contents

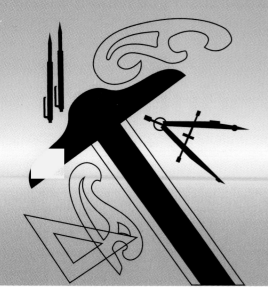

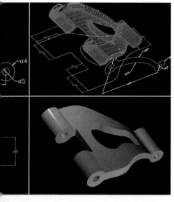

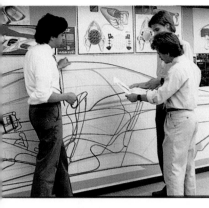

Many technical drawings are required to accurately produce this vehicle. Would you like to be a member of a design team which comes up with new and exciting ideas such as this? (Buick)

Unit 1
Why Study Drafting?

After studying this unit, you should be able to:
◆ Identify many fields of drafting.
◆ Cite why drafting is called a universal language.
◆ Explain why drawings are often the best way to describe or show our ideas.
◆ Describe, in a limited way, the technological changes in the ways drawings are made and stored.

Drafting is a form of graphic communication. It is concerned with the preparation of the drawings needed to develop and manufacture a product, Figure 1-1. It is a very important part of modern industry because drawings are often the only way to explain or show our ideas, Figure 1-2.

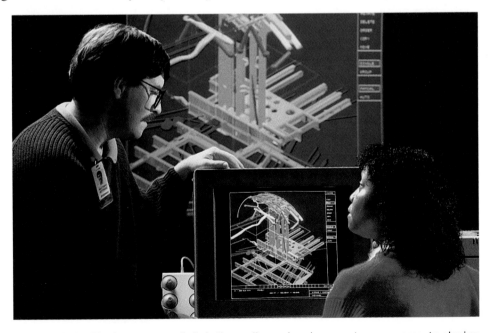

Figure 1-1. Engineers use digital, three-dimensional computer programs to design parts and systems for the Boeing aircraft, such as the 777. Parts can be viewed from any angle and cross sections can be taken from the solid description of the parts. The software system, known as CATIA (Computer-Aided Three-dimensional Interactive Application), allows engineers from many disciplines to work at the same time on the 777 design, leading to improved designs and avoiding costly rework. Designers can "pre-assemble" the airplane on the computer screen to detect problems before costly tools and parts are made. (The Boeing Company)

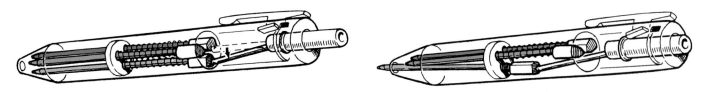

Figure 1-2. Drawings are often the best way to explain or show ideas. Think how difficult it would be to explain how this simple three-color pen operates using only the written word.

Drafting is frequently called a "universal language." Like other languages, symbols (lines and figures) that have specific meaning are used. The symbols accurately describe the shape, size, material, finish, and fabrication or assembly of a product. These symbols have been standardized over most of the world. This makes it possible to interpret or understand drawings made in other countries, Figure 1-3.

Drafting is also the "language of industry." Industry uses this precise language because the drawings must communicate the information the designer had in mind to those who produce the product.

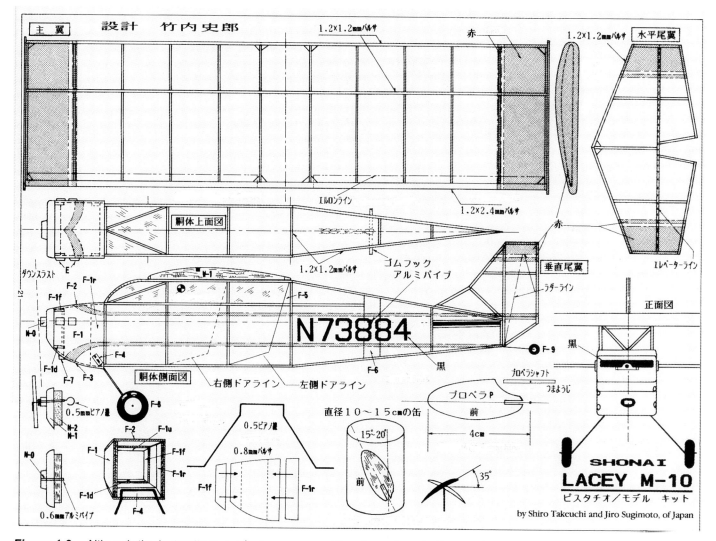

Figure 1-3. Although the instructions are in Japanese and the dimensions in the metric system, it would be possible for you to construct this model airplane. (Bill Hannan, *Peanuts & Pistachios*)

Drawings can be made manually (by hand) on paper or film using drafting tools such as drawing board, pencil or pen, angles, compass, etc. Drawings are also produced using Computer-Aided Design and Drafting (CADD) systems, Figures 1-4 and 1-5. When the design is completed, high-speed plotters or printers turn out hard (paper) copy showing the part or design, Figure 1-6. (Computer graphics and CADD is explained in more detail in Unit 24.)

Figure 1-4. Example from industry showing engineers and drafters working on CADD stations. (Emerson Electronics)

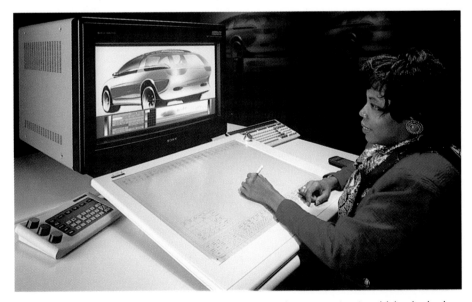

Figure 1-5. Computer-generated graphics use the computer to aid in designing and engineering a product. (Ford Motor Company)

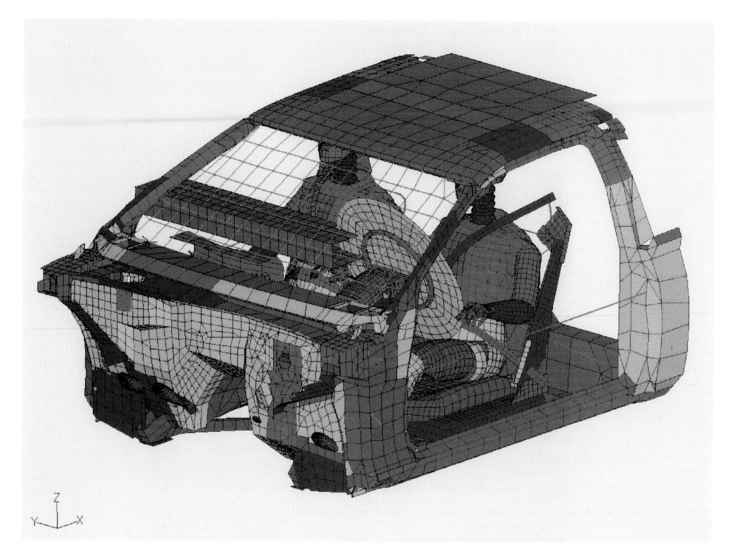

Figure 1-6. Computer design can be converted into hard (paper) copy using a high-speed plotter or printer. This computer drawing shows engineering research on airbag deployment during a simulated crash. This helps in the design of safer vehicles. (Ford Motor Company)

It would be difficult in today's world to name an occupation that does not require the ability to read and understand graphic information such as drawings, charts, or diagrams. You use drawings when you construct a model or electronic kit, Figure 1-7.

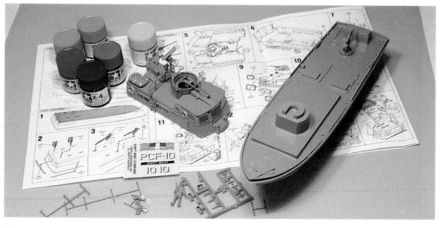

Figure 1-7. You use drawings when you construct models. The instructions for building this boat are in pictorial form.

Many specialized fields of drafting have been developed—aerospace, architectural, automotive, electrical and electronic, printed circuitry design, and topographical.

Basically, they all use the same drafting equipment and employ similar drafting techniques. (Computer graphics may require specialized programs.) However, the finished product of each field varies greatly, Figures 1-8, 1-9, and 1-10. You will learn the skills and techniques of the drafter as you study this book.

Figure 1-8. An aerospace drafter must have the technical knowledge to carry out the ideas of the aerospace engineer. (Bombardier Inc., Canadair of Canada)

Figure 1-9. The drafter conveys or communicates with drawings all the information required to construct this motel.

Figure 1-10. Engineers discussing the design of this photocopy machine. The drafters who produced the drawings to manufacture this product needed a working knowledge of many technical areas. The design of this equipment involved optics, printed circuits, plastics, metals, and manufacturing processes. While designers and engineers furnished the basic information, drafters put their ideas into production form. (Eastman Kodak Company)

Drafting Vocabulary

Aerospace
Architectural
Computer graphics
Computer-aided drafting
  and design (CADD)

Diagram
Drafting
Manually
Manufacture

Plotters
Printers
Symbols
Topographical

Test Your Knowledge—Unit 1

Please do not write in the book. Place your answers on another sheet of paper.

1. What does the term drafting mean?
2. Drawings are often used because _____.
   A. they are easy to make
   B. they are the best means available to explain or show many ideas
   C. people who cannot read can understand drawings

3. Drafting is called _____.
   A. a picture language
   B. the language of industry
   C. a universal language
4. There are very few occupations that do not require the ability to read and understand graphic information. Name five occupations that require this skill.
5. What does the term CADD mean?
6. When a CADD designed problem is completed, it is converted into hard (paper) copy on a(n) _____ or a(n) _____.
7. List five specialized fields of drafting.

Outside Activities

1. Prepare a list with two columns. In the first column write in the names of as many occupations that you can think of that require the ability to read and understand graphic information. In the second column, place the names of occupations that do not require this skill.
2. Secure samples of drawings used by the following industries:
   A. Aerospace.
   B. Building construction.
   C. Structural.
   D. Manufacturing.
   E. Map making.
   F. Electrical and electronic.
3. Make a collection of pictures (magazine clippings, photographs, drawings, etc.) that show products made by industries listed in activity 2. (DO NOT cut them from library books and magazines.)
4. Visit a local architect who designs residences. After discussing a project in work, prepare a report on the steps normally followed when designing a home for a client. Prepare your questions carefully before you make your visit.
5. Visit a local surveyor and make a report on the work he or she does. Borrow samples of completed work for a bulletin board display.
6. Obtain and display copies of drawings made in a foreign country.

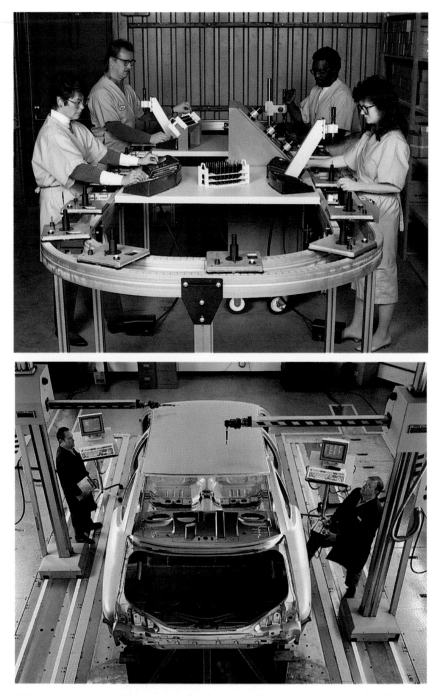

Clear, concise drawings are required to manufacture a product, regardless of its complexity. Many drafting and design tasks are involved in creating the drawings. (Enidine, Inc.; Ford Motor Company)

Unit 2
Careers in Drafting and Design

After studying this unit, you should be able to:

◆ Identify many career possibilities related to the fields of drafting and design.

◆ Cite many of the skills needed to prepare for the various fields that use drafting.

◆ Identify the traits of leadership.

◆ Summarize the role of leadership in society and how it relates to drafting technology.

◆ Describe what to expect when entering the world of work.

◆ Identify sources of information about careers in drafting technology.

It would be difficult to name an industry that does not use drawings. These drawings may be in the form of conventional production drawings, Figure 2-1, instructional booklets, charts, graphs, or maps. It typically takes more than 27,000 drawings to manufacture an automobile. The field of drafting provides employment for over one million men and women. The work of other millions requires them to be able to read and interpret drawings. See Figure 2-2.

Drafting Occupations

Job titles and duties will vary from one company to another. However, the following drafting occupations are typical of those found in industry.

Figure 2-1. Skilled drafters are always in demand. This photo, taken at a National VICA (Vocational Industrial Clubs of America) conference, shows a couple competitors working in their trade. Competition is keen because the students are highly skilled in drafting procedures and techniques. How do you think you would fare in such a competition? (VICA)

Figure 2-2. Quality control specialists must be able to read and understand drawings. Here, a laser measuring system is being set up and calibrated. It will be used to inspect the fit of car doors. (Ford Motor Company)

Drafters make working plans and detailed drawings. They prepare them from specifications and information received verbally, from sketches, and from notes.

The drafter usually starts out as a ***trainee drafter*** where he or she redraws or repairs damaged drawings. The trainee may revise engineering drawings or make simple detail drawings under the direct supervision of a senior drafter.

During the training period, trainee drafters are often enrolled in formal classes such as CADD, as shown in Figure 2-3, mathematics, electronics, or manufacturing processes. These classes may be held within the company or at a local technical school or community college.

Upon completion of the training period, the trainee drafter usually advances to ***junior drafter.*** Similar positions are known as: ***detailer, detail drafter,*** and ***assistant drafter.***

A junior drafter calls for the preparation of detail and working drawings of machine parts, electrical/electronic devices, or structures, from rough design drawings. It may also require preparation of simple assembly drawings, charts, or graphs. The junior drafter must be able to prepare simple calculations made according to established drafting room procedures.

From junior drafter, the next step forward is ***drafter.*** The drafter applies independent judgment in the preparation of original layouts with intricate details. He or she must have an understanding of machine shop practices, the proper use of materials, and be able to make extensive use of reference books and handbooks.

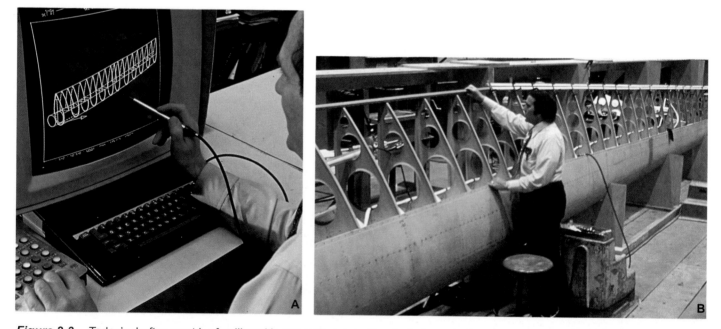

Figure 2-3. Today's drafter must be familiar with conventional drafting techniques before he or she can work with computer-aided design and drafting equipment. A—Here a drafter is using CAD to design the blade of a wind-powered electric generating unit. B—Completed and partially completed wind turbine blades that were shown in the design stage in Figure 2-3A. These blades require extreme care in their design because they will be used in a location where there is often 50 miles per hour difference in the wind flow between the top and bottom blades when they are in a vertical position. (Lockheed-California)

With experience, the drafter will become a *senior drafter* and be expected to do complex original work.

In time, the senior drafter can become a **lead drafter** or **chief drafter.** Such a person is responsible for all work done by the department.

Most drafters specialize in a particular field of technical drawing: aerospace, architectural, structural, etc. Regardless of the field of specialization, drafters should be able to draw rapidly, with accuracy and neatness. They will also need a working knowledge of computer graphics.

Firms employing computer-aided design and drafting usually establish job applications for computer graphics drafters/specialists. A few job titles include **CAD drafter, CAD/CAM specialist, computer graphics specialist,** and **computer graphics technician.** These specialists seldom prepare drawings manually on "the board." Instead, their "drawings" are computer-generated with hard (paper) copy produced on a plotter.

All drafters must have a thorough understanding of mathematics, science, materials, and manufacturing processes in their areas of specialization. The manufacturing industries, Figure 2-4, employ large numbers of drafters. Others are employed by

Figure 2-4. Many drafters were required to prepare drawings for the conversion of a Boeing 747 commercial jet to a Space Shuttle transporter. The transporter is used to return the orbiter to Cape Kennedy after a west coast landing. (NASA)

architectural and engineering firms and local, state, and federal government.

Industrial Designers

The work of the *industrial designer* influences most every item used in our daily living, whether it is the design of a small CD player or a giant jet plane. Many designers are on the staffs of major automobile manufacturers, Figure 2-5.

Figure 2-5. A thorough knowledge of advanced technical practices is required to place a present-day automobile into production. (General Motors Corp.)

In general, the chief function of the industrial designer is to simplify and improve the operation and appearance of industrial products. Design simplification usually means fewer parts to wear or malfunction. Appearance plays an important role in the sale of a product. The industrial designer must be aware of changing customs and tastes. He or she must know why people buy and use different products.

It is recommended that prospective designers have an engineering degree and a working knowledge of engineering and manufacturing techniques and materials. Artistic ability is necessary.

Tool Designers

Tools and devices needed to manufacture industrial products are designed by the *tool designer.* The tool designer originates the designs for cutting tools, special holding devices (fixtures), jigs, dies, and machine attachments that are necessary to manufacture the product. He or she must be familiar with machine shop practices, be an accomplished drafter, and have a working knowledge of algebra, geometry, and trigonometry. It is also important to be familiar with computer-aided design and computer-aided manufacturing (CAD/CAM), and robotics, Figure 2-6.

As industrial technology expands and more automated machinery is utilized, there will be a constantly increasing demand for competent tool designers.

Teachers

Teaching is a satisfying and challenging profession. Teachers of industrial technology, vocational, and technical education are in a fortunate position because they have a freedom not found in many other professions. Teachers are needed in schools such as yours, as well as in industry, Figure 2-7.

Four years of college training are needed and some industrial experience is highly recommended.

Figure 2-6. Design engineers develop the tools and devices needed to manufacture industrial and commercial products. Here engineers are determining the special tooling that must be designed to machine these castings. (American Foundrymen's Society, Inc.)

Figure 2-7. Teaching is a challenging and rewarding profession. (Amp, Inc.)

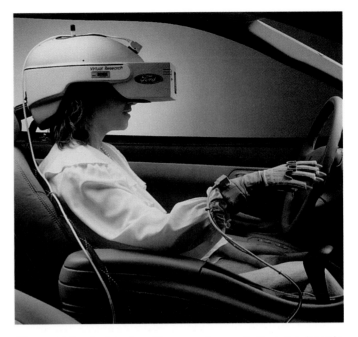

Figure 2-8. A virtual reality computer-controlled simulator is used by this engineer to judge the ergonomics of a proposed car. Ergonomics is the science concerned with matching a design to the human body. In this example, ergonomics is concerned with matching the car's design, seating comfort, and location of controls to the human body. The simulator realistically portrays vehicle operation. This information is important in helping designers and engineers design car interiors for driver and passenger comfort and safety. (Ford Motor Company)

Engineers

The *engineer* usually specializes in one of the recognized engineering specialties—aeronautical, industrial, marine, structural, civil, electrical/electronic, and metallurgy to name but a few. See Figure 2-8. Engineers provide technical and, in many instances, managerial leadership in industry and government. Depending upon the area of specialization, engineers may be responsible for the design and development of new products and processes, plan structures and highways, or work out new ways of transforming raw materials into salable products.

Laws provide for licensing engineers whose work may affect life, health, and property. A professional engineering license usually requires graduation from an approved engineering college, four years of experience, and passing an examination. Some states will accept experience in place of a college degree provided the required tests are passed.

Architects

In general, *architects* plan and design all kinds of structures. However, they may specialize in specific fields of architecture such as private homes, industrial buildings, schools, or commercial buildings.

When planning a structure, an architect first consults the client on the purpose of the building, size, location, cost range, and other requirements. Upon completion and approval of preliminary design drawings, detailed working drawings and specification sheets are prepared. As construction progresses, the architect usually makes periodic inspections to determine whether the plans are being followed and construction details are to specifications.

Most architects are licensed. This requires a bachelor's degree from an approved college program, several years of experience working for an architectural firm (similar to the internship of a physician), and passing a special examination. In many cases, a master's degree for advanced or specialized study is obtained.

Modelmakers

Industry makes extensive use of models, mock-ups, and prototypes for engineering and planning purposes. Their preparation is often the responsibility of the engineering drafting group although professional **modelmakers** are frequently used. See Figure 2-9. Modelmakers must be able to read and understand drawings. They must also have skills in working metal, wood, and plastics.

Figure 2-10. Illustration of a two-stage-to-orbit launch vehicle. The smaller stage of the launch vehicle is shown as it separates from the first-stage supersonic aircraft that began the boost toward space. It was drawn by a technical illustrator using engineering information and specifications. (Boeing Defense & Space Group)

Figure 2-9. In order to interpret the designer's or engineer's plans accurately, modelmakers must be able to read and understand drawings. (General Motors Corp.)

Technical Illustrators

Technical illustration is a process of preparing artwork for industry. The **technical illustrator** prepares pictorial matter for engineering, marketing, or educational purposes. See Figure 2-10. Technical illustrators must have both technical and artistic abilities. The technical ability is necessary to understand the mechanical aspects of the job. The artistic ability will show the building or product in three dimensions. A technical illustrator must be able to make accurate sketches and finished drawings according to industry standards.

Leadership in Drafting Technology

So far in your study of this unit, you have learned that drafting technology requires highly skilled people. However, the nature of industry is that strong and dynamic leadership is required at all levels if it is to be competitive with other technologically oriented countries.

Leadership is the ability to be a leader. A **leader** is a person who is in charge or in command. The quality of leadership usually determines whether an organization will be successful or be a failure, Figure 2-11.

What Makes a Good Leader?

Most good leaders usually have the following traits (qualities) in common:

Vision. Knows what must be done. Continually looks for positive ways to reach these goals.

Communication. Can communicate in such a way that will encourage the assistance and cooperation of others.

Figure 2-11. Good leadership brought the latest manufacturing processes like this master computer that continually monitors 58 welding robots in use and helps keep the company profitable. (Chrysler Corp.)

Persistence.   Willing to work long hours to achieve success.

Organizational Qualities.   Knows how to organize and direct the activities of the group.

Responsibility.   Accepts responsibility for actions but readily gives credit to others when they deserve or warrant it.

Delegates Authority.   Not afraid to make assignments that will help others within the group to take on leadership roles.

Getting Leadership Experience

How can you get leadership experience? The Industrial Technology program or other areas in your school, such as clubs, offer leadership training to students willing to take advantage of the opportunities. Many programs provide leadership experience through student technology education organizations such as the *Technology Student Association (TSA)* and the *Vocational Industrial Clubs of America (VICA).* These organizations develop leadership and personal abilities as they

Figure 2-12. Student organizations like the Technology Student Association (TSA) and Vocational Industrial Clubs of America (VICA) offer many opportunities for students to develop leadership skills. (VICA)

relate to the industrial-technical world. Contact the national offices for information, Figure 2-12.

What to Expect as a Leader

As a leader, you will be expected to set a good example. The responsibility of leadership also means that you must get all members of the group involved in activities. This requires encouragement and tact. A good leader is not expected to have answers for every problem the group encounters. Members should be encouraged to recommend possible solutions. It is up to the leader to decide which to use.

Leadership is not easy or without difficulties. Decisions must be made that some group members may not like. People and even friends may have to be reprimanded or dismissed (fired) for being incompetent (doing poor work) or for trying to take advantage of the group.

If you think you have the traits of a strong leader (or believe you can develop them) and can handle the unpleasant tasks that are part of leadership, by all means look into joining groups where it will be possible for you to become a leader.

What to Expect When You Enter the World of Work

You will be very disappointed if you think that graduation means the end of training and study. Advancing technology, Figure 2-13, means constantly developing new ideas, materials, processes, and manufacturing techniques that in

Figure 2-13. Computer-aided design (CAD) helped develop this advanced concept aircraft. Advanced technology such as this is creating jobs that did not previously exist. (Beechcraft)

turn are creating occupations and jobs that did not previously exist. It has been said that young graduates will be employed in the average of five different jobs during their lifetime, and three of them do not now exist!

To hold your job and advance in it, you will have to study to keep up-to-date with the knowledge and new skills needed by modern technology.

Industry expects a fair day's work for a fair day's pay. High manufacturing costs and competition from foreign made products have made this necessary. You will be expected to start work on time, and be on the job each workday. Sick leave was originated to help workers and, therefore, should not be abused.

Do your assigned work and never knowingly turn out a piece of substandard or faulty work, Figure 2-14. Take pride in what you do. YOU must assume the responsibility for your actions. Industry is always on the lookout for bright young people who are not afraid to work and assume responsibility.

Where to Get Information about Careers in Drafting Technology

Information on careers can be found in many sources. The guidance office and industrial technology department in your school are the two good starting places. Almost all community colleges have *career information centers.* They can furnish any needed occupational and career information.

Look in the *Occupational Outlook Handbook* published by the United States Department of Labor and available through the Government Printing Office. The Department of Labor also publishes the *Dictionary of Occupational Titles,* which has job descriptions of most occupations. Both books can usually be found in the school library or guidance office.

The local *State Employment Service* (it may have a different name in your state) is another excellent source of occupational and career information. They will also be able to tell you about local and statewide employment opportunities.

Figure 2-14.   Never knowingly turn out a piece of substandard work. Your work should be such that you can take pride in it. How would you feel if the work you did resulted in the disaster shown above?

Drafting Vocabulary

Architects
Assistant drafter
CAD drafter
CAD/CAM specialist
Career information centers
Chief drafter
Communication
Computer graphics
   specialist
Computer graphics
   technician
Detail drafter
Detailer

Dictionary of
   Occupational Titles
Drafters
Engineer
Industrial designer
Junior drafter
Lead drafter
Leader
Leadership
Modelmakers
Occupational Outlook
   Handbook
Organizational qualities

Persistence
Responsibility
Senior drafter
Teaching
Technical illustrator
Technology Student
   Association (TSA)
Tool designer
Trainee drafter
Vision
Vocational Industrial
   Clubs of America
   (VICA)

Test Your Knowledge—Unit 2

Please do not write in the book. Place your answers on another sheet of paper.

1. Drafters make _____.
2. The drafter usually starts as a(n) _____ and advances to a(n) _____, and then to a(n) _____.
3. A good drafter should be able to draw with accuracy and _____.
4. List the job classification relating to drafting and design occupations using the computer.
5. The industrial designer's main job is to _____.
6. _____ is a profession that offers a freedom not usually found in other professions.
7. _____ typically plan and design all kinds of structures and buildings. Most states require them to be _____.
8. The engineer usually specializes in one of the many branches of the profession. List four types of engineering.
9. The modelmaker makes models, _____ or _____ to show the engineer's or designer's plans.
10. The _____ _____ must have a combination of technical and artistic abilities.

Match each word or phrase in the left column with the correct sentence in the right column.

11. Leadership
12. Leader
13. Quality of leadership
14. Unpopular decisions

A. One of the difficulties of leadership.
B. Determines whether an organization will be a success or be a failure.
C. The ability to be a leader.
D. A person who is in charge or in command.

Outside Activities

1. Invite a representative from the local State Employment Service Office to visit your class, and discuss employment opportunities in the drafting occupations.
2. Make a study of the Help Wanted columns in your daily newspaper for a period of two weeks. Prepare a list of drafting and related jobs available, salaries offered, and the minimum requirements for securing the jobs. How often are additional benefits such as insurance, hospitalization, etc., mentioned in the ads?

   Note the length of time (number of days) each ad is run in the column. Calculate the average number of days jobs are listed.
3. Summarize the information on the drafting occupations given in the Occupational Outlook Handbook (a government publication) and make it available to the class. With your teacher's permission, contact the Technology Student Association and Vocational Industrial Clubs of America for information on how to organize a technology or vocational club in your school. (Conduct a student survey to see if there is interest in forming such a club.) TSA and VICA can be contacted at the following addresses:

Technology Student Association
1914 Association Drive
Reston, VA 22091

Vocational Industrial Clubs of America
P.O. Box 3000
Leesburg, VA 22075

Unit 3
Sketching

After studying this unit and sketching the assigned problems, you should be able to:

◆ Demonstrate sketching skills and techniques.

◆ Apply the "alphabet of lines" to a drawing.

◆ Sketch basic geometric shapes.

◆ Enlarge or reduce the drawn size of an object by the graph method.

Sketching is one of many drafting techniques. It is a quick way to show an idea that would be difficult to describe with words alone. Industry uses many sketching techniques to develop engineering concepts (ideas). See Figures 3-1 and 3-2.

Figure 3-1. Sketching is a quick way to show ideas that would be difficult to describe with words alone. A—Sketch used in a sports car study, which was one of the many used. While the car will never be produced, the front end of the sketch (slightly modified) has already found its way into production. B—The "Mighty Mouse" is an ultralight, single-person, high-mileage vehicle for use within the city. It was used in a study to reduce air pollution in cities.

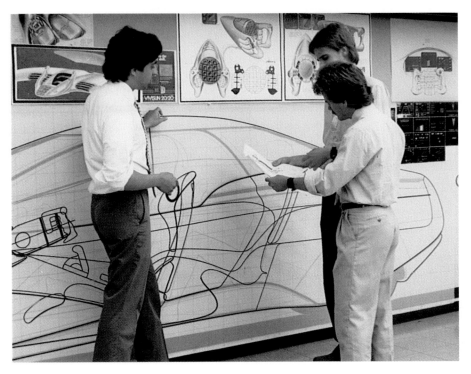

Figure 3-2. The auto industry uses narrow adhesive-backed tape to study ideas for future vehicles. Tape "sketches" are made full size and can be easily modified or changed. (Pontiac Div., General Motors Corp.)

A good sketch shows the shape of the object. It also provides dimensions and may include special instructions on how the object is to be made or finished.

Sketching does not require a great deal of equipment. Properly sharpened F, 2H, or HB grade pencils and sheets of standard-size paper (8 1/2 x 11 in.) are well-suited for this purpose. See Figure 3-3. The paper can be plain or cross sectioned (graph or squares). A good eraser is also needed.

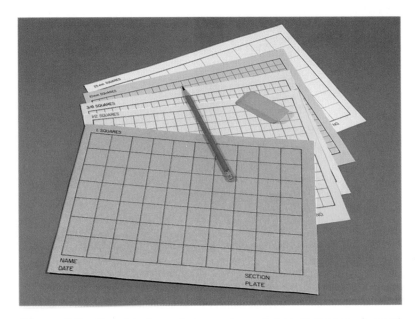

Figure 3-3. Preprinted graph paper is useful in sketching. A good eraser is also a necessity. The colors indicate different size grids.

In sketching, a line is drawn by making a series of short strokes, Figure 3-4. It is recommended that light (thin) construction lines be drawn first. Corrections will then be easier to make. For right-handed people, horizontal lines are drawn from left to right. In addition, vertical lines are drawn from the top down. For left-handed people, the direction of drawing is opposite.

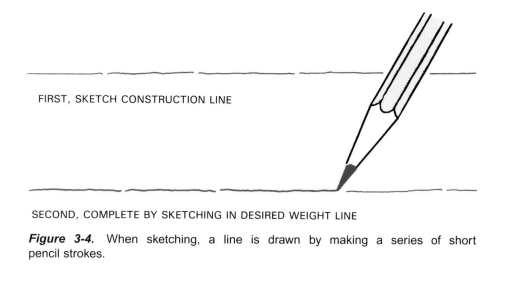

FIRST, SKETCH CONSTRUCTION LINE

SECOND, COMPLETE BY SKETCHING IN DESIRED WEIGHT LINE

Figure 3-4. When sketching, a line is drawn by making a series of short pencil strokes.

The instructions that follow tell and show you how to sketch basic geometric shapes. You will find that even the most complex drawings are made using a combination of these basic geometric shapes.

Alphabet of Lines

A drafter uses lines of various weights (thicknesses) to make a drawing. Each line has a special meaning, Figure 3-5. Contrast between the various line weights or thicknesses help to make a drawing easier to read. It is essential that you learn this ***alphabet of lines***.

Construction and Guide Line

Construction lines are used to lay out drawings. ***Guide lines*** are used when lettering to help you keep the lettering uniform in height. These lines are drawn lightly using a pencil with the lead sharpened to a long conical point.

Border Line

The ***border line*** is the heaviest (thickest) line used in sketching. First, draw light construction lines as a guide. Then, go over them using a pencil with a heavy rounded point to provide the border lines.

Object Line

The ***object line*** is a heavy line, but slightly less in thickness than the border line. The object line indicates visible edges. In sketching object lines, use a pencil with a medium lead and a rounded point.

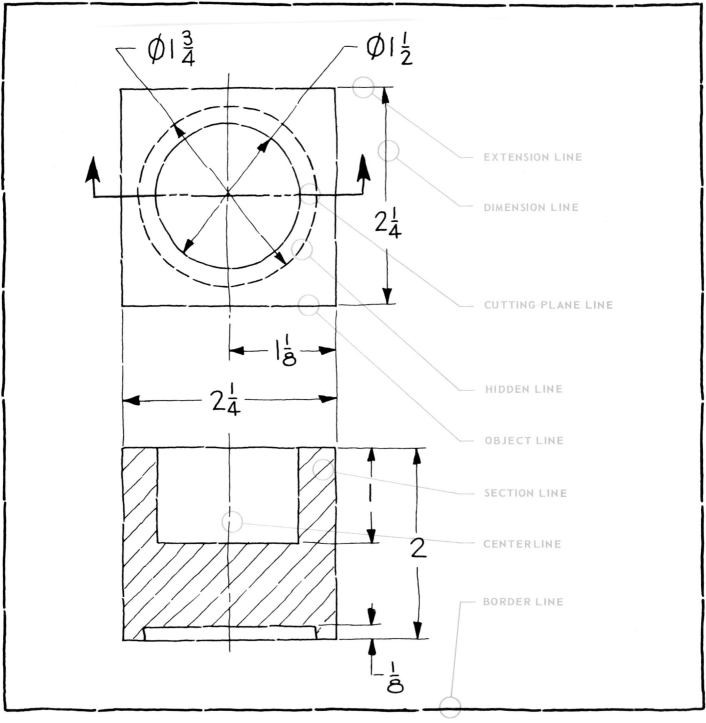

Figure 3-5. The alphabet of lines as used in sketching. Note the different thicknesses.

Hidden Line   — — — — — — — — — — — — — — — — —

Hidden lines are used to indicate or show the hidden features of a part. The hidden line is made up of a series of dashes (1/8 in.) with spaces (1/16 in.) between the dashes.

Dimension Line   ←————————— 2 —————————→

Dimension lines generally terminate (end) with arrowheads at the ends. They are usually placed between two extension lines. A break is made, usually in the center, to place the dimension. The dimension line is placed from 1/4 in. to 1/2 in. away from the drawing. It is a fine line and is drawn using a pencil sharpened to a long conical point.

Extension Line   ———————————————

Extension lines are the same weight as dimension lines. These lines indicate points from which the dimensions are given. The extension line begins 1/16 in. away from the view and extends 1/8 in. past the last dimension line.

Centerline   ——— — ——— — ——— — ———

Centerlines are made up of alternate long (3/4 in. to 1 1/2 in.) and short (1/8 in.) dashes with 1/16 in. spaces between. These are drawn about the same weight as dimension and extension lines, and are used to locate centers of symmetrical objects.

Cutting-Plane Line   ——— — — ——— — — ——— — —

A *cutting-plane line* indicates where an object has been cut to show interior features. Two types are used: 1/4 in. dashes with 1/16 in. spacing; a long dash (3/4 in. to 1 1/2 in.), then two short dashes (1/8 in.) with 1/16 in. spacing. Draw the cutting-plane line slightly heavier than an object line, using a pencil with a rounded point.

Section Line

Section lines are used when drawing inside features of an object to indicate the surfaces exposed by the cutting-plane line. Section lines are also used to indicate general classifications of materials. These lines, light in weight, are drawn with a pencil sharpened to a long conical point.

Phantom Line   ——— —— ——— —— ———

Phantom lines are used to show alternating positions of a moving part, repeated details, or the path of motion of an object. The line weight is the same as centerlines. A phantom line consists of dashes 3/4 in. to 1 1/2 in. long, separated by a set of two short (1/8 in.) dashes.

How to Sketch a Horizontal Line

1. Mark off two points spaced a distance equal to the length of the line to be drawn. The points should be parallel to the top or bottom edge of your paper.

2. Move your pencil back and forth and connect these points with a construction line.

3. Start from the left point and sketch an object line to the right point. This line is sketched over the construction line.

How to Sketch a Vertical Line

1. Mark off two points spaced a distance equal to the length of the line to be drawn. The two points should be parallel to the right or left edge of the sheet. Move your pencil back and forth and connect these points with a construction line.

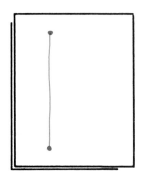

2. Start from the top point and sketch down and over the construction line to draw the desired line.

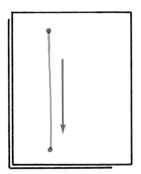

3. Vertical lines can also be sketched by rotating the paper into a horizontal position and proceeding as explained in **How to Sketch a Horizontal Line**.

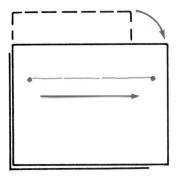

How to Sketch an Inclined Line

1. Mark off two points at the desired angle. Connect these points with a construction line.

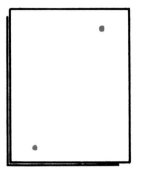

2. Sketch the desired weight line over the construction line. Sketch in the directions illustrated. Sketch up when the line inclines to the right. Sketch down when the line inclines to the left.

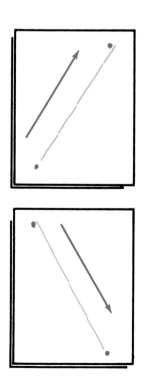

3. Inclined lines can also be sketched by rotating the sheet so the points are in a horizontal position. Sketch the line as previously described.

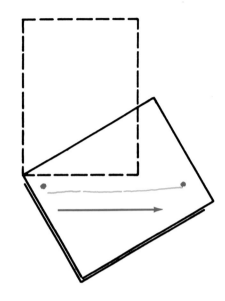

4. For some sketching problems, it may be easier to rotate the paper so the points are in a vertical position. Proceed as explained in **How to Sketch a Vertical Line**.

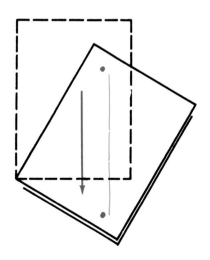

How to Sketch Squares and Rectangles

1. Sketch a horizontal line and a vertical line (axes).

3. Sketch construction lines through the desired points.

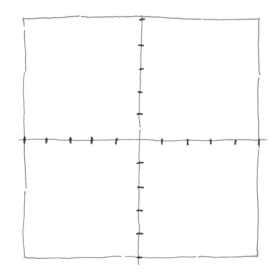

2. Begin at the intersection of these lines and lay out equal units on both lines in each direction. For example: If you want to draw a 2 1/2 in. square, you would estimate a unit of 1/4 in. and mark off five of these units on the vertical axis above and below the horizontal axis. Lay out the horizontal axis in the same manner.

4. Go over the construction lines forming the square to produce the desired weight line.

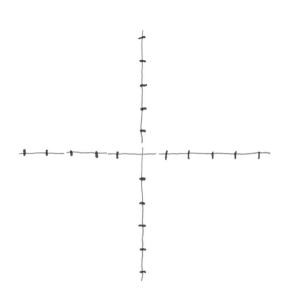

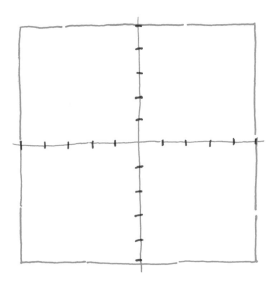

5. Rectangles are sketched in the same way except that you will have more units on one axis (line) than the other axis (line).

How to Sketch Angles

1. Sketch vertical and horizontal construction lines. These lines will form a 90 degree or *right angle.*

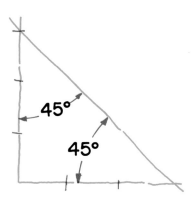

2. A 45 degree angle is sketched by marking off equal number of units on both lines. Connect the last unit of each line. This will form a 45 degree angle with the vertical and the horizontal lines.

3. To sketch 30 and 60 degree angles, mark off three units on one line and five units on the other line. Connecting the last unit on each line will give the required angles.

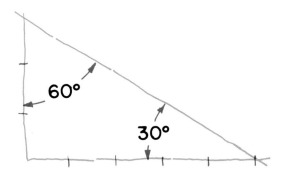

4. Other angles may be drawn by sketching an angle and subdividing this into the approximate number of degrees required. Example: Dividing a 30 degree angle into thirds will give a 10 degree angle.

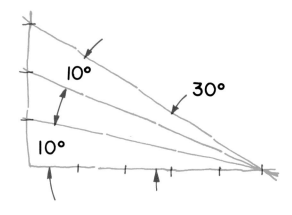

5. Another method used to develop angles in sketching is to sketch a quarter circle and divide the resulting arc into the desired divisions. Example: Dividing the arc into three parts will give 30 and 60 degree angles.

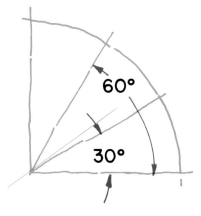

How to Sketch Circles

1. Sketch vertical, horizontal, and inclined axes.

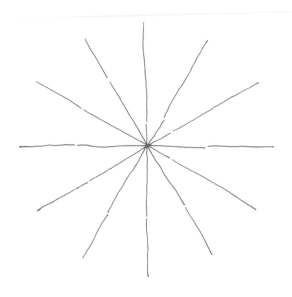

2. Mark off units equal to the radius of the required circle on each axis.

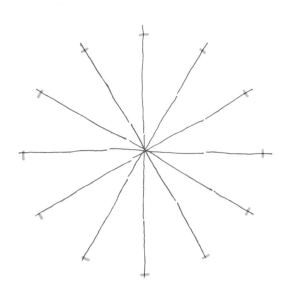

3. The radius units can be quickly and accurately located by marking off the desired radius on a piece of paper and using the paper as a measuring tool.

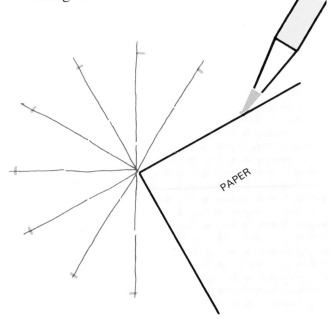

4. Sketch a construction line through the points. When satisfied with the construction line, fill it in with a line of the desired weight.

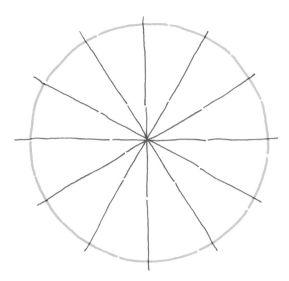

How to Sketch an Arc

1. Sketch a right (90 degree) angle. Use construction lines.

2. Units equal to the length of the desired radius are marked on each leg of the angle. Connect these points with a construction line.

3. Divide this line into two equal parts. Starting from the point where the legs of the angle intersect, sketch a line through the dividing point of the diagonal line.

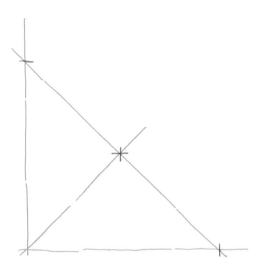

4. Mark off a point half way between the diagonal line and the intersection of the legs of the angle. Sketch an arc through the three points as shown.

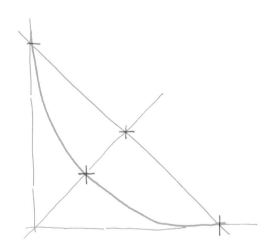

How to Sketch an Ellipse

1. Sketch horizontal and vertical lines as shown. Mark off equal size units on the centerlines to construct a rectangle with the dimensions equal to the major axis (the long axis) and the minor axis (the small axis) of the desired ellipse.

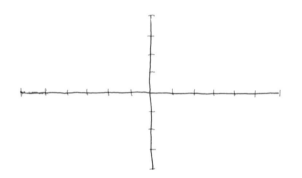

2. Construct the rectangle by sketching construction lines through the outer points.

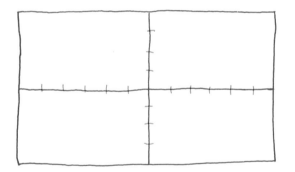

3. Lightly sketch arcs tangent to the lines that form the rectangle.

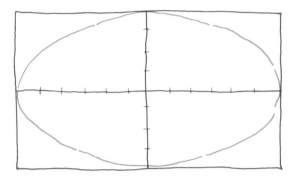

4. When you are satisfied with the shape of the ellipse, complete it by going over the construction lines with lines of the desired weight.

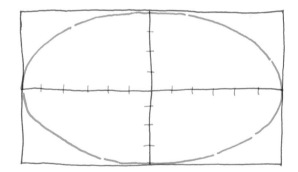

How to Sketch a Hexagon

1. Sketch vertical and horizontal centerlines, and inclined lines at 30 and 60 degrees. Construct a circle with a diameter equal to the distance across the flats of the required hexagon. Use construction lines.

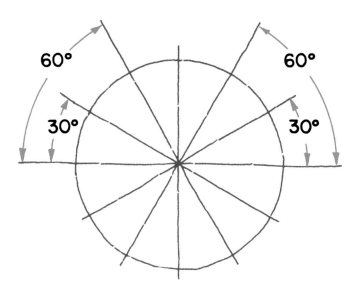

3. Sketch inclined parallel lines at 60 degrees and tangent to the circle at the point where the 30 degree inclined line intersects the circle.

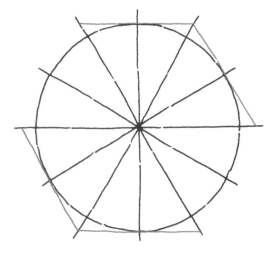

4. Complete the hexagon and go over the construction lines to produce the proper weight line.

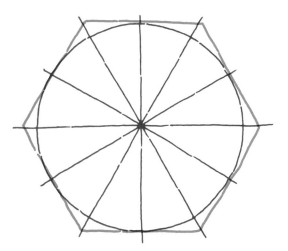

2. Sketch horizontal parallel lines at right angles (90 degree) to the vertical centerline. The lines are tangent to the circle at these points.

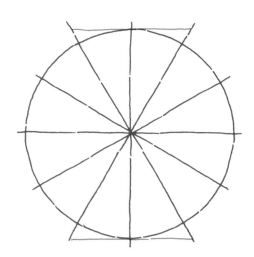

How to Sketch an Octagon

1. Sketch vertical and horizontal centerlines and inclined lines at 45 degrees. Construct a circle with a diameter equal to the distance across the flats of the required octagon. Use construction lines.

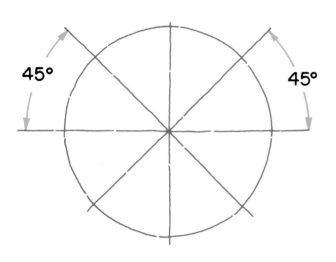

2. Sketch parallel lines tangent to the circle where the horizontal and vertical centerlines intersect the circle.

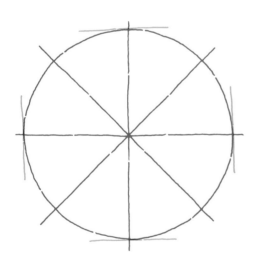

3. Sketch inclined parallel lines at 45 degree and tangent to the circle at the point where the 45 degree inclined lines intersect the circle.

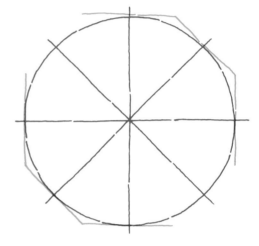

4. Complete the octagon and go over the construction lines to produce the desired weight line.

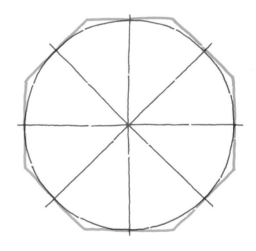

Sheet Layout for Sketching

1. Sketch a 1/2 in. border around the edges of the paper. Use a construction line. The sheet should be 8 1/2 in. by 11 in. It may be plain or graph paper. Sketch in guide lines as shown in Figure 3-6.

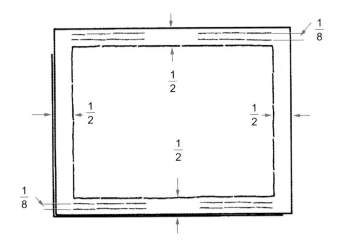

Figure 3-6. Sketching border and lettering guide lines. Since no rule is used, these dimensions are only approximate.

2. The edge of your drawing board or desk may be used as a guide in sketching the border and guide lines, Figure 3-7. Place the pencil in a fixed position and move your fingers along the edge of the drawing board or desk.

Figure 3-7. The edge of your drawing board can be used as a guide when sketching border lines. Note how the pencil is held.

3. Sketch a border line over the construction lines, letter in information as shown, Figure 3-8, or as specified by your instructor.

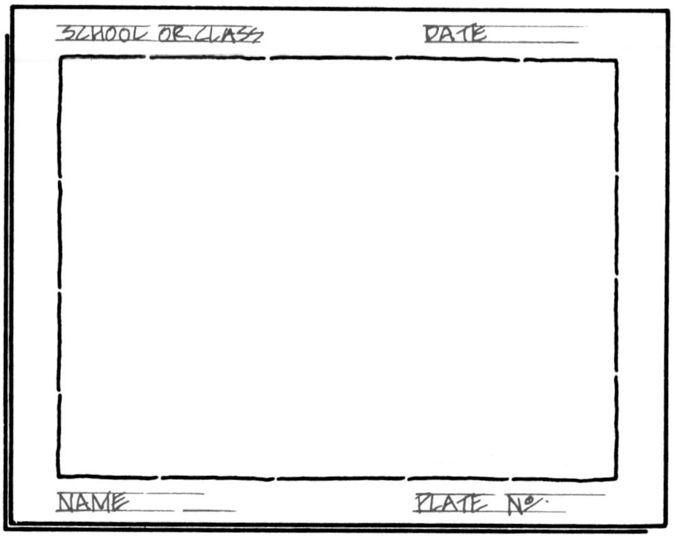

Figure 3-8. Lettering in the required information.

4. Take your time and sketch in the border and information carefully and neatly.

Enlarging or Reducing by the Graph Method

Drawings can be reduced or enlarged easily and quickly using this technique. The original drawing is blocked off into squares, Figure 3-9. The squares are numbered as shown. The horizontal numbers are referred to as X-coordinates and the vertical numbers are Y-coordinates. After deciding how much larger or smaller the new drawing is to be made, draw squares of the new size on a blank sheet of paper. Add coordinate numbers to the new squares. Using the design in the original drawing as a guide, sketch the design into the larger or smaller blank squares.

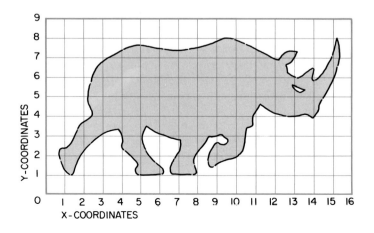

Figure 3-9. Drawing to be enlarged has been blocked in.

For example, a drawing of the design is to be enlarged to twice its original size. The original drawing is marked off into 1/4 in. squares. Square size will vary depending on the size of the original design. Number the squares as shown in Figure 3-9. Another sheet is made up with squares that are twice the size of the 1/4 in. squares, or 1/2 in. squares. Identify the large squares in the same manner as the smaller squares. Then, sketch in the details freehand, Figure 3-10.

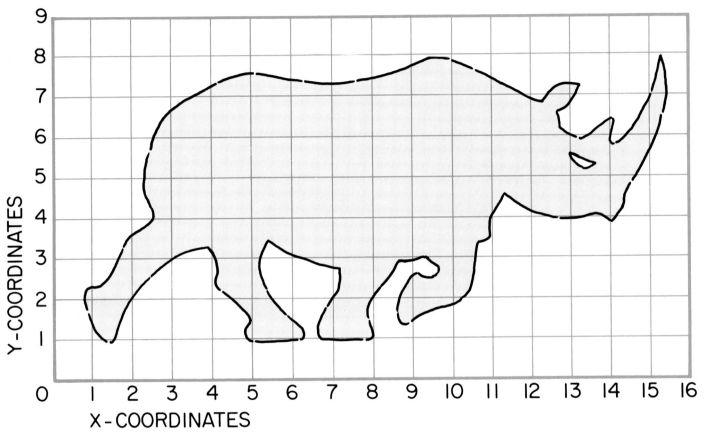

Figure 3-10. Enlarged drawing sketched into larger squares.

Computer-aided design and drafting (known as CADD) uses a similar system of coordinates in a grid pattern to enlarge, reduce, or duplicate original drawings. You will learn more about computer graphics in Unit 24.

Drafting Vocabulary

Accuracy	Dimension line	Octagon
Alphabet of lines	Dimensions	Parallel
Arc	Ellipse	Radius
Axis	Extension line	Right angle
Border line	Geometric	Section line
Centerline	Guide line	Sketching
Conical	Hexagon	Symmetrical
Construction line	Hidden line	Tangent
Coordinates	Horizontal	Terminate
Cutting-plane line	Inclined	Vertical
Diagonal	Intersection	
Diameter	Object line	

Test Your Knowledge—Unit 3

Please do not write in the book. Place your answers on another sheet of paper.

1. In sketching, a line is drawn by making a series of _____ _____.
2. The heaviest line used in sketching is the _____ line.
3. Drawings can be _____ or _____ easily by using the graph method.
4. When sketching an inclined line, sketch up when a line inclines to _____ and down when a line inclines to _____.
5. Extension lines are the same weight as _____ lines.
6. In sketching, horizontal lines are drawn from _____ _____ _____; vertical lines from the _____ _____.
7. Dimension lines generally terminate in _____ at the ends.

Sketching plays an important role in developing ideas. Several sketches of an end table are shown on the left. The photograph on the right shows the completed end table.

PROBLEMS

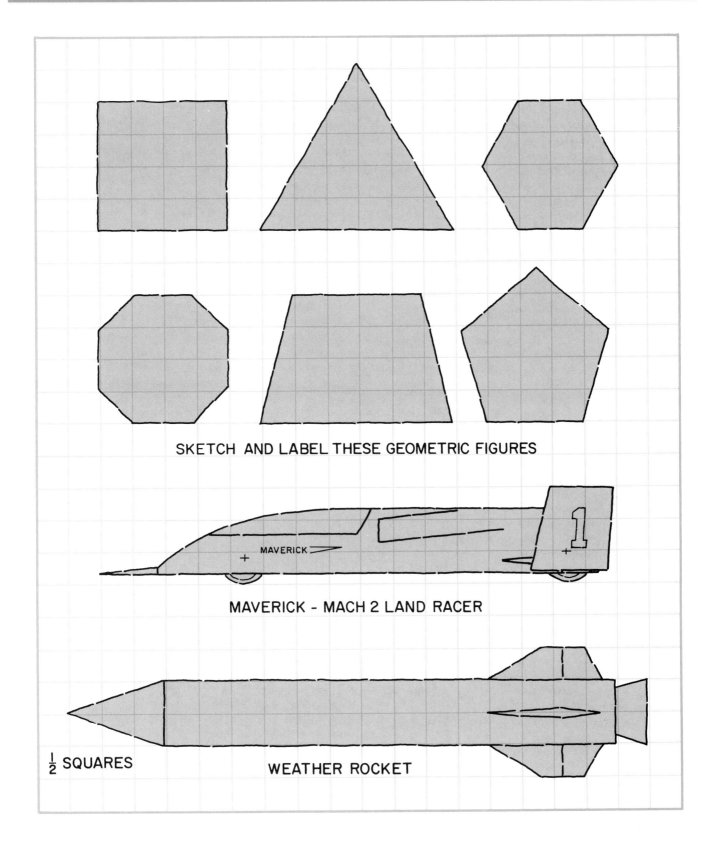

SKETCH AND LABEL THESE GEOMETRIC FIGURES

MAVERICK - MACH 2 LAND RACER

$\frac{1}{2}$ SQUARES   WEATHER ROCKET

Problem Sheet 3-1.   SKETCHING. Practice your sketching techniques using these objects. You may enlarge or reduce by using the graph method.

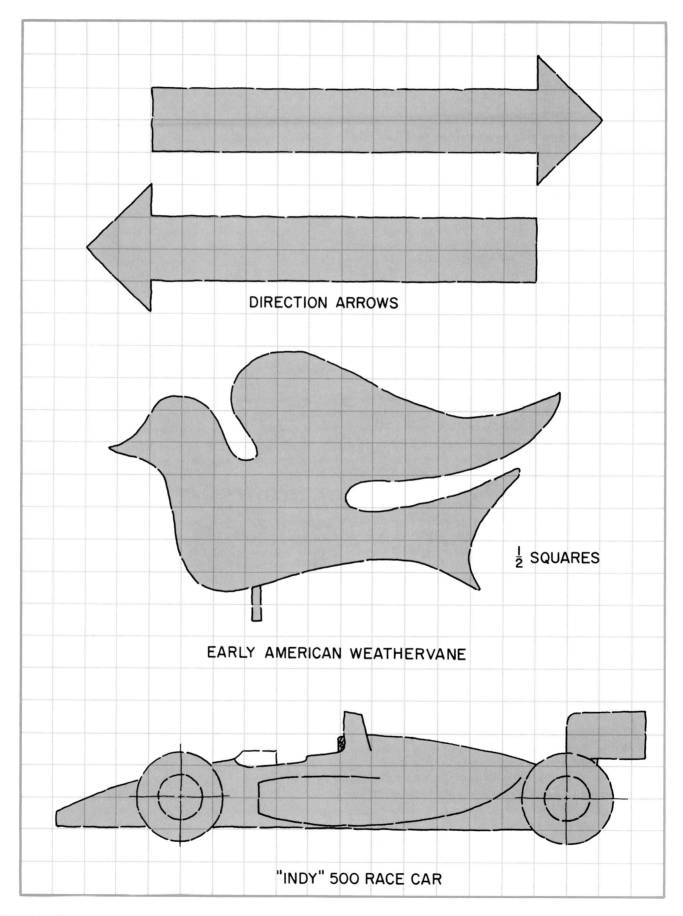

DIRECTION ARROWS

$\frac{1}{2}$ SQUARES

EARLY AMERICAN WEATHERVANE

"INDY" 500 RACE CAR

Problem Sheet 3-2.   SKETCHING. Practice your sketching techniques using these objects. You may enlarge or reduce by using the graph method.

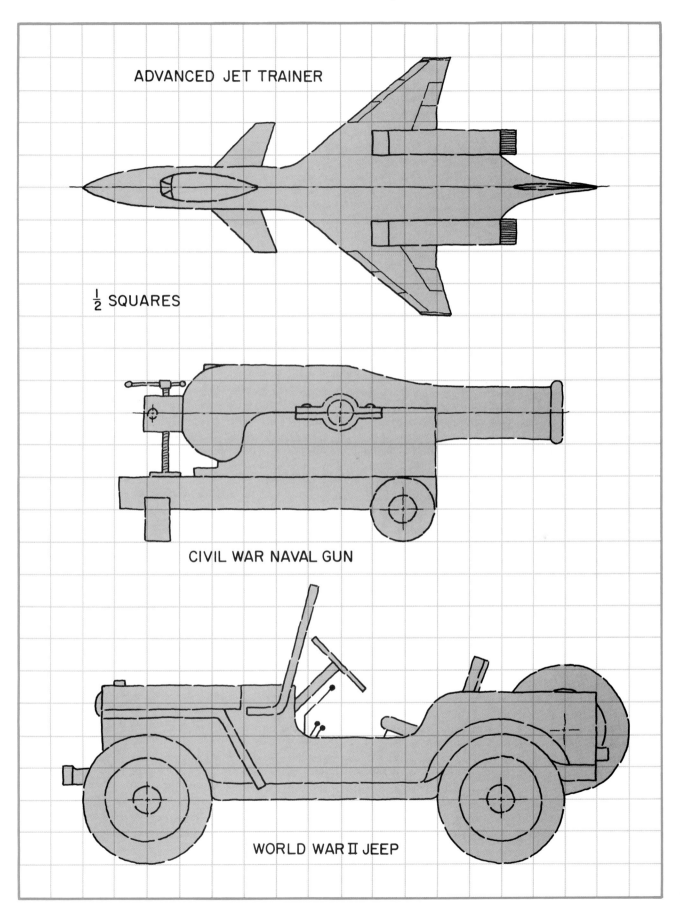

ADVANCED JET TRAINER

$\frac{1}{2}$ SQUARES

CIVIL WAR NAVAL GUN

WORLD WAR II JEEP

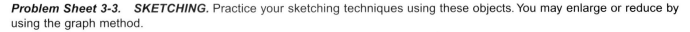

Problem Sheet 3-3.   SKETCHING. Practice your sketching techniques using these objects. You may enlarge or reduce by using the graph method.

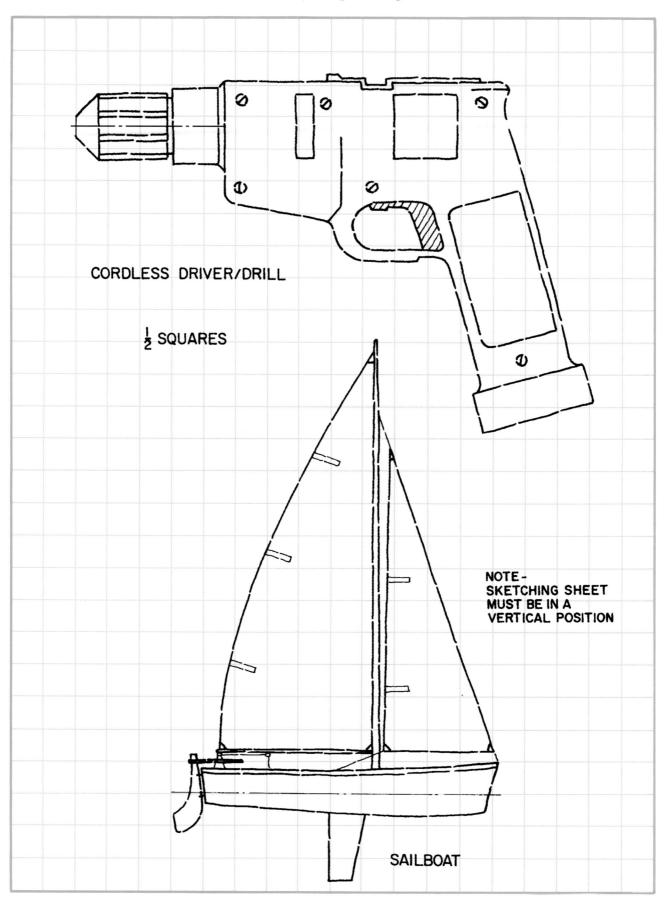

CORDLESS DRIVER/DRILL

½ SQUARES

NOTE-
SKETCHING SHEET
MUST BE IN A
VERTICAL POSITION

SAILBOAT

Problem Sheet 3-4. SKETCHING. Practice your sketching techniques using these objects. You may enlarge or reduce by using the graph method.

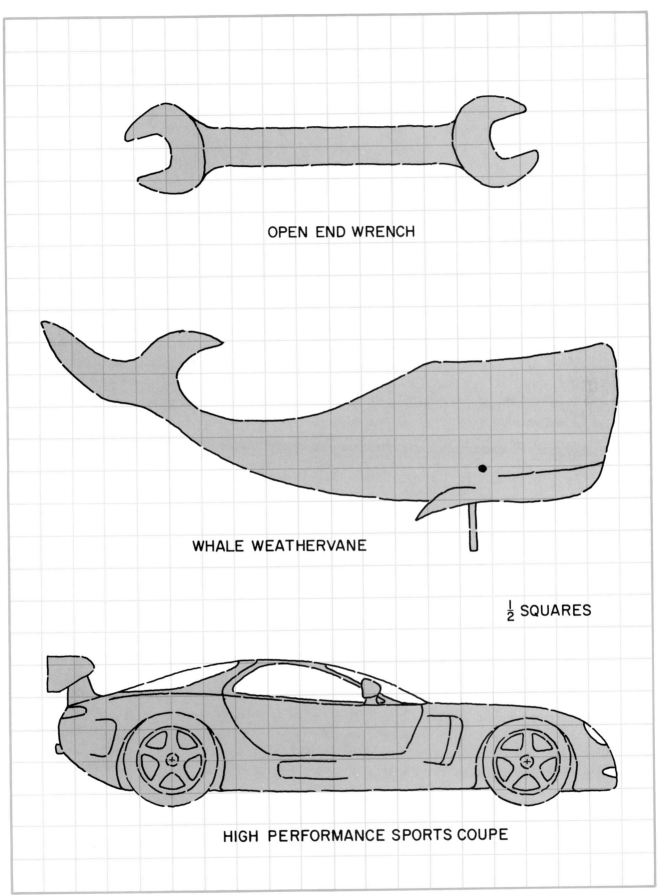

OPEN END WRENCH

WHALE WEATHERVANE

$\frac{1}{2}$ SQUARES

HIGH PERFORMANCE SPORTS COUPE

Problem Sheet 3-5. SKETCHING. Practice your sketching techniques using these objects. You may enlarge or reduce by using the graph method.

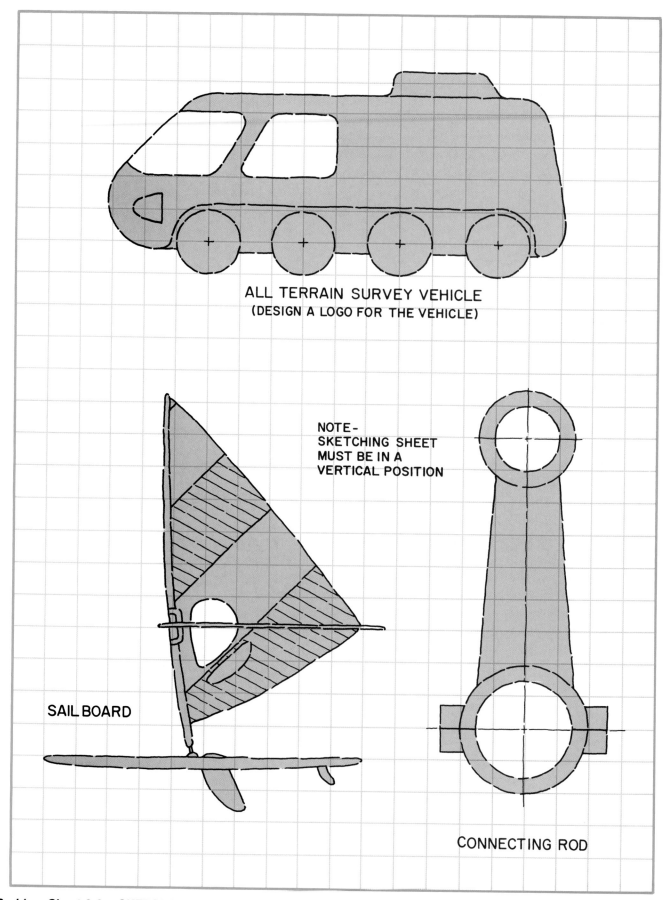

ALL TERRAIN SURVEY VEHICLE
(DESIGN A LOGO FOR THE VEHICLE)

NOTE–
SKETCHING SHEET
MUST BE IN A
VERTICAL POSITION

SAIL BOARD

CONNECTING ROD

Problem Sheet 3-6.   SKETCHING. Practice your sketching techniques using these objects. You may enlarge or reduce by using the graph method.

Drafting Equipment

After studying this unit, you should be able to:

◆ Identify basic drafting equipment.

◆ Use drafting equipment in a safe and efficient manner.

◆ Explain the terms CAD and CADD.

◆ Explain why the principles of drafting are common to both traditional drafting and CAD.

Drafting tools must be in good condition to make first-rate drawings. Typical equipment found in many schools is shown in Figure 4-1. It is always a good idea to check your drafting tools' condition before you use them.

The *drawing board* provides the smooth, flat surface needed for drafting. The tops of many drafting tables are designed for this purpose. Drawing boards are manufactured in a variety of sizes. The majority of them are made from selected, seasoned basswood.

A right-handed drafter will use the left edge of the board as the working edge; a left-handed drafter uses the right edge. The working edge should be checked periodically for straightness.

Drafters often tape a piece of heavy paper or special vinyl board to cover the working face of the

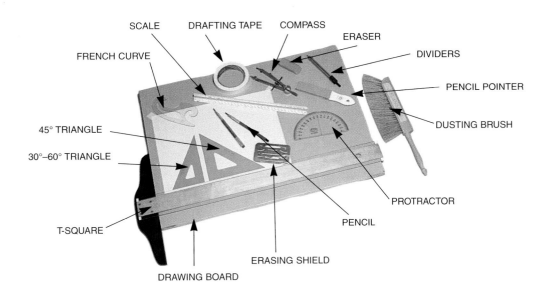

Figure 4-1. Type of drafting equipment found in many school drafting rooms.

drawing board to protect its surface. The vinyl surface is easily cleaned.

Horizontal lines are drawn with the **T-square**. It also supports triangles when they are used to draw vertical and inclined lines.

A T-square consists of two parts, the head and the blade or straightedge. The head is usually fixed solidly to the blade. However, a T-square with a protractor head and adjustable blade is available.

Clear plastic strips inserted in the blade edge of many T-squares make it easier to locate reference points and lines. The blade must never be used as a guide for a knife or other cutting tool since this may affect the "trueness" of the edge.

If accurate line work is to be done, it is essential that the T-square be held firmly against the working edge of the board.

It is recommended that the blade be left flat on the board or stored suspended from the hole in its end. This will keep warping or bowing of the blade to a minimum.

Triangles

When supported on a T-square blade, the 30-60 degree and 45 degree *triangles* are used to draw vertical and inclined lines, Figure 4-2. They are

made of transparent plastic and are available in a number of different sizes.

To prevent warping, a triangle should be left flat on the drawing board when not being used.

When drawing vertical lines, rest the triangle solidly on the T-square blade while holding the T-square head firmly against the working edge of the board. Angles of 15 and 75 degrees can be drawn by combining the triangles as shown in Figure 4-3.

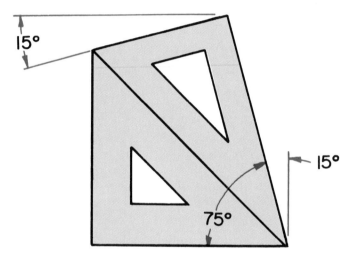

Figure 4-3. Drawing 75 degree and 15 degree angles. Note how the two triangles are combined to produce the required angles.

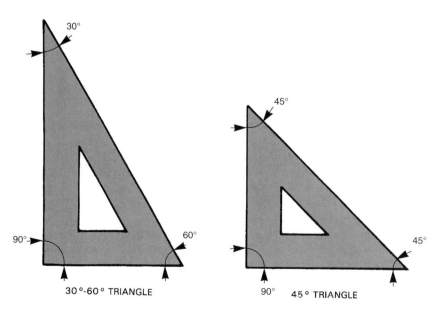

Figure 4-2. Triangles must be used with a T-square in order to give you the noted angles.

Drafting Machines

Industry makes considerable use of ***drafting machines***, Figure 4-4. This device replaces both the T-square and triangles. The straightedges can be adjusted to any angle using the built-in protractor blade. Drafting machines are often used in place of T-squares and triangles in school drafting rooms.

Figure 4-4. Typical drafting machine. Note the adjustable protractor blade.

Compass

In drafting, circles and arcs are drawn with a ***compass***, Figure 4-5. The tool is held as shown in Figure 4-6. For best results, the lead should be adjusted so that it is about 1/32 in. (0.5 mm) shorter than the needle. Both legs will be the same length when the needle penetrates the paper. Fit the compass with lead that is one grade softer than the pencil used to make the drawing. The lead must be kept sharp.

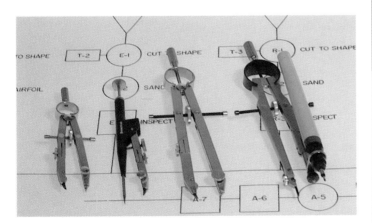

Figure 4-5. Compasses (from left to right). Small bow compass. Drop bow compass. Big bow compass. Big bow compass with inking attachment and pen.

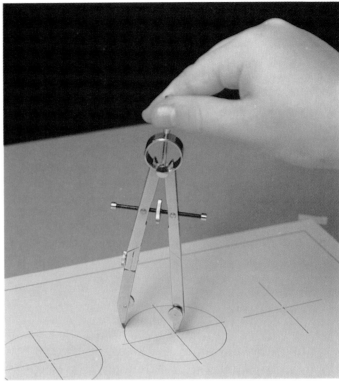

Figure 4-6. The correct way to hold a compass when drawing a circle.

Several compass attachments are available. A ruling pen is substituted for the pencil "leg", Figure 4-7. The technical pen, shown in Figure 4-5, is now commonly used for inking tasks. An extension is used to draw large circles. The ***beam compass*** will do the job better, Figure 4-8.

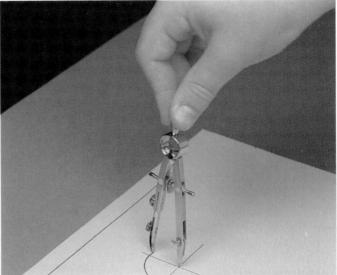

Figure 4-7. Inking a circle using a compass. Some compass models replace the inking "leg" with a technical pen.

Figure 4-8. When drawing large circles, it is easier to use a beam compass rather than a compass with an extension.

To set a compass to the size of the circle radius, first draw a straight line on a piece of scrap paper and then measure off the required distance equal to the radius. Set the compass on this line. Avoid setting a compass on a scale. "Sticking" the compass needle into the scale will eventually destroy the scale's accuracy.

Dividers

Distances are subdivided and measurements are transferred with *dividers*, Figure 4-9. Careful adjustment of the divider points is necessary.

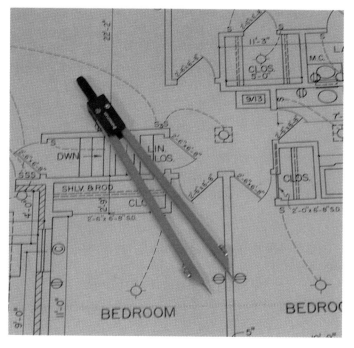

Figure 4-9. Dividers.

Safety Note: Be careful where and how you place dividers or compass after use. It is very painful to accidentally run the point into your hand when reaching for the tool.

Pencil Pointer

It is not necessary to resharpen a drawing pencil every time it starts to dull. It can be repointed quickly with a *pencil pointer*, Figure 4-10. Use the wooden pencil sharpener only when the point becomes very blunt, or when it breaks.

BEFORE   AFTER

Figure 4-10. Mechanical pencil pointers sharpen the lead, but do not remove any wood. Always rotate the pointer in a clockwise direction.

Keep the pencil pointer clear of the drawing area when repointing a pencil. The graphite dust will smudge your paper when you attempt to remove eraser crumbs.

Many types of pencil pointers are available. The sandpaper pad, Figure 4-11, may be found in the school drafting room. A piece of styrofoam cemented to the back of the pad can be used to remove graphite dust from the newly pointed pencil.

Erasers

Many shapes and kinds of *erasers* are manufactured for use in the drafting room, Figure 4-12. The type of material being drawn upon—paper,

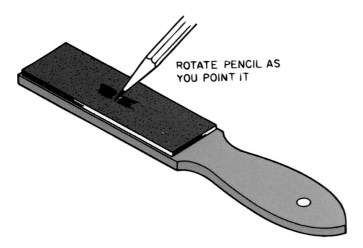

Figure 4-11. Using a sandpaper pad to point a drafting pencil. Do not hold your pencil over the drawing board when pointing the lead.

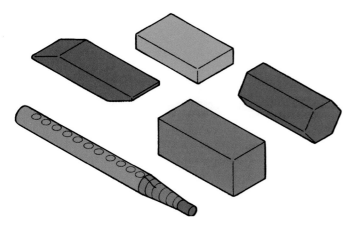

Figure 4-12. Examples of typical erasers used in the drafting room.

film, or vellum—will determine the type of eraser to be used.

Note! Always brush away eraser crumbs before starting to draw again.

Erasing Shield

Small errors or drawing changes can be erased without soiling a large section of the drawing if an *erasing shield* is employed, Figure 4-13. Place a shield opening of the proper shape and size over the area to be changed and then erase. The erasure is made without touching other parts of the drawing.

One widely used shield is made from stainless steel. It is very thin, wear-resistant, and does not stain or "smudge" drawings.

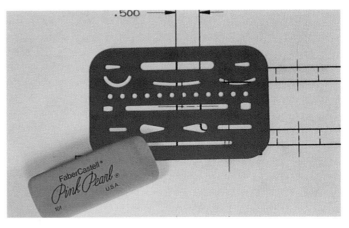

Figure 4-13. An erasing shield is used to protect a portion of the drawing while you are erasing another area.

Fasteners

Two preferred methods of attaching drafting media to a drawing board are shown in Figure 4-14. Drafting tape is recommended as the most desirable. It does not puncture the paper or affect the surface of the drawing board.

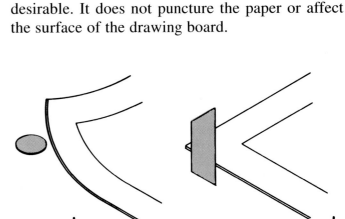

PREFERRED

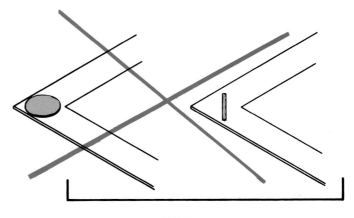

AVOID

Figure 4-14. Two-sided drafting tape dots or drafting tape are preferred methods of attaching paper to the drawing board. Staples and tacks should not be used.

Adhesive tape is sometimes used to attach large drawing sheets. Continued use may affect the working surface of the drawing board.

Thumb tacks and staples are to be avoided. These quickly destroy the smooth surface of a drawing board.

Dusting Brush

No matter how careful you are, some erasing crumbs and dirt particles will collect on the drawing area. They should be removed with a *dusting brush*, Figure 4-15. *Do not remove them with your hand.* Using your hand will cause smudges and streaks.

Figure 4-16. Protractor used in the drafting room.

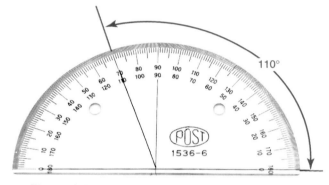

Figure 4-17. Measuring angles with a protractor.

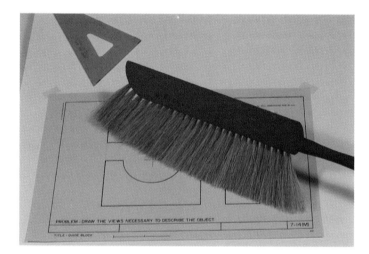

Figure 4-15. Removing erasure crumbs from the drawing surface with a dusting brush.

French Curves

Curved lines that are not exactly circular in form are drawn with a *French curve*, Figure 4-18. After the curved line is carefully plotted, it is drawn with a French curve as shown in Figure 4-19.

Protractor

The *protractor* shown in Figure 4-16 is employed to measure and lay out angles on drawings. Protractors are usually made from clear plastic and may be either circular or semicircular in shape. The degree graduations are scribed or engraved around the circumference of a protractor.

When measuring or laying out an angle, place the centerlines of the protractor at the point (apex) of the required angle as shown in Figure 4-17. Read or mark the angle from the graduations on circumference of the tool.

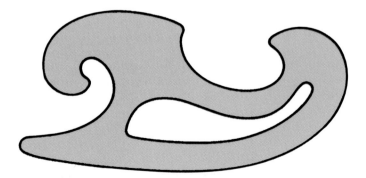

Figure 4-18. French curve, also known as an irregular curve. It is available in many sizes and configurations.

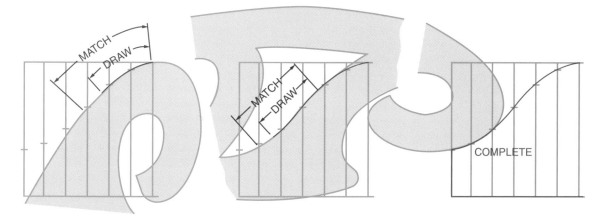

Figure 4-19. Using a French curve to draw an irregular (not a true circle) curve.

The curves are made of transparent acrylic plastic. They range in size from a few inches to several feet in length and may be purchased individually or as a set. They are also known as *irregular curves*.

Templates

Templates are available in an almost unlimited range of shapes, Figure 4-20. Made of thin, transparent plastic, they contain openings of different sizes and shapes. Most templates allow for the thickness of the pencil lead or pen tip.

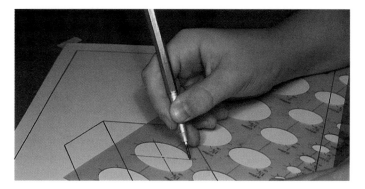

Figure 4-21. Templates enable the drafter to do normally time-consuming jobs with ease and accuracy.

Drafting Media

Drawings are made on many different materials—paper, tracing vellum, mylar film, etc. A heavyweight opaque paper that is buff (light yellow), pale green, or white in color is used in many school drafting rooms.

While most papers take pencil lines well, they are often difficult to erase because the pencil point makes a depression in the paper when a line is drawn. Plan your work carefully to prevent mistakes.

Industry makes much use of tracing vellum and film because reproductions or prints must be made of all drawings.

The bulk of the drawings used by industry are put on standard size drawing sheets. This makes them easier to file and identify. A listing of standard sheet sizes is shown in Figures 4-22 and 4-23.

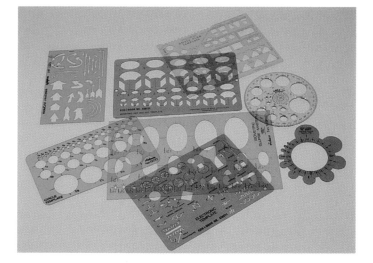

Figure 4-20. A few of the many types of templates used in a drafting room.

Templates enable a drafter to do normally time-consuming jobs with ease and accuracy, Figure 4-21.

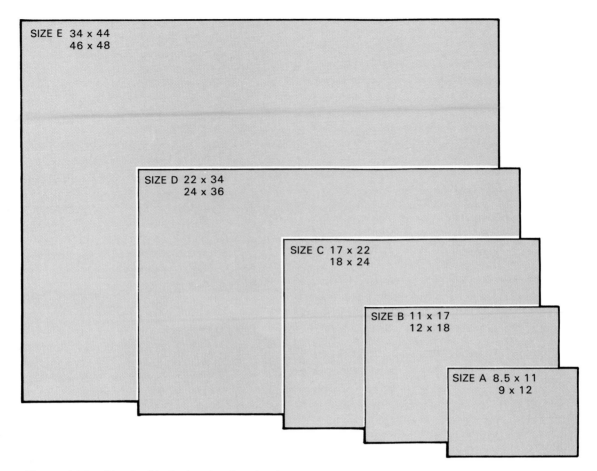

Figure 4-22.   Standard inch-size drawing sheets.

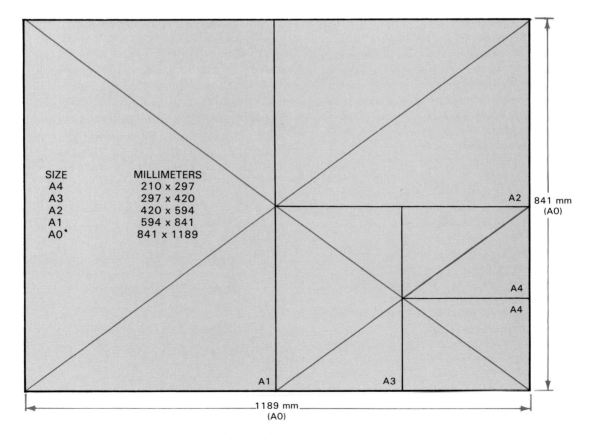

Figure 4-23.   Standard metric-size drawing sheets.

Designers, drafters, surveyors, engineers, and architects make considerable use of commercially prepared graph paper to make preliminary design studies, Figure 4-24. Isometric grid paper makes it very easy to convert an orthographic drawing into an isometric drawing, Figure 4-25.

Many types and sizes of grid patterns are available commercially in the form of prepunched loose leaf sheets (8 1/2 in. x 11 in.), other sheet sizes, and in roll form. See Figure 4-26. The grid lines are printed in many colors—green, orange, black, and in a pale blue that will not reproduce when a drawing made on it is run through a photocopy machine.

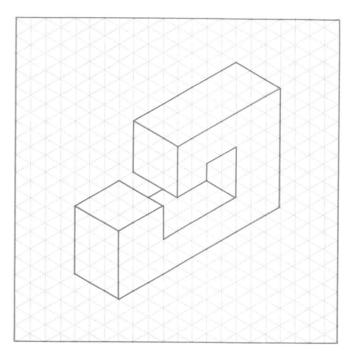

Figure 4-25. An isometric drawing made on graph paper designed for that purpose.

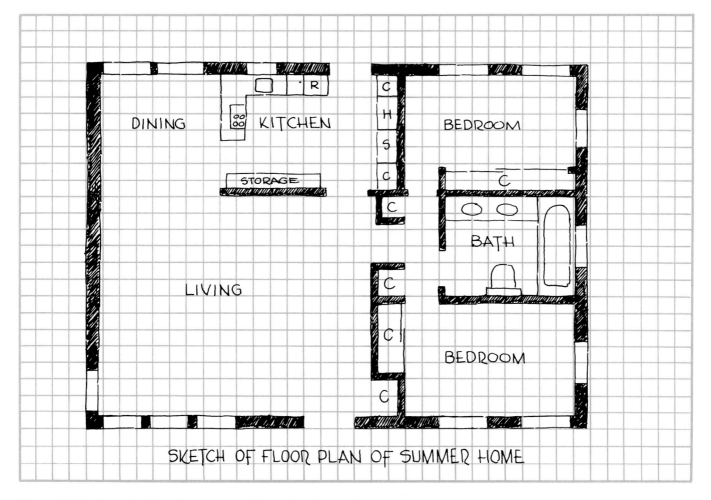

SKETCH OF FLOOR PLAN OF SUMMER HOME

Figure 4-24. Preprinted graph paper speeds up sketching. A floor plan sketched on graph paper. Each square equals one foot.

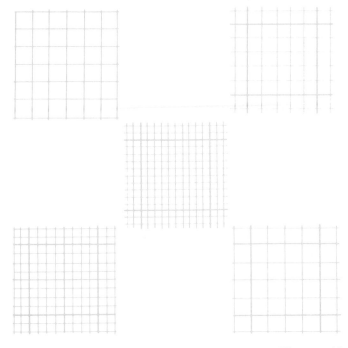

Figure 4-26. Graph paper is available in many different grid (square) sizes.

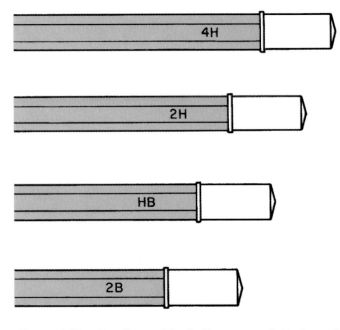

Figure 4-27. Pencils used in drafting are available in a wide range of hardness.

Pencils

Since many drawings are prepared with a pencil, it is important that the proper pencil be selected. The drawing media employed will determine the type pencil for best results. A conventional lead pencil is satisfactory with most papers and tracing vellums, while a pencil with plastic lead is necessary if the drawing is made on film.

The drafter can select from 17 grades of pencils that range in hardness from 9H (very hard) to 6B (very soft), Figure 4-27. Many drafters use a 4H or 5H pencil for lay out work and an H or 2H pencil to darken the lines and to letter. In general, use a pencil that will produce a sharp, dense black line because this type line reproduces best on prints.

Avoid using a pencil that is too soft. It will wear rapidly, smear easily, and soil your drawing. Also, the line will be "fuzzy" and will not produce usable prints.

A conical-shaped pencil point is preferred for most general-purpose drafting. To sharpen the pencil, cut the wood away from the unlettered end. Sharpen with a knife or mechanical sharpener and point the lead with a pencil pointer, Figure 4-28.

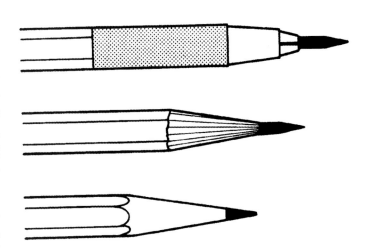

Figure 4-28. Pencil points. A—Sharpened with a mechanical pencil pointer. B—Sharpened with a knife and pointed on a sandpaper pad. C—Sharpened with a regular pencil sharpener.

A semiautomatic pencil, Figure 4-29, is usually preferred to a wood pencil. With this type pencil it is not necessary to cut away wood to expose the lead. The extended lead is shaped on a pencil pointer.

Scales

A scale can be used to make measurements that are full-size or larger, or smaller than full-size.

Figure 4-29. A variety of semiautomatic drafting pencils. Only the lead needs to be replaced.

The size of the object and drawing sheet size will determine the scale of a drawing.

Scales are in constant use on the drawing board because almost every line on a mechanical drawing (a drawing made with instruments) must be a measured length, Figure 4-30. Accurate drawings require accurate measurements.

Due to the diversity of work that must be drawn, scales used by drafters are made in many shapes, lengths, and measurement graduations. They may be made of wood, plastic, or a combination of both materials. Graduations are printed on inexpensive scales, and are machine engraved (much more accurate) on the more costly ones.

Architect's Scale. All of the scales represent one foot. The scales are 1/8, 3/32, 3/16, 1/4, 3/8, 1/2, 3/4, 1, 1 1/2, and 3. The scale of a particular section is marked on the face at the end of each edge of the measuring device. Each division represents one foot (12 inches). The first division is used like a foot rule. For example, the first division of the 3/4 scale is divided into 24 parts. Each graduation equals 1/2 in.

One edge of this triangular scale is divided into 1/16 in. graduations.

Engineer's Scale. Used mostly where large reductions are required. It is divided into 10, 20, 30, 40, 50, and 60 units per inch.

Mechanical Drafter's Scale. Most commonly divided into the following graduations—full-size, and 3/4, 1/2, 1/4, and 1/8 in. to one foot.

Metric Scale. This scale is now a required tool in the drafting room. Common divisions are in cen-

timeters (cm) divided into millimeters (mm) and centimeters divided into half-millimeters (0.5 mm). Metric scales are also available in a number of enlargement ratios (2:1, 3 1/3:1, etc.) and reductions (1:2, 1:3, etc.).

Computer Drafting Equipment

Much design and drafting work is done with computers, Figure 4-31. Prints will be made on high speed plotter from information generated on a computer, Figure 4-32. A number of large firms

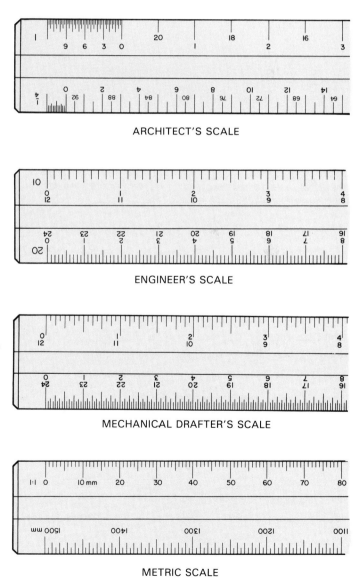

ARCHITECT'S SCALE

ENGINEER'S SCALE

MECHANICAL DRAFTER'S SCALE

METRIC SCALE

Figure 4-30. A few of the various types of the scales used in a drafting room.

Figure 4-31. Many drafting rooms use computers to develop the drawings needed to manufacture the many products we use. When needed, prints are made on high-speed plotters from information generated on the computer. (Amp, Inc.)

Figure 4-32. High-speed plotter. (Calcomp)

small companies utilize computer-aided design and drafting (known as CAD or CADD).

The basic CAD system uses a *mouse* for inputting data and information, a *CRT* (cathode ray tube) to display the drawing, a *disk drive* or some other device to store and retrieve data, and the

processor which is the heart of the computer. To make the CAD system function, *software* is used to program the unit. The software allows the operator to design, draw, alter, erase, move, etc., any item on the screen. The selection of software is a major consideration in the operation of CADD.

The principles of drafting are common to both traditional drafting and CAD. Drafting technicians must have a working knowledge of basic drafting practices, standards, and procedures (the same ones you are now learning) before being able to use CAD to its fullest potential.

CAD and other computer graphics related to industry will be explained in more detail in Unit 24—Computer-Aided Drafting and Design.

Drafting Vocabulary

Architect's scale	Erasers	Ratio
Beam compass	Erasing shield	Reproduction
Circular	French curve	Scales
Circumference	Graduations	Semiautomatic
Compass	Irregular curves	Semicircular
CRT (cathode ray tube)	Keyboard	Software
Disk drive	Mechanical drafter's scale	Templates
Dividers	Metric scale	Transparent
Drafting machines	Pencil pointer	Triangles
Drawing board	Processor	T-square
Dusting brush	Protractor	Vinyl
Engineer's scale		

Test Your Knowledge—Unit 4

Please do not write in the book. Place your answers on another sheet of paper.

1. A drawing board provides the _____ necessary for drafting.
2. A piece of heavy paper or a special vinyl board cover is sometimes attached to the working surface of the board to _____.
   A. provide a drawing surface
   B. protect its surface
   C. permit the pencil to draw better
3. The T-square is used to draw _____.
4. Vertical and angular lines are drawn with _____.
5. A compass is used in drafting to draw _____ and _____.
6. A pencil is repointed with a(n) _____.
7. The erasing shield is used to _____.
8. List the two preferred methods for attaching a drawing sheet to the board. Which methods cause damage to the board and are to be avoided?
9. Angles can be measured and laid out on drawings with a(n) _____.
10. A(n) _____ H or _____ H pencil is recommended for layout work.
11. A(n) _____ H or _____ H pencil is recommended for darkening the lines and for lettering.
12. Why are scales important to drafters?
13. List the four types of scales described in the text.
14. What drafting concepts are common to both traditional board drafting and computer-aided drafting?

Outside Activities

1. Examine the drafting equipment assigned to you and make a record of its condition. Notify your teacher if any of your tools need repair.
2. Prepare a bulletin board display on computer-aided drafting (CAD).
3. Using catalogs of available drafting equipment, make a comparison of the cost of a fully equipped drafting room using drafting boards, drafting machines, triangles, etc., relative to the cost of computer-aided design and drafting equipment.

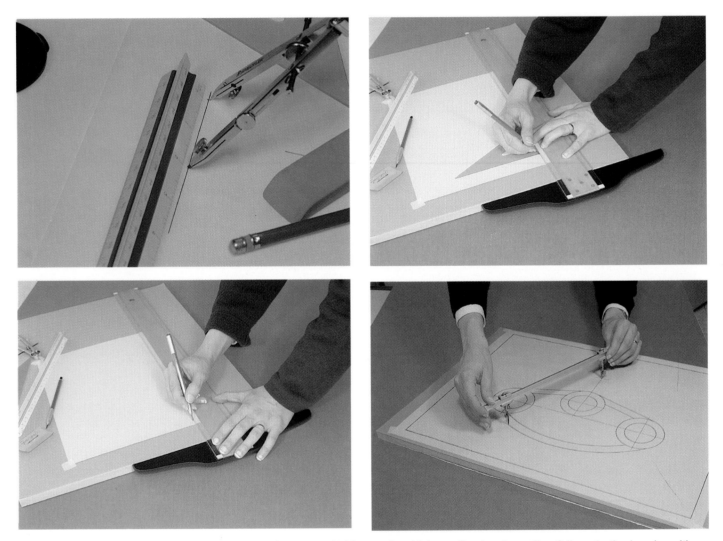

Basic drafting techniques, such as those shown here, are vital in creating high-quality drawings. Carefully note the hand positions in each of these photographs.

Unit 5
Drafting Techniques

After studying this unit and completing the assigned problems, you should be able to:

◆ Demonstrate basic drafting skills.

◆ Use basic drafting skills and techniques when solving drawing problems.

◆ Identify the various drafting scales.

◆ Make accurate measurements using drafting scales.

You were introduced to the *alphabet of lines* in the unit on sketching. The lines you sketched can be drawn more uniformly and accurately with drafting instruments. See Figure 5-1.

To understand the *language of industry,* it is necessary that you know the characteristics of the various lines and the correct way to use them in a drawing. Figure 5-2 shows an example of the correct use of hidden lines. In drafting room language, the characteristics and uses of lines are known as *line conventions.*

Each type or kind of line has a specific meaning. It is most important that each line is drawn properly, is opaque, and is uniform in width throughout its entire length. See Figure 5-3.

Alphabet of Lines

Construction and Guide Lines (very thin)

Construction lines are drawn very lightly. They are used to block in drawings, while guide lines are used for lettering. They may be erased, if necessary, after they have served their purpose.

Border Line (very thick)

The *border line* is the heaviest weight line used in drafting. It varies from 1/32 in. to 1/16 in. depending upon the size of the drawing sheet.

Object Line (thick)

The *object line* , also known as a *visible line,* is used to outline the visible edges of the object being drawn. They should be drawn so that the views stand out clearly on the drawing. All of the visible object lines on the drawing should be the same weight (thickness).

Dimension Line (thin)

The *dimension line* is usually capped at each end with arrowheads and is placed between two extension lines. With few exceptions, it is broken with the dimension placed at midpoint between the arrowheads. The dimension line is a light line a bit heavier than the construction line. It is placed 1/4 to 1/2 in. away from the drawing.

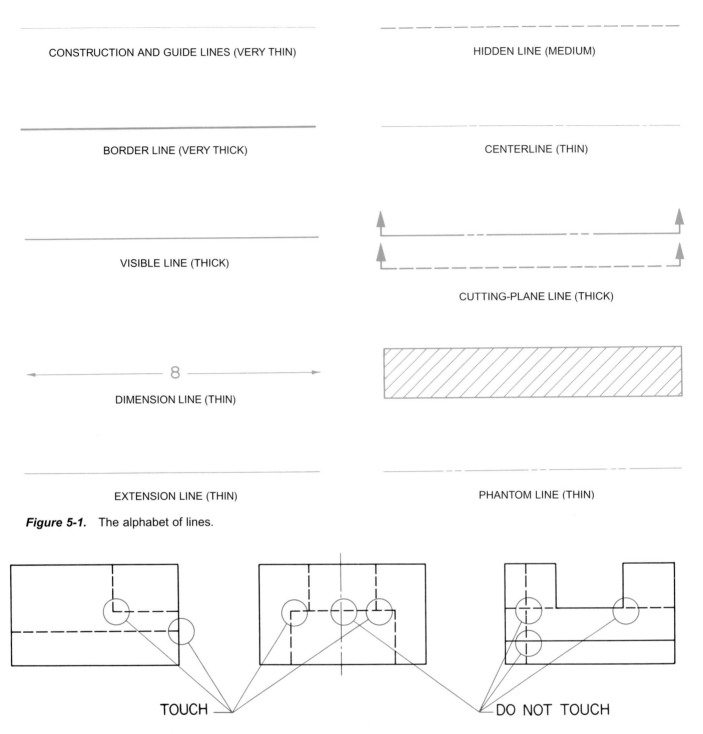

Figure 5-1. The alphabet of lines.

CONSTRUCTION AND GUIDE LINES (VERY THIN)

HIDDEN LINE (MEDIUM)

BORDER LINE (VERY THICK)

CENTERLINE (THIN)

VISIBLE LINE (THICK)

CUTTING-PLANE LINE (THICK)

8

DIMENSION LINE (THIN)

EXTENSION LINE (THIN)

PHANTOM LINE (THIN)

TOUCH

DO NOT TOUCH

Figure 5-2. Correct use of hidden lines.

Extension Line (thin)

The *extension line* is the same weight as the dimension line. It extends the dimension beyond the outline of the view so that the dimension can be read easily. The line starts about 1/16 in. beyond the object and extends about 1/8 in. past the last dimension line.

Hidden Line (medium)

The *hidden line* is used to show the hidden features of the object. It is drawn medium weight, slightly larger than the object line and is composed of short lines approximately 1/8 in. long separated by spaces approximately 1/16 in. They may vary slightly according to the size of the drawing.

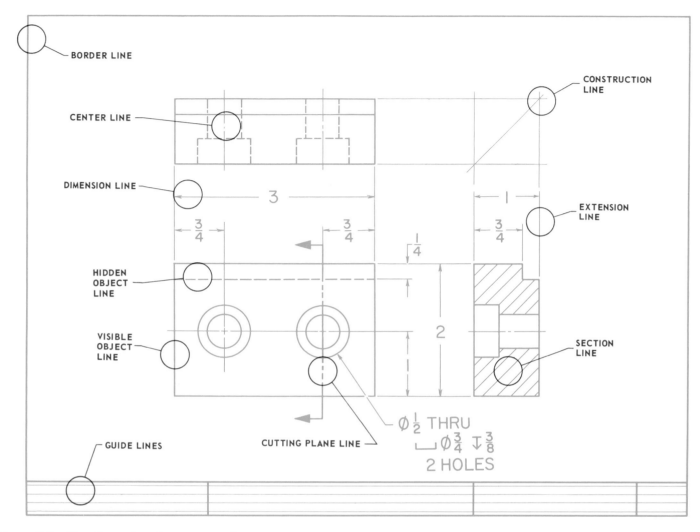

Figure 5-3. How the various lines are used.

Hidden object lines should always start and end with a dash in contact with the object line. See Figure 5-2 for the correct uses of the hidden line.

Centerline (thin)

The *centerline* is used to indicate the center of symmetrical objects. It is a fine dark line composed of alternate long (3/4 in.) and short (1/8 in.) dashes with 1/16 in. spaces between the dashes. The centerline should extend uniformly only a short distance beyond the circle or view. It starts and ends with long dashes and should not cross at the spaces between the dashes.

Cutting-Plane Line (thick)

The *cutting-plane line* is a heavy line. It is used to indicate where the sectional view will be taken. Two forms are recommended for general use. The first form is composed of a series of long (3/4 to 1 1/2 in.) and short (1/8 in. with 1/16 in. space) dashes. The second form is composed of equal dashes about 1/4 in. long with 1/16 in. spacing. The cutting-plane line will be further explained in Unit 10 on Sectional Views.

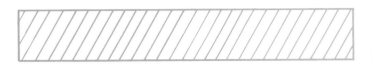

Section Line (thin)

Section lines are used when drawing the inside features of the object. They indicate material cut by the cutting-plane line, and also indicate the general classification of the materials. The lines are fine dark lines.

Phantom Line (thin)

The *phantom line* is used to indicate alternate positions of moving parts and for repeated detail like threads and springs. It is a thin dark line made of long dashes (3/4 to 1 1/2 in.) long, alternated with pairs of short dashes 1/8 in. in length, with 1/16 in. spaces.

How to Measure

Almost every line on a mechanical drawing must be a measured line. If your drawings are to be made accurately, *you* must be able to make accurate measurements.

Measurements are made in the drafting room with *scales*, Figure 5-4. The term *scale* means both the device or tool for making measurements, and the size to which the drawing is made.

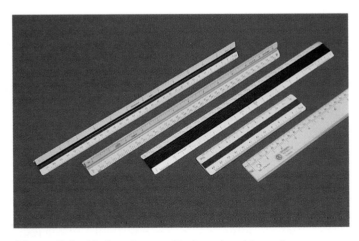

Figure 5-4. Various types of inch and metric scales.

Scales have graduations on the edges that show lengths used to indicate larger units of measure (such as 1/4 in. equals one foot). While several different shapes of scales are available, Figure 5-5, triangular-shaped scales are most widely found in the school drafting room.

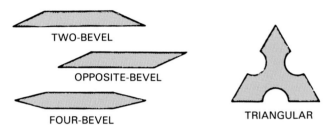

Figure 5-5. Scales are available in a number of different shapes.

A *scale clip* provides a means to lift a triangular scale and keeps the desired scale edge in an upright position, Figure 5-6.

Figure 5-6. The scale clip keeps the desired scale edge in an upright position. No time is wasted looking for the scale edge you are using.

There are four types of scales in common usage. They are classified as *architect's, engineer's, mechanical drafter's* and *metric scales,* Figure 5-7.

The triangular architect's scale (most often used in school drafting rooms) has six faces, Figure 5-8 (left). Each face is graduated differently.

You will find one face where the inch divisions are each divided into sixteen (16) parts. Each divi-

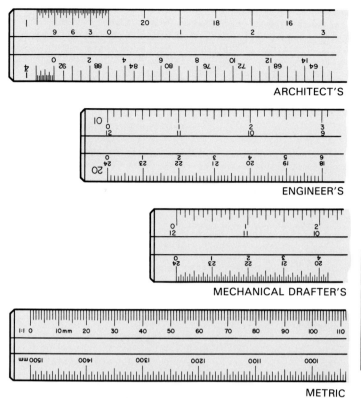

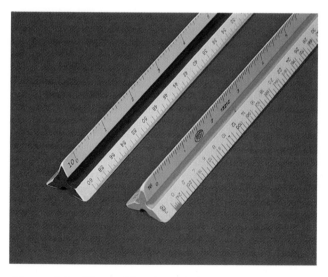

Figure 5-7. Scales in common usage in school drafting rooms.

Figure 5-8. The architect's scale is on the right and the engineer's scale is on the left.

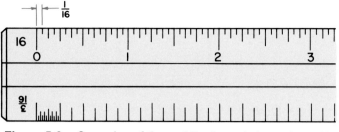

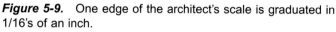

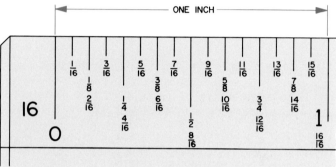

Figure 5-9. One edge of the architect's scale is graduated in 1/16's of an inch.

Figure 5-10. The divisions of the 1/16's edge of the architect's scale are numbered for learning experiences.

A section of the scale is shown in Figure 5-11. How many of the measurements can *you* read correctly?

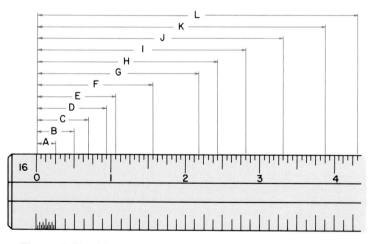

Figure 5-11. How many can you answer correctly? On a piece of paper write the letters A to L. After each letter, write the correct measurement. Reduce fractions to the lowest terms.

sion is equal to one-sixteenth (1/16) of an inch, Figure 5-9. To read the scale, imagine that the one-sixteenth (1/16) divisions are numbered as shown in Figure 5-10. At first you may find it easier to count the spaces when you measure. However, after some practice this should not be necessary.

After you can read and make measurements accurately and quickly to one-sixteenth of an inch, note the remaining faces on the scale. Each is

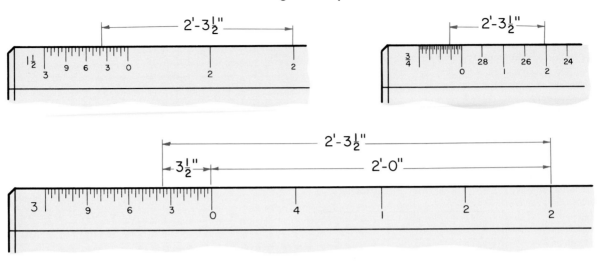

Figure 5-12. Graduations on the scale edges indicate larger or smaller scale units of measure.

divided to represent one foot (12 in., 1'-0") of actual measurement reduced to a particular length.

There are two scales on each face and each is marked to show scale divisions of 3/32 and 3/16; 1/8 and 1/4; 3/8 and 3/4; and 1 1/2 and 3. For example, the face marked with a 3 means the foot (12 in.) has been reduced to 3 in. A drawing made using this scale would be one-quarter (1/4) actual size. See Figure 5-12.

Using a Metric Scale

There are no architect's or engineer's scales in the metric system. Most drawings will be dimensioned in millimeters. Full-size drawings will require a 1:1 ratio scale. Use a 1:2 scale when drawings are reduced to half size, Figure 5-13. The preferred metric scales are shown in Figure 5-14. For architectural drafting, reduction scales

COMMON DRAFTING SCALES

CUSTOMARY INCH		NEAREST ISOMETRIC EQUIVALENT (mm)
1:2500	(1 in. = 200 ft.)	1:2000
1:1250	(1 in. = 100 ft.)	1:1000
1:500	(1/32 in. = 1 ft.)	1:500
1:192	(1/16 in. = 1 ft.)	1:200
1:96	(1/8 in. = 1 ft.)	1:100
1:48	(1/4 in. = 1 ft.)	1:50
1:24	(1/2 in. = 1 ft.)	1:20
1:12	(1 in. = 1 ft.)	1:10
1:4	Quarter size (3 in. = 1 ft.)	1:5
1:2	Half-size (6 in. = 1 ft.)	1:2
1:1	Full-size (12 in. = 1 ft.)	1:1

Figure 5-14. Preferred metric scales recommended by ISO (International Standards Organization).

above 1:10 are most often used. See Unit 23 for more details.

Most metric drawings you will make are measured with a 1:1 metric scale. Each division is 1.0 mm, and the numbered lines are 10, 20, 30, etc., Figure 5-15.

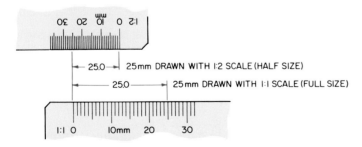

Figure 5-13. Graduations on a metric scale also indicate larger or smaller scale units of measure.

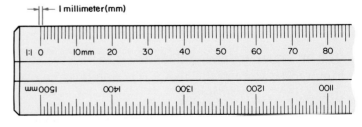

Figure 5-15. Section of 1:1 metric scale. Each division is equal to 1.0 mm.

To make a measurement of 52.5 mm, for example, it is a simple matter to come out to the 50 division, then add 2.5 mm more, Figure 5-16.

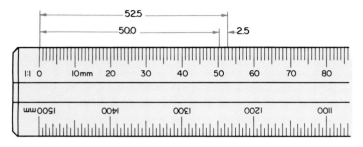

Figure 5-16. Making a measurement of 52.5 mm.

A section of a 1:1 scale is shown in Figure 5-17. How many measurements can you read correctly?

Making Measurements

To make a measurement, observe the scale from directly above. Mark the desired measurement on the paper by using a light perpendicular line made with a sharp pencil, Figure 5-18.

Keep the scale clean. Do not mark on it or use it as a straightedge.

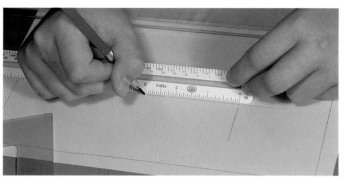

Figure 5-18. When making a measurement, observe the scale from directly above. Mark the desired measurement on the paper using a light perpendicular line made with a sharp pencil.

How to Draw Lines with Instruments

Care must be taken when drawing lines of the same type. Lines must be uniform in width and darkness throughout their entire length.

When lines are drawn using instruments, hold the pencil perpendicular to the paper, and inclined at an angle of about 60 degrees in the direction the line is being drawn. To keep the lines uniform in weight, especially if a long line is being drawn, rotate the pencil as you draw. Rotating the pencil will keep the point sharp.

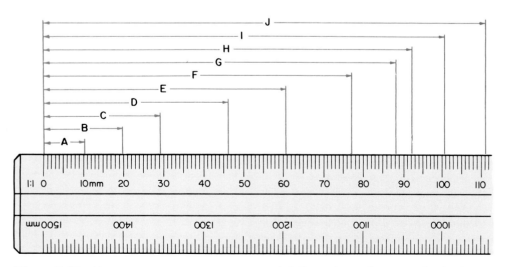

Figure 5-17. How many can you answer correctly? On a piece of paper, write the letters A to J. After each letter, write the correct answer. Ask your instructor to check your answers.

Horizontal lines are drawn with a T-square, Figure 5-19. The lines are drawn from left to right. Hold the T-square head firmly against the *left* edge of the drawing board (left-handed drafters will use the *right* edge and draw lines from right to left). See Figure 5-20.

Figure 5-21. Vertical lines are drawn using a triangle. The lines are drawn from the bottom to the top of the sheet. Incline the pencil in the direction of travel.

Figure 5-19. Use the T-square to draw horizontal lines. Note how the T-square head is held against the edge of the drawing board. Incline the pencil about 60 degrees in the direction the line is being drawn.

direction of the slope. Lines that incline to the left are drawn more easily from the top downward, Figure 5-22. Those that incline to the right should be drawn from the bottom upward, Figure 5-23.

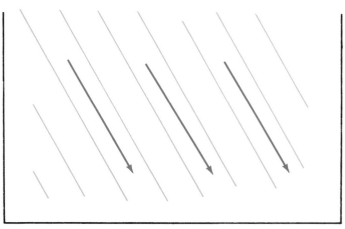

Figure 5-22. Lines that incline to the left are drawn from the top down.

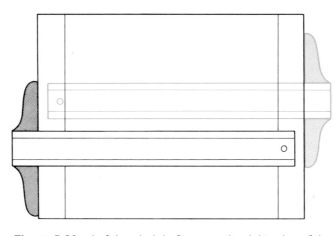

Figure 5-20. Left-handed drafters use the right edge of the drawing board.

Vertical lines are drawn using a triangle and are drawn from bottom to the top of the sheet, Figure 5-21. The base of the triangle must rest on the blade of the T-square. Use your left hand to hold the triangle in place.

Inclined lines are lines that are not vertical or horizontal. The drawing procedure depends on the

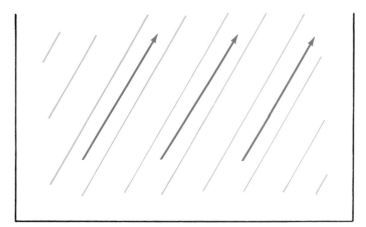

Figure 5-23. Lines that incline to the right are drawn from the bottom up.

How to Draw a Line Perpendicular to a Given Line Using Instruments

To draw a line perpendicular to a given line using instruments, place the hypotenuse (long edge) of any triangle parallel to the given line, Figure 5-24. Support the triangle on the T-square or another triangle. Rotate the first triangle around the 90 degree corner to draw a line perpendicular to the given line.

Another technique used to draw a line perpendicular to a given line requires that you place either leg of the triangle parallel to the given line, Figure 5-25. Support this triangle's hypotenuse side on the T-square or another triangle. Slide the first triangle on the supporting triangle or T-square, then draw the line. It will be perpendicular to the given line.

How to Erase

When drawing, every effort should be made to prevent mistakes. However, even the best drafter must occasionally make changes on a drawing. This will require erasing.

Some suggestions to follow when erasing:

1. Keep your hands and instruments clean. This will help to keep "smudges" to a minimum.

2. Use an *erasing shield* whenever possible, Figure 5-26. Select an opening that will expose

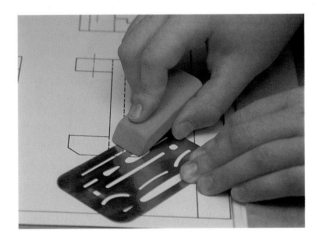

Figure 5-26. An erasing shield protects the area around where the correction or erasure is being made.

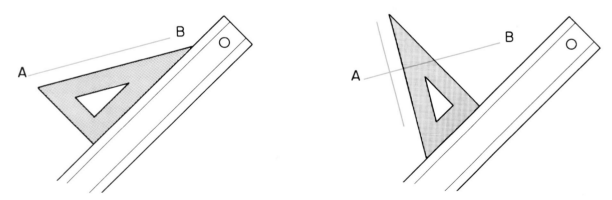

Figure 5-24. How to draw a line perpendicular to a given line using a T-square and triangle. Be sure to hold the base (T-square) firmly in place when the triangle is moved.

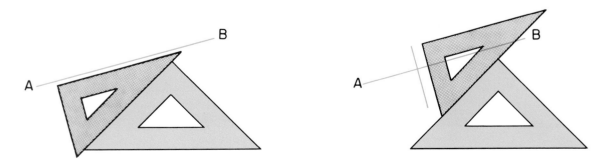

Figure 5-25. How to draw a line perpendicular to a given line using two triangles.

only the area to be erased. The shield will protect the rest of the drawing while the erasure is made.

3. Clean all eraser crumbs from the board immediately after making an erasure. Remove them with a brush (preferred) or clean cloth, *not your hands.*

4. Hold the paper with your free hand when erasing. This will keep the paper from wrinkling.

5. Erasing will remove the lead but will not take the pencil grooves from the paper. Avoid deep, wide grooves by first blocking in all views with light construction lines.

How to Use a Compass

In general drafting work, circles and arcs are drawn with a compass. Care must be taken so that the line drawn with the compass is the same weight as the line produced with the pencil. To accomplish this, it is recommended that the compass lead should be one or two grades softer than that of the pencil.

Sharpen the lead and adjust the point as shown in Figure 5-27. Do not forget to adjust the point after each sharpening.

Figure 5-28. Adjust the compass to the desired radius size on a measured line. Never set it to size directly on the scale.

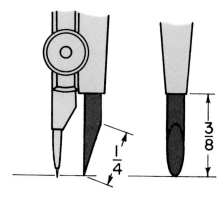

Figure 5-27. Sharpening and adjusting the compass lead. Readjust point each time the lead is sharpened.

To *set a compass*, draw a line on scrap paper that is equal in length to the desired radius. Adjust the compass on this line, Figure 5-28. Avoid setting a compass on the scale. The point will eventually ruin the division lines on the scale.

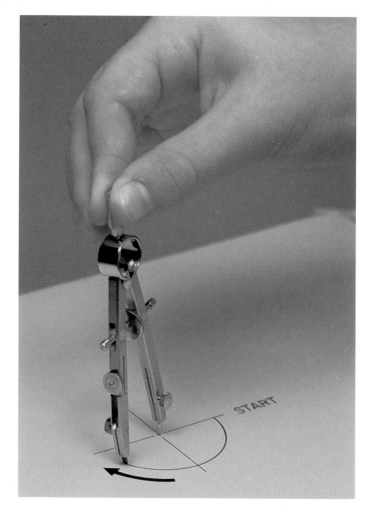

Figure 5-29. Drawing a circle. Note that the compass is inclined in the direction of rotation. Start drawing the circle on a centerline if one is available.

To ***draw a circle***, rotate the compass in a clockwise direction, Figure 5-29. Incline the tool in the direction of rotation. Start and complete the circle on a centerline. When drawing a series of concentric circles (circles with the same center), draw the smallest circle first.

Attaching a Drawing Sheet to the Board

The drawing sheet should be attached to the board with drafting tape. Tape is preferred because it does not damage the board. Before attempting to attach the paper, remove all eraser crumbs.

To attach the paper, place the sheet on the board as shown in Figure 5-30. Left-handed drafters should use the right-hand portion of the board.

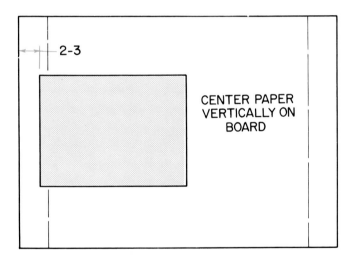

Figure 5-30. Locating the drawing sheet on the board.

Place the T-square on the board with the head firmly against the left edge. Slide it up until the top of the blade is in line with the top edge of the drawing sheet, Figure 5-31. Position the sheet so the top edge is parallel with the T-square blade. Fasten the sheet to the board. Larger sheets may also require fasteners on the bottom corners.

The following procedure is recommended when positioning and attaching A-size (8 1/2 × 11

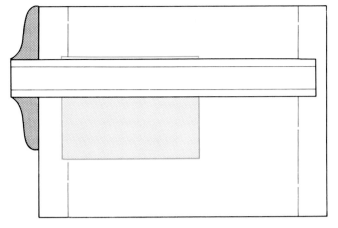

Figure 5-31. Aligning the top of the drawing sheet with T-square.

in.) and A4 (210 × 297 mm) size drawing sheets. Position the sheet on the board as shown in Figure 5-30.

Place the T-square head firmly against the left edge of the board. Align the drawing sheet by sliding the T-square blade until it contacts the bottom edge of the paper, Figure 5-32. Align the sheet with this edge. Fasten the sheet to the board.

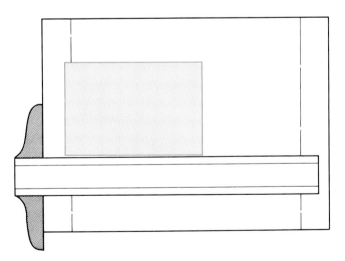

Figure 5-32. The drawing sheet can also be aligned on the board by placing the bottom edge of the sheet on the T-square blade.

Lightweight paper, like tracing vellum, is slightly more difficult to attach to the board. It has a tendency to wrinkle. This paper is aligned using the sequence shown in Figure 5-33.

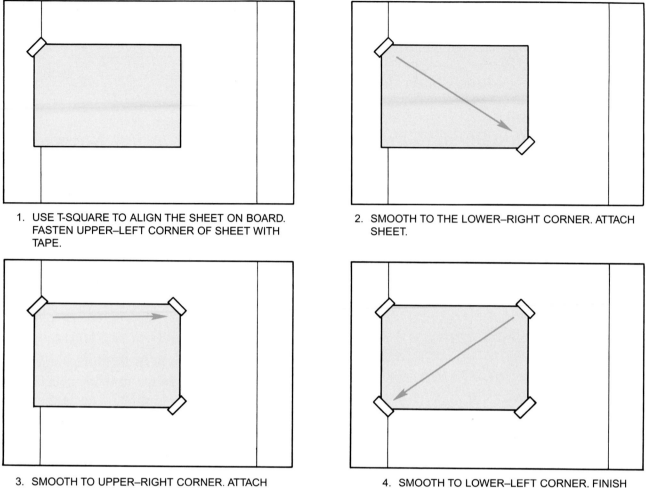

1. USE T-SQUARE TO ALIGN THE SHEET ON BOARD. FASTEN UPPER–LEFT CORNER OF SHEET WITH TAPE.

2. SMOOTH TO THE LOWER–RIGHT CORNER. ATTACH SHEET.

3. SMOOTH TO UPPER–RIGHT CORNER. ATTACH SHEET.

4. SMOOTH TO LOWER–LEFT CORNER. FINISH ATTACHING SHEET TO BOARD.

Figure 5-33. Sequence recommended for attaching lightweight papers (such as tracing vellum) to the board.

Figure 5-34. Preprinted drawing sheets save a great deal of time for the drafter.

Drafting Sheet Format

Most drafting rooms use a standard format in the layout of their drawing sheets. In general, the format consists of the border, title block, and standards notes, Figure 5-34. A border is included to define the drawing area of the sheet. The title block and standard notes provide information that is necessary for the manufacture and/or assembly of the object described on the drawing sheet. A preprinted title block is shown in Figure 5-35.

Most industrial firms employ standard drawing sheet sizes. Drawings made on standard sheet sizes are easier to file. They also present less difficulty when prints are made from them.

With few exceptions, the drawings in this text should be drawn on 8 1/2 × 11 in., or 11 × 17 in. size sheets. The small sheet is known as an A-size sheet, and the large sheet as a B-size sheet. Corresponding metric sheet sizes would be A4 (210 × 297 mm) and A3 (297 × 420 mm) size sheets.

Plan your work carefully. Avoid using a large sheet when a smaller size sheet will do.

Recommended Drawing Sheet Formats

The following drawing sheet formats are suggested. The first is similar to the sketching sheet format, Figure 5-36. It should be used when class time is limited.

A more formal sheet format is prepared as follows. Put a 1/2 in. (12.5 mm) border on the sheet, Figure 5-37. Use a short light pencil stroke, not a dot, as the guide for locating the border line. The

UNLESS OTHERWISE SPECIFIED DIMENSIONS ARE IN INCHES TOLERANCES ON FRACTIONS ± 1/64 DECIMALS ± 0.010 ANGLES ± 1°	DRAWN BY	WALKER INDUSTRIES	
MATERIAL	DATE	TITLE	
	CHK'D		
	HEAT TREATMENT	SCALE	DRAWING NO.
		SHEET	

Figure 5-35. Preprinted title block. When it is used, only the border has to be drawn.

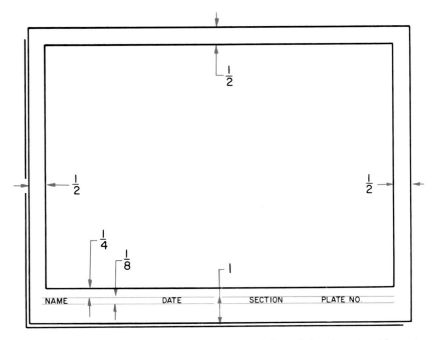

Figure 5-36. How to start laying out a more informal drawing sheet format.

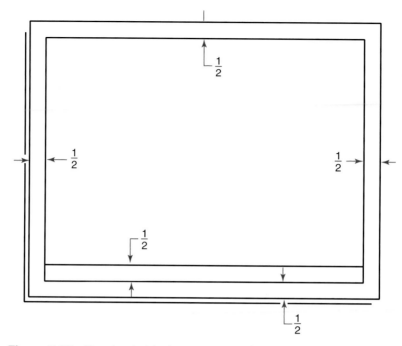

Figure 5-37. How to start laying out a more formal drawing sheet format.

short light guide mark should be covered when the border line is drawn. Allow another 1/2 in. (12.5 mm) for the title block. Divide the title block as shown in Figures 5-38 and 5-39. Guide lines for the title block are drawn, Figure 5-40. They should be drawn very lightly. Letter in the necessary information, Figure 5-41.

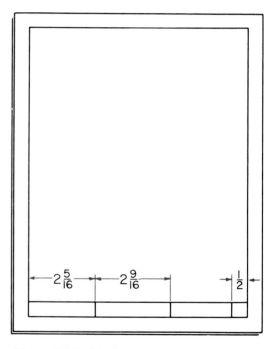

Figure 5-39. Vertical drawing sheet format.

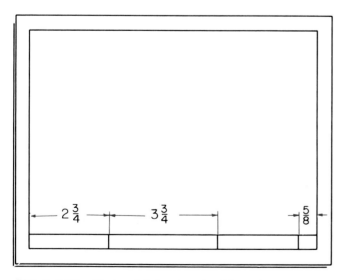

Figure 5-38. Horizontal drawing sheet format.

It is recommended that modern duplicating equipment be used to prepare preprinted drawing sheets. The sheet format and the title block should be designed to meet the specific requirements of the drafting department and the industry. Prescribed tolerances can be printed on the sheets. Preprinted drafting sheets will save a great deal of time by eliminating repetitive and time-consuming drawing operations.

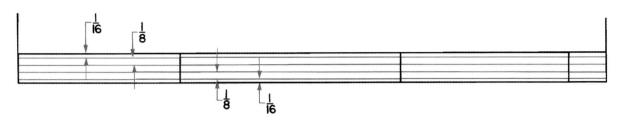

Figure 5-40. Layout of guide lines needed for lettering.

| NAME OF YOUR SCHOOL | TITLE OF DRAWING | YOUR NAME   SECTION   DATE | DRAWING NUMBER |
| SCHOOL LOCATION | | SCALE   CHECKED BY | 5-3 |

Figure 5-41. Suggested information to be lettered on drawing sheets for problems in this text.

Drafting Vocabulary

Alphabet of lines
Architect's scale
Border line
Centerline
Characteristic
Concentric
Construction lines
Cutting-plane line
Dimension

Dimension line
Divisions
Engineer's scale
Extension line
Graduations
Hidden line
Hypotenuse
Line conventions
Mechanical drafter's

Metric scale
Object line
Perpendicular
Phantom line
Preprinted
Scale
Scale clip
Section lines
Visible line

Test Your Knowledge—Unit 5

Please do not write in the book. Place your answers on another sheet of paper.

1. Identify these lines:

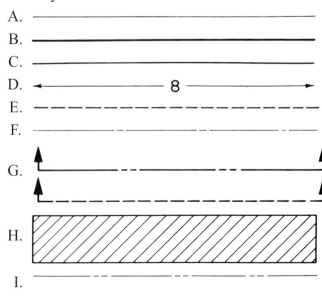

2. In drafting room language, the characteristics of the lines in question 1 and their correct use are known as _____ _____ _____.
3. Why is it important for a drafter to be able to measure accurately?
4. In drafting, the term "scale" has two meanings. What are they?
5. Prepare sketches which show four different shapes of scales available.
6. List three types of scales in common use.
7. In drafting, horizontal lines are drawn using the _____.
8. Vertical lines are drawn using _____.
9. When erasing, the _____ is often used to protect surrounding areas.
10. Circles and arcs are drawn with a _____.
11. The recommended method of attaching drawing sheets is _____ _____.

Outside Activities

1. Secure samples of drawing sheet formats used by industry. After studying these industrial examples, design a sheet format for the school's drafting department.
2. Obtain a compass and demonstrate to the class the proper way to sharpen the lead, adjust the length of the point, set the compass to the proper radius, and draw a circle.

Problems

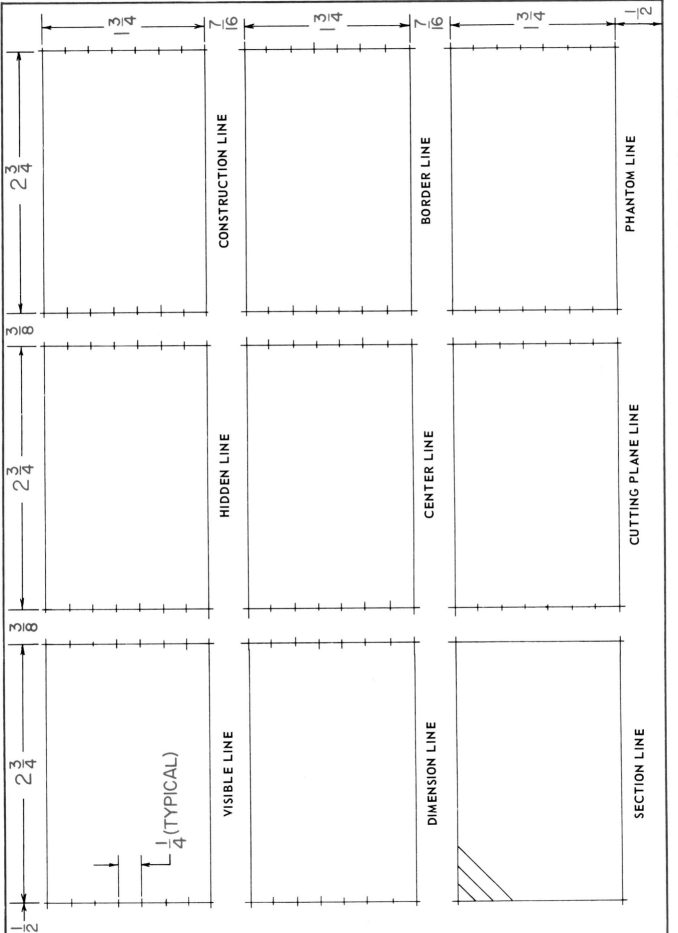

Problem Sheet 5-1. ALPHABET OF LINES. Duplicate this drawing on a separate sheet of paper. Construct the nine different lines called for.

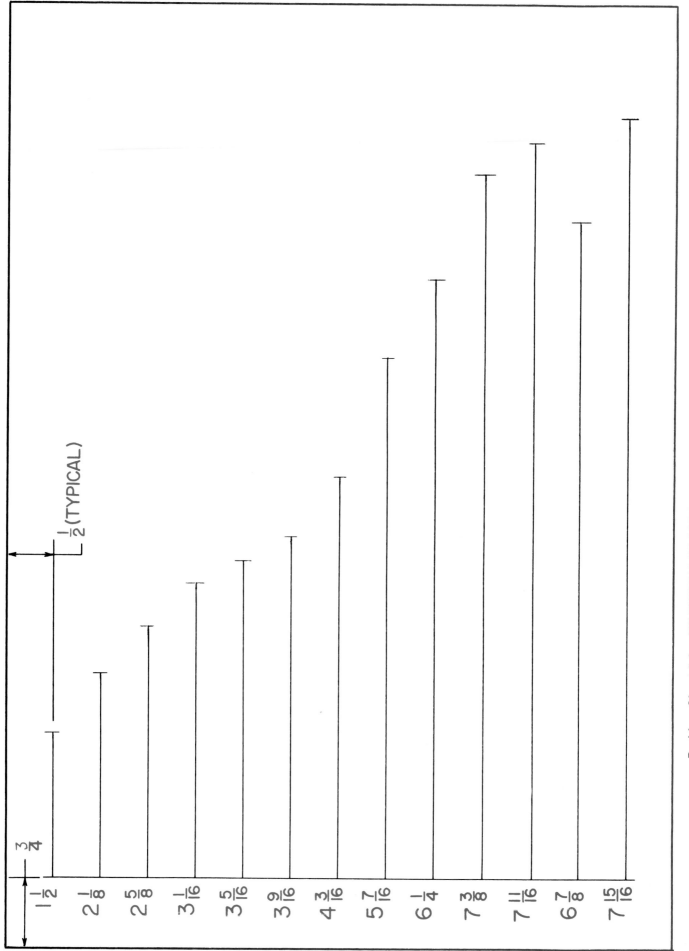

$\dfrac{3}{4}$

$\dfrac{1}{2}$ (TYPICAL)

$1\dfrac{1}{2}$

$2\dfrac{1}{8}$

$2\dfrac{5}{8}$

$3\dfrac{1}{16}$

$3\dfrac{5}{16}$

$3\dfrac{9}{16}$

$4\dfrac{3}{16}$

$5\dfrac{7}{16}$

$6\dfrac{1}{4}$

$7\dfrac{3}{8}$

$7\dfrac{11}{16}$

$6\dfrac{7}{8}$

$7\dfrac{15}{16}$

Problem Sheet 5-2. MEASURING PRACTICE. Duplicate this drawing on a separate sheet of paper.

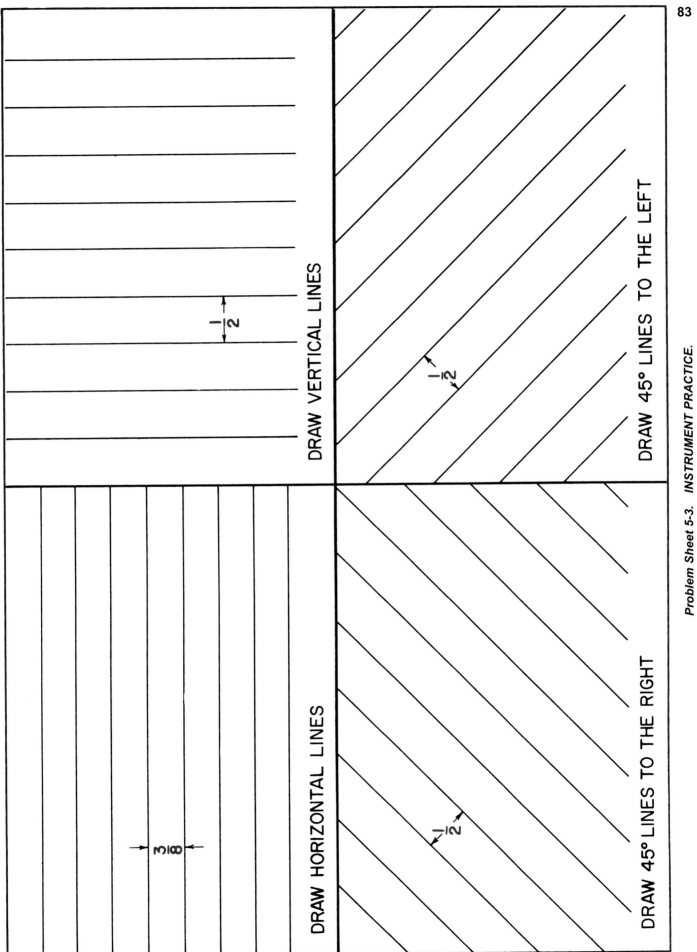

DRAW VERTICAL LINES

DRAW 45° LINES TO THE LEFT

DRAW HORIZONTAL LINES

DRAW 45° LINES TO THE RIGHT

$\frac{1}{2}$

$\frac{1}{2}$

$\frac{3}{8}$

$\frac{1}{2}$

Problem Sheet 5-3. INSTRUMENT PRACTICE.

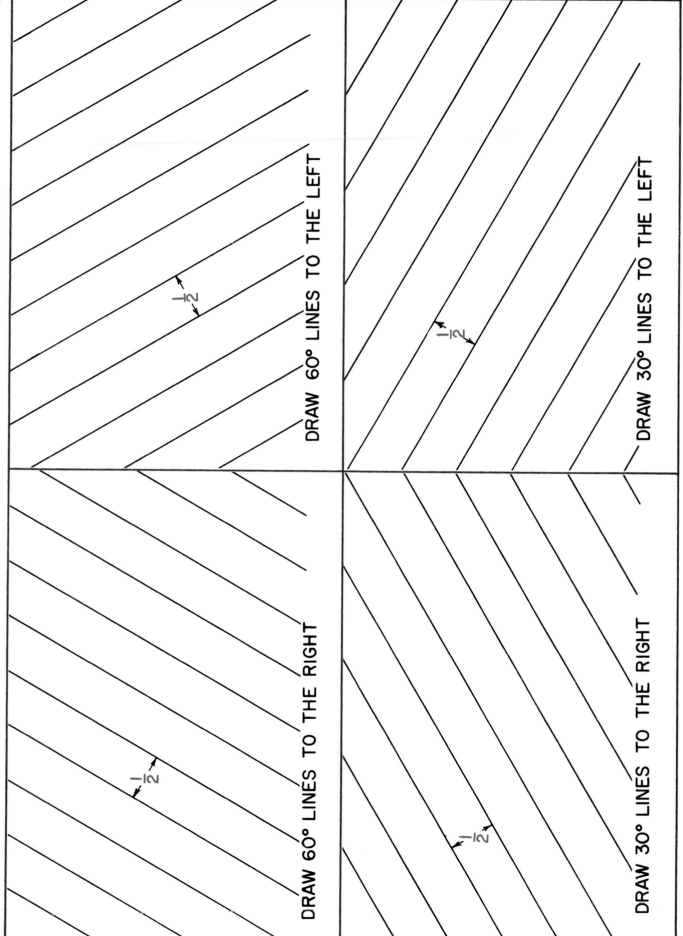

84

DRAW 60° LINES TO THE LEFT

$\frac{1}{2}$

DRAW 30° LINES TO THE LEFT

$\frac{1}{2}$

DRAW 60° LINES TO THE RIGHT

$\frac{1}{2}$

DRAW 30° LINES TO THE RIGHT

$\frac{1}{2}$

Problem Sheet 5-4.   INSTRUMENT PRACTICE.

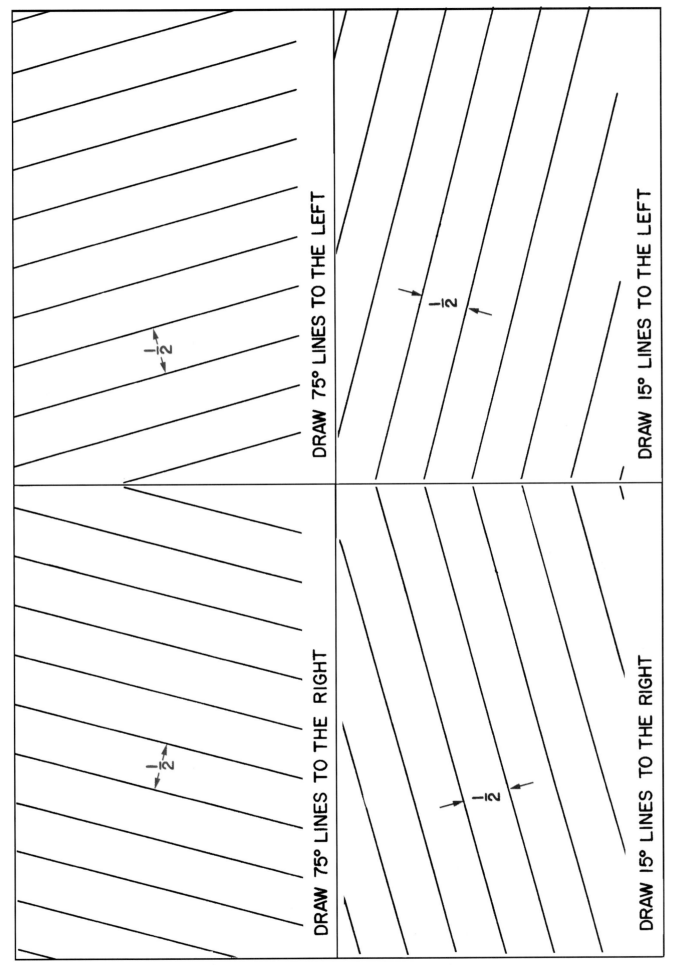

DRAW 75° LINES TO THE LEFT

DRAW 75° LINES TO THE RIGHT

DRAW 15° LINES TO THE LEFT

DRAW 15° LINES TO THE RIGHT

Problem Sheet 5-5. INSTRUMENT PRACTICE.

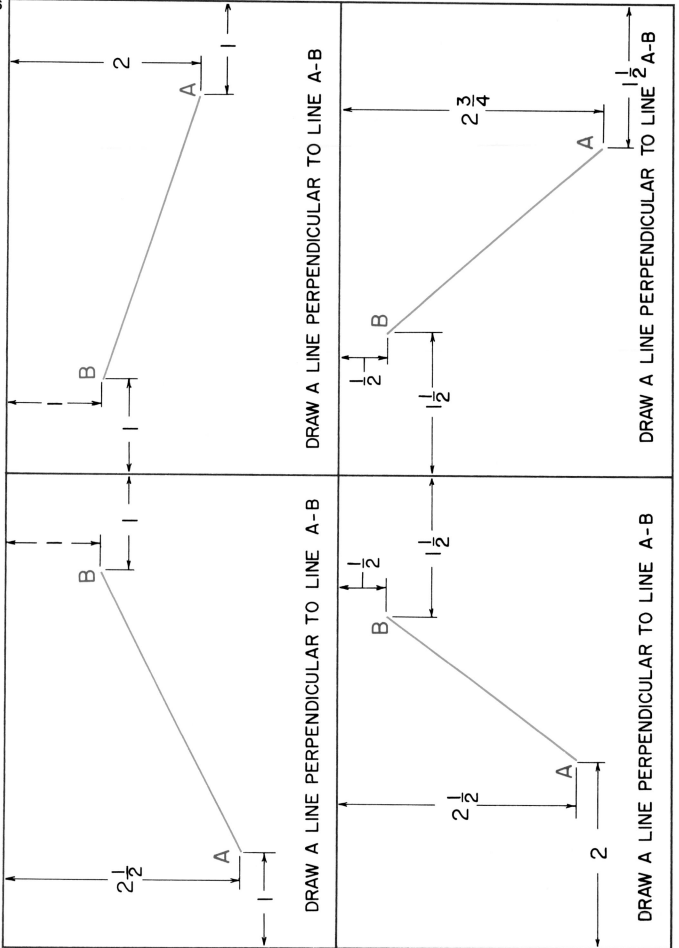

DRAW A LINE PERPENDICULAR TO LINE A-B

DRAW A LINE PERPENDICULAR TO LINE A-B

DRAW A LINE PERPENDICULAR TO LINE A-B

DRAW A LINE PERPENDICULAR TO LINE A-B

Problem Sheet 5-6. INSTRUMENT PRACTICE.

Basic Geometric Construction

After studying this unit and completing the assigned problems, you should be able to:

Identify basic geometric shapes.

Bisect a line, arc, and angle.

Transfer an angle and construct an equilateral triangle.

Draw a square with a side given.

Construct a pentagon, a hexagon, and an octagon.

Draw an arc tangent to two lines.

Draw an arc tangent with a straight line and a given arc and draw tangent arcs.

Divide a line into a given number of equal divisions.

Construct an ellipse by several methods.

Whether we realize it or not, we see geometry in use every day, Figure 6-1. The design of the aircraft that flies overhead and the automobile that passes on the street is based on geometrics. Buildings and bridges utilize squares, rectangles, triangles, circles, and arcs in their construction. Every drawing employed to manufacture a structure or product is composed of one or more geometric shapes.

Geometry is the basis of all computer-generated drawings, Figure 6-2. The first stage in developing a design is to generate a mesh or wireframe model of the design using basic geometric shapes.

It is important that you acquire the ability to visualize and draw the basic geometric shapes presented in this Unit, Figure 6-3. This will aid you in solving drafting problems. It will also provide you an opportunity to improve your skill with drafting instruments.

Solving Basic Geometric Problems

Size does *not* enter into the solution of most geometric problems. For that reason, no dimensions are given for the solution of the basic geometric problems. Most of them require little space in their solution; therefore, several problems may be included on a single drawing sheet. A suggested sheet layout is shown in Figure 6-4.

Drawing sheets can be made more interesting and attractive if the geometric figures are constructed with colored pencils.

Figure 6-1. Geometric shapes are found in the design of most products and structures. Some are more obvious than others. How many geometric figures can you identify in the above photos? (Lockheed Advanced Development Company and Champion Spark Plug Company)

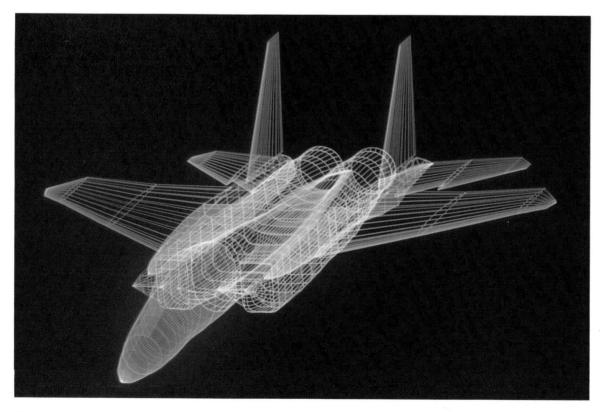

Figure 6-2. Geometry is the basis of all computer-generated drawings. This is a "wireframe" model of a proposed aircraft. (Evans & Sutherland)

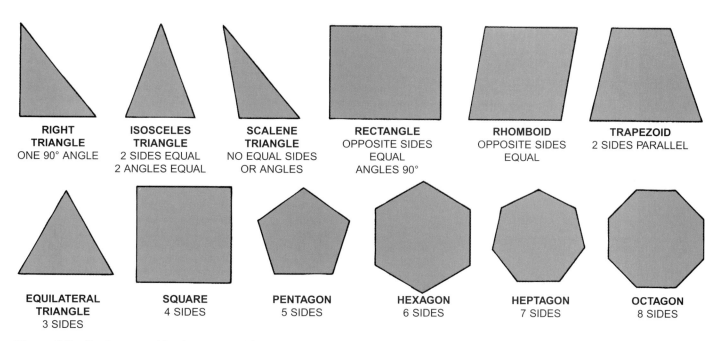

Figure 6-3. Basic geometric shapes on which drawings are based.

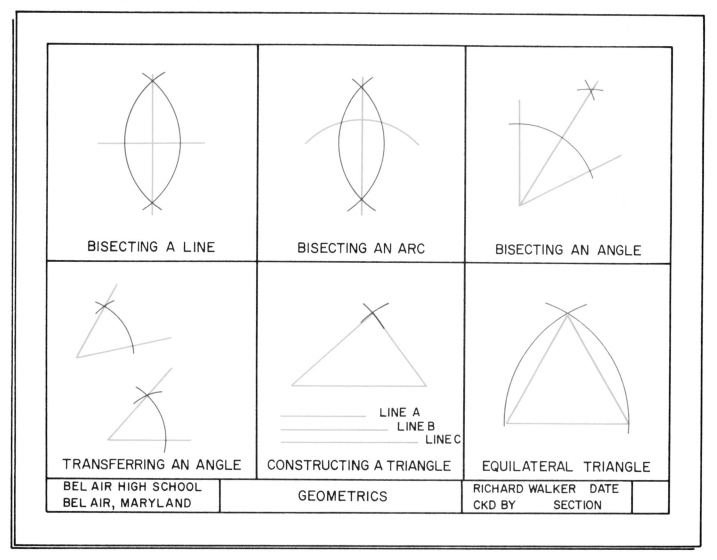

Figure 6-4. Suggested sheet layout for drawing basic geometric problems. Several figures may be placed on a single sheet.

How to Bisect or Find the Middle of a Line

Using this method, the bisecting line will be at a right angle (90 degrees) to the given line.

1. Let line A-B be the line to be bisected.
2. Set your compass to a distance larger than one-half the length of the line to be bisected. Using this setting as the radius and the end of the line at A as the center point, draw arc C-D. Using the same compass setting but the end of the line at B as the center point, draw arc E-F.
3. Draw a line through the points where the arcs intersect. This line will be at a right angle (90 degrees or perpendicular) to and bisect the original line A-B.

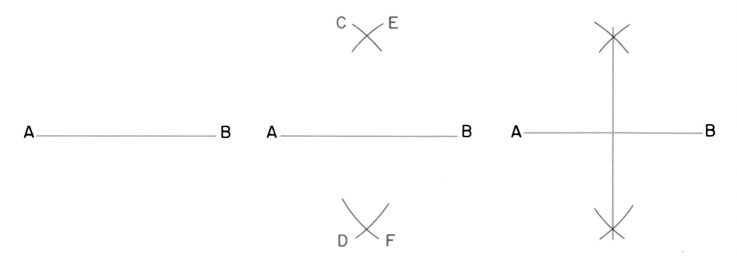

How to Bisect an Arc

1. Let arc A-B be the arc to be bisected.
2. Set your compass to a distance larger than one-half the length of the arc to be bisected. Using this setting as the radius and the end of the arc at A as the center point, draw arc C-D. Using the same compass setting but the end of the arc at B as the center point, draw arc E-F.
3. Draw a line through the points where the arcs intersect. This line will bisect the original arc A-B.

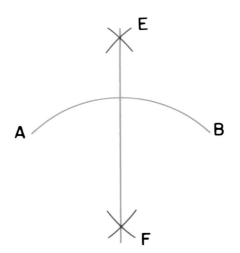

How to Bisect an Angle

1. Let the angle formed by lines A-B and B-C be the angle to be bisected.
2. With B as the center, draw an arc intersecting the angle at D and E.
3. Using a compass setting greater than one-half D-E as centers, draw intersecting arcs. A line through this intersection and B will bisect the angle.

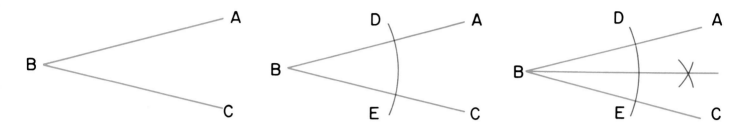

How to Transfer or Copy an Angle

1. Let the angle formed by lines A-B and B-C be the angle to be transferred or copied.
2. Locate the new position of the angle and draw line A'-B'.
3. With B as the center point, draw an arc of any convenient radius on the given angle. This arc intersects the given angle at D and E.
4. Using the same radius and B' as the center draw arc D'-E'.
5. Set your compass equal to D-E. With point D' as a center and D-E as the radius, strike an arc which intersects the first arc at point E'.
6. Draw a line through the intersecting arcs to complete the transfer of the given angle.

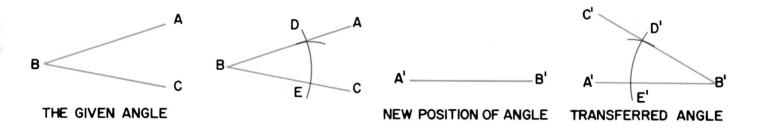

THE GIVEN ANGLE   NEW POSITION OF ANGLE   TRANSFERRED ANGLE

How to Construct a Triangle from Given Line Lengths

1. Let lines A, B, and C be the sides of the required triangle.
2. Draw a line that is equal in length to line A.
3. Set your compass to a length equal to line B. Use one end of line A as a center and strike an arc. Reset your compass to a length equal to line C and with the other end of line A as a center, strike another arc.
4. Connect the ends of line A to the points where the two arcs intersect.

LINE C

LINE B

LINE A

How to Construct an Equilateral Triangle

An *equilateral triangle* is one having all sides equal in length and all angles are equal.

1. Let line A-B be the length of the sides of the triangle.
2. With A as the center and with the compass setting equal to the length of line A-B, strike the arc B-C. Using the same compass setting, but with B as the center, strike the arc A-D. These arcs intersect at E.
3. Complete the triangle by connecting A to E and B to E.

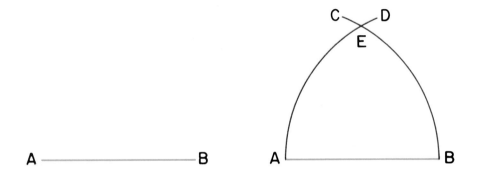
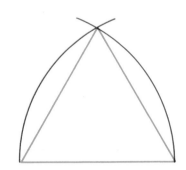

How to Draw a Square with the Diagonal Given

1. Draw a circle with a diameter equal to the length of the diagonal.
2. Connect the points where the centerlines intersect the circle.

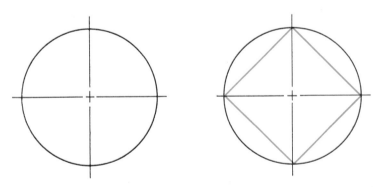

How to Draw a Square with the Side Given

1. Draw a circle with a diameter equal to the length of the side.
2. Draw tangents at a 45 degree to the centerlines.

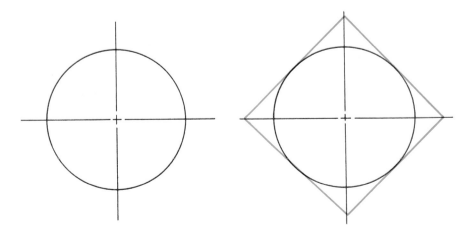

How to Construct a Pentagon or Five-Point Star

1. Draw a circle. Let A-C and B-D be the centerlines and O be the point where the centerlines intersect.

2. Bisect the line (radius) O-B. This will locate point E.

3. With E as the center and with the compass set to the radius E-A, strike the arc A-F.

4. The distance A-F is one-fifth the circumference of the circle. Set your compass or dividers to this distance and circumscribe the circle starting at A. Connect these points as a pentagon or as a five-point star.

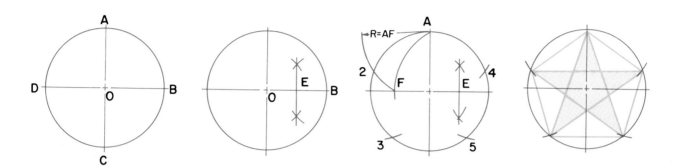

How to Construct a Hexagon (First method)

1. Draw a circle.

2. With a compass setting equal to the radius of the drawn circle:

   A. With point A as the center, swing an arc to locate points B and F. With point D as the center, swing an arc and locate points C and E.

   B. Start at point A and move around the circle locating points B, C, D, E, and F in sequence.

3. Connect the points with a straightedge to complete the hexagon.

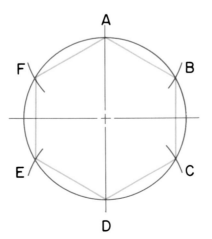

How to Construct a Hexagon (Second method)

1. Draw a circle.
2. Use the 30-60 degree triangle and draw construction lines at 60 degrees to the vertical centerline and tangent to the circle.
3. Draw two vertical construction lines tangent to the circle. Fill in the construction lines with visible object lines to complete the hexagon.
4. The same basic technique can be used to draw (inscribe) a hexagon inside the circle.

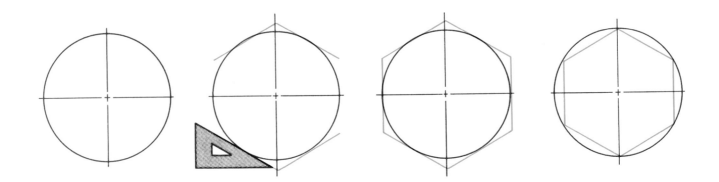

How to Draw an Octagon Using a Circle

1. Draw a circle with a diameter equal to the distance across the flats of the desired octagon.
2. Draw the vertical and horizontal lines tangent to the circle. Use construction lines.
3. Complete the octagon by drawing the 45 degree angle lines tangent to the circle. Fill in the construction lines with object lines.

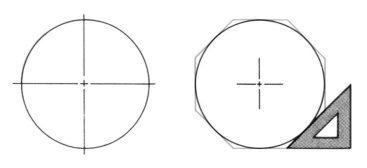

How to Draw an Octagon Using a Square

1. Draw a square with the sides equal in length to the distance across the flats of the required octagon. Draw the diagonals A-C and B-D. The diagonals intersect at O.

2. Set your compass to radius A-O and with corners A, B, C, and D as centers, draw arcs that intersect the square at points 1, 2, 3, 4, 5, 6, 7, and 8.

3. Complete the octagon by connecting point 1 to 2, 2 to 3, 3 to 4, 4 to 5, 5 to 6, 6 to 7, 7 to 8, and 8 to 1 with object lines.

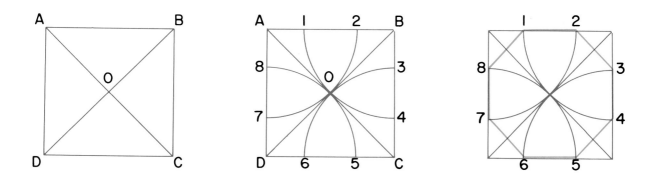

How to Draw an Arc Tangent to Two Lines at a Right Angle

1. Let A-B and B-C be the lines that form the right angle.

2. Set your compass to the radius of the required arc and using B as the center, strike arc D-E. With D and E as centers and with the same compass setting, draw the arcs that intersect at O.

3. With O as the center and with the compass at the same setting, draw the required arc. It will be tangent to lines A-B and B-C at points D and E.

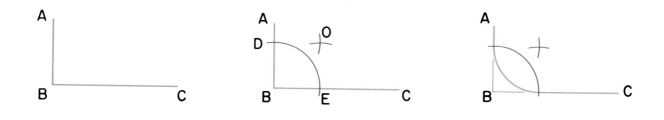

How to Draw an Arc Tangent to Two Straight Lines

1. Let lines A-B and C-D be the two straight lines.
2. Set your compass to the radius of the arc to be drawn tangent to the two straight lines and with points near the ends of the lines A-B and C-D as centers, strike two arcs on each line.
3. Draw straight construction lines tangent to the arcs.
4. The point where the two lines intersect (O) is the center for drawing the required arcs.

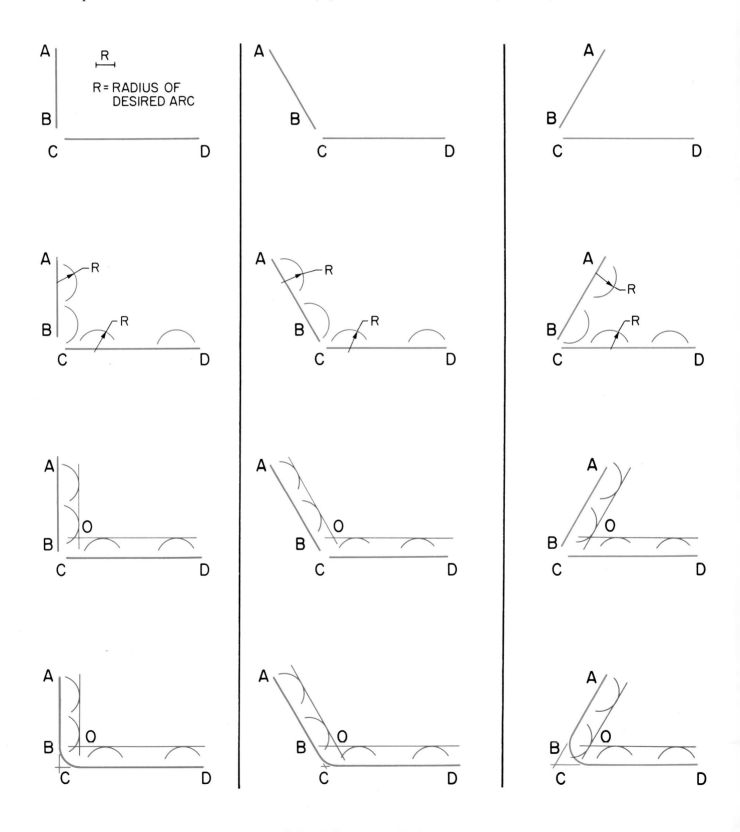

How to Draw an Arc Tangent with a Straight Line and a Given Arc

1. Draw the given arc r and straight line a-b in proper relation to one another.
2. Draw line A-B a distance equal to the radius (R) of the desired arc from and parallel to the straight line a-b.
3. Draw arc C-D by setting your compass to a radius equal to R + r. This arc intersects with line A-B at point O.
4. Using point O as the center and with the compass set to the desired radius R, draw the required arc. It will be tangent to the given arc and to the straight line a-b. Fill in construction lines with object lines.

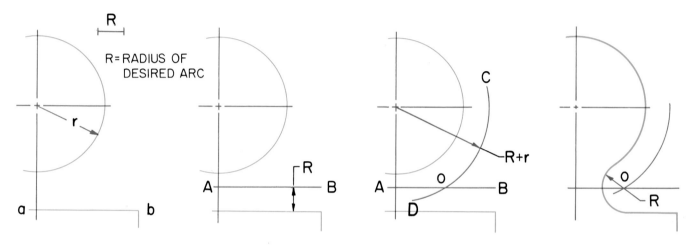

How to Draw Tangent Arcs

1. Draw the two arcs (ra and rb) that are to be joined by the tangent arc. Let O and X be the centers of these arcs.
2. Set the compass to a distance equal to the radius R + ra. Using X as the center draw arc A-B. Reset the compass to a distance equal to the radius R + rb. Using O as the center draw arc C-D. These arcs intersect at Y.
3. Set the compass to the radius of the required arc R. With Y as the center draw the desired arc. This arc will be tangent to the given arcs.

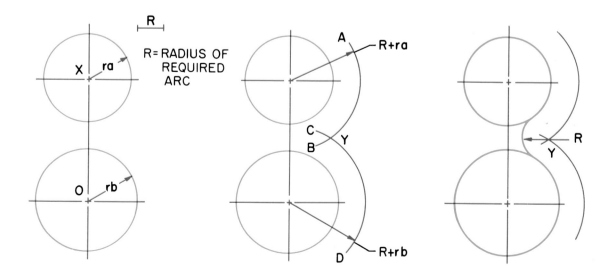

How to Divide a Line into a Given Number of Equal Divisions

1. Problem: Divide line A-B into five (5) equal parts.
2. With the T-square and triangle draw vertical line B-C. Use construction lines.
3. Locate the scale with one point at A. Adjust the scale until a multiple of the divisions required (in this case five 1 in. divisions) lies between A and vertical line B-C.
4. Make vertical points at each of the five 1 in. divisions. Project vertical lines from these points parallel to line B-C. These vertical lines divide line A-B into five equal parts.

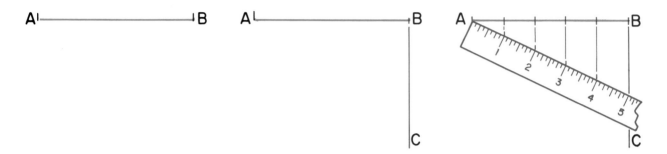

How to Draw an Ellipse Using Concentric Circles

1. Draw two concentric circles. The diameter of the large circle is equal to the length of the large axis of the desired ellipse. The diameter of the small circle is equal to the length of the small axis of the desired ellipse.
2. Divide the two circles into twelve equal parts. Use your 30-60 degree triangle.
3. Draw horizontal lines from the points where the dividing lines intersect the small circle. Vertical lines are drawn from the points where the dividing lines intersect the large circle.
4. Connect the points where the vertical and horizontal lines intersect with a French curve.

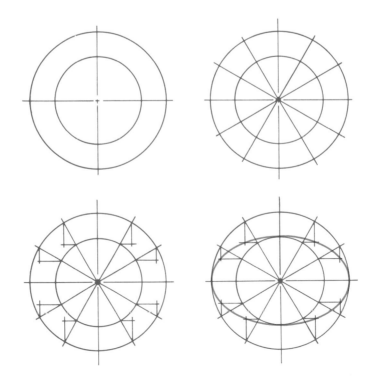

How to Draw an Ellipse Using the Parallelogram Method

This method is satisfactory for drawing large ellipses.

1. Let line A-B be the major axis and line C-D be the minor axis of the required ellipse.
2. Construct a rectangle with sides equal in length and parallel to the axes.
3. Divide A-O and A-E into the same number of equal parts.
4. From C draw a line to point 1 on line A-E. Draw a line from D through point 1 on line A-C. The point of intersection of these two lines will establish the first point of the ellipse. The remaining points in this section are completed as are similar points in the other three sections (quadrants).
5. Connect the points with a French curve to complete the ellipse.

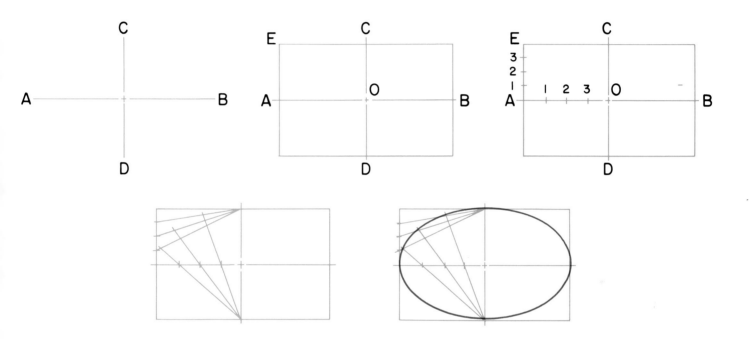

How to Construct an Ellipse Using the Four-Center Approximation Method

1. Given major axis A-B and minor axis C-D intersecting at point O.
2. Draw diagonal C-B. With point O as the center and O-C as the radius, strike an arc. The arc will intersect line O-B at E.
3. With the radius E-B, and using C as the center, strike an arc that intersects line C-B at F.
4. Construct a perpendicular bisector of F-B and extend it to intersect the major and minor axes at G and H.
5. Points G and H are the centers for two of the arcs needed to construct the ellipse. With O as the center, use a compass to locate points J and K. They are symmetrical with points G and H.
6. Draw a line from H extending through J and from K through G and J.
7. Using G and J as centers, strike arcs G-B and J-A.
8. With H and K as centers, strike arcs H-C and K-D.

These four arcs will be tangent to each other and form a four-center approximate ellipse.

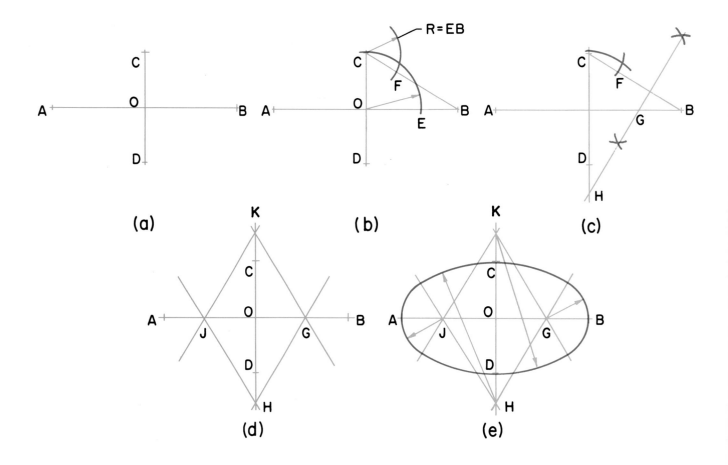

Drafting Vocabulary

Arc	Equilateral triangle	Parallelogram
Axis	Geometric	Pentagon
Bisect	Hexagon	Quadrant
Circle	Intersect	Radius
Circumscribe	Length	Rectangle
Concentric	Major axis	Square
Diagonal	Minor axis	Tangent
Diameter	Octagon	Triangle
Ellipse	Parallel	Vertical

Outside Activities

1. Prepare a bulletin board display that illustrates the use of geometric shapes in buildings and bridges.

2. Develop a list of everyday items that make use of geometric shapes. For example: Nut and bolt heads are round, square, and hexagonal.

3. Secure a photo or a drawing of a modern aircraft. Place a sheet of tracing vellum over it and sketch in the various geometric shapes used in its design.

4. Do the same using a photo or drawing of a late-model automobile.

5. Carefully examine a bicycle. Make a list of all the geometric shapes used in the design and construction of the bicycle. Be sure to look at all of the component parts.

Problems

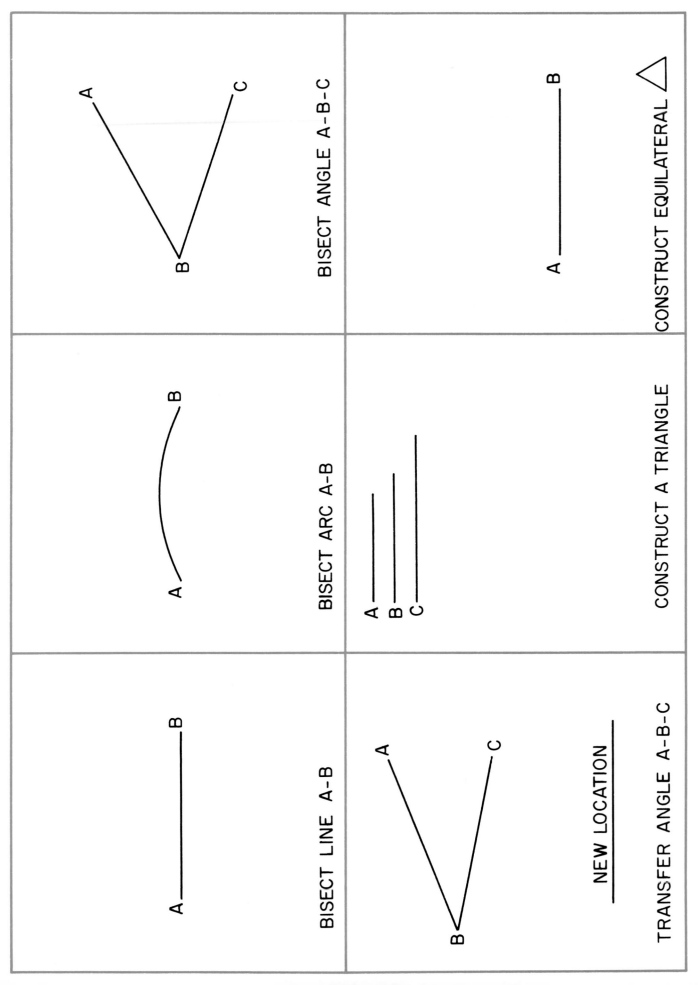

BISECT LINE A-B

BISECT ARC A-B

BISECT ANGLE A-B-C

TRANSFER ANGLE A-B-C

NEW LOCATION

CONSTRUCT A TRIANGLE

CONSTRUCT EQUILATERAL △

Problem Sheet 6-1. GEOMETRICS.

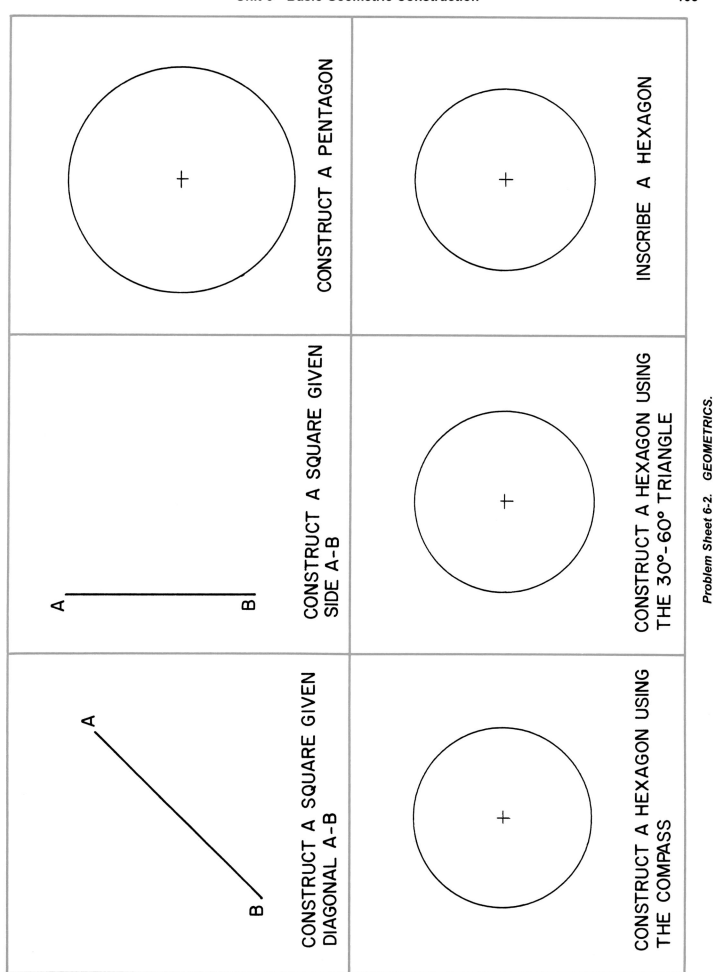

CONSTRUCT A PENTAGON

INSCRIBE A HEXAGON

CONSTRUCT A SQUARE GIVEN
SIDE A-B

CONSTRUCT A HEXAGON USING
THE 30°-60° TRIANGLE

CONSTRUCT A SQUARE GIVEN
DIAGONAL A-B

CONSTRUCT A HEXAGON USING
THE COMPASS

A

B

A

B

Problem Sheet 6-2.   GEOMETRICS.

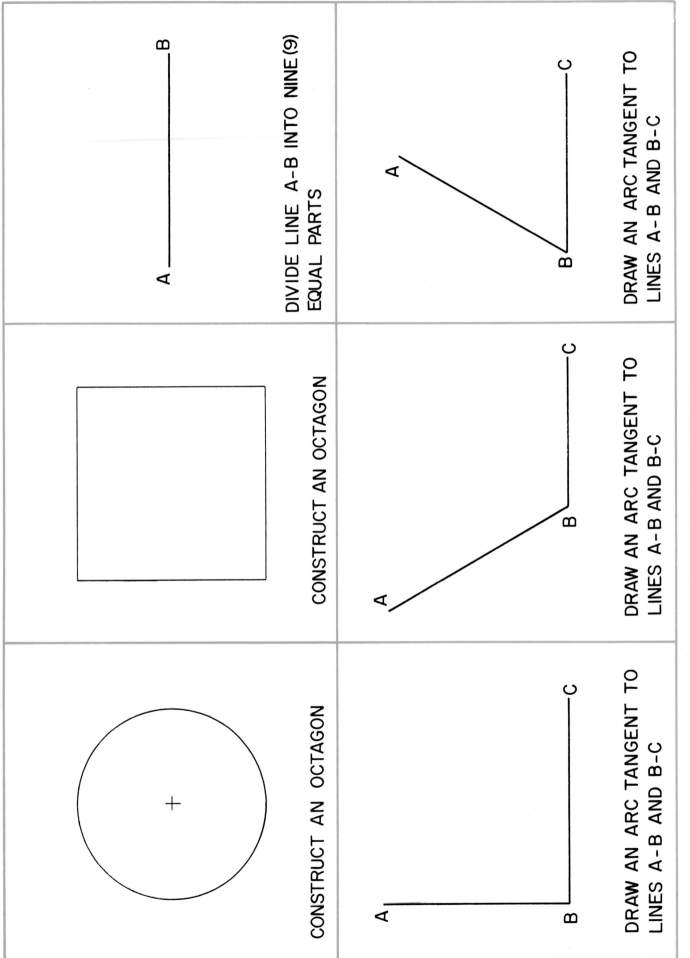

A ————————— B

DIVIDE LINE A-B INTO NINE (9) EQUAL PARTS

CONSTRUCT AN OCTAGON

CONSTRUCT AN OCTAGON

DRAW AN ARC TANGENT TO LINES A-B AND B-C

DRAW AN ARC TANGENT TO LINES A-B AND B-C

DRAW AN ARC TANGENT TO LINES A-B AND B-C

Problem Sheet 6-3. GEOMETRICS.

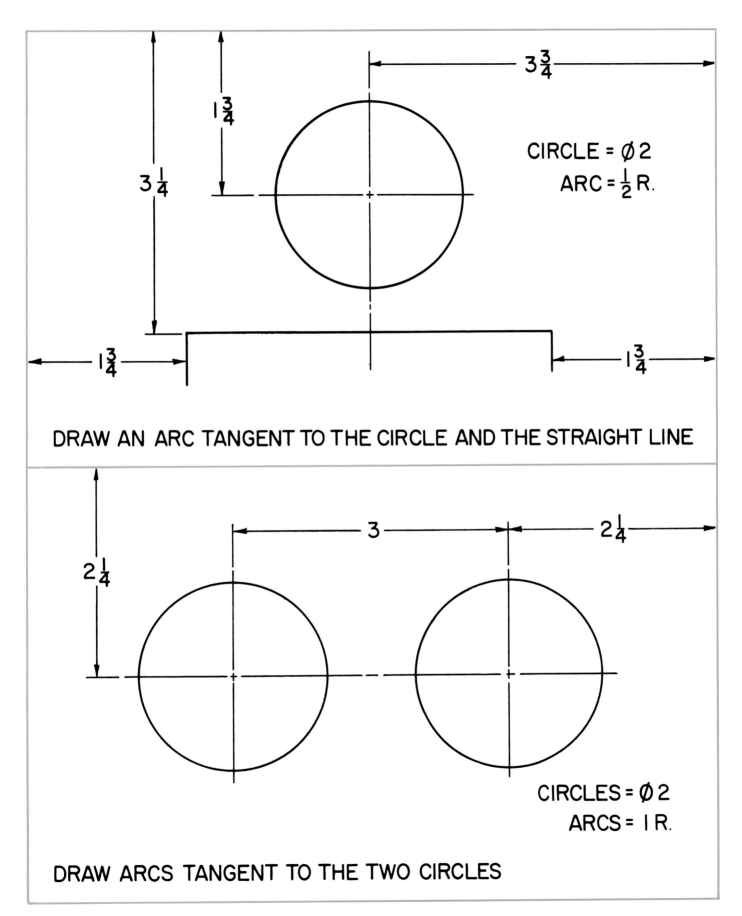

$3\frac{3}{4}$

$1\frac{3}{4}$

$3\frac{1}{4}$

CIRCLE = $\emptyset\,2$

ARC = $\frac{1}{2}$ R.

$1\frac{3}{4}$ $1\frac{3}{4}$

DRAW AN ARC TANGENT TO THE CIRCLE AND THE STRAIGHT LINE

$2\frac{1}{4}$ 3 $2\frac{1}{4}$

CIRCLES = $\emptyset\,2$

ARCS = 1 R.

DRAW ARCS TANGENT TO THE TWO CIRCLES

Problem Sheet 6-4.   Use a vertical sheet layout. Divide the work area as shown. Dimensions do not have to be put on your finished drawing.

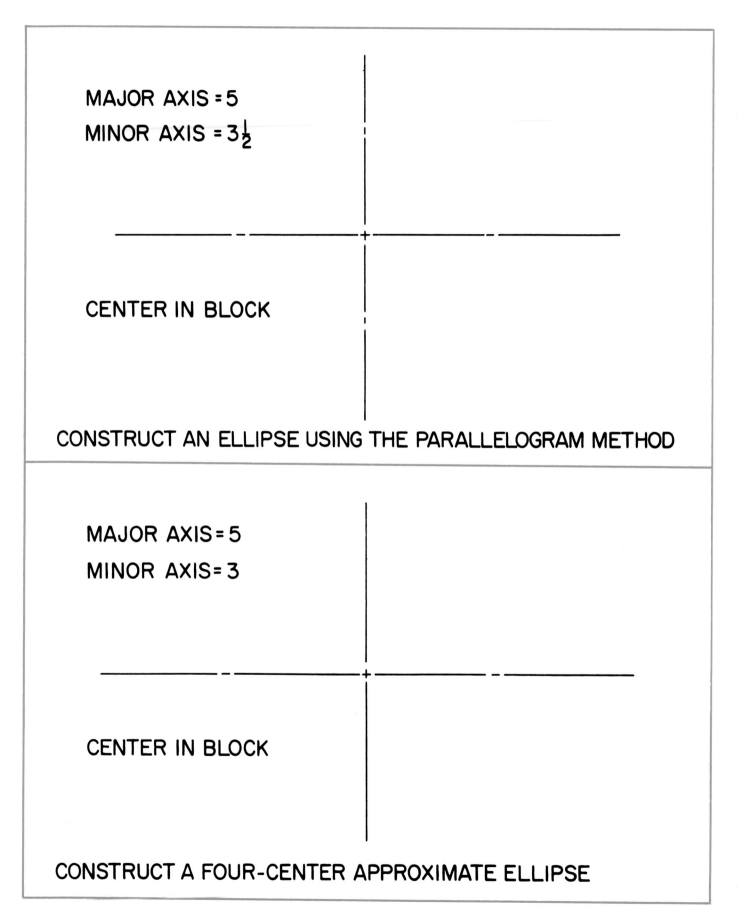

MAJOR AXIS = 5
MINOR AXIS = 3½

CENTER IN BLOCK

CONSTRUCT AN ELLIPSE USING THE PARALLELOGRAM METHOD

MAJOR AXIS = 5
MINOR AXIS = 3

CENTER IN BLOCK

CONSTRUCT A FOUR-CENTER APPROXIMATE ELLIPSE

Problem Sheet 6-5. Use a vertical sheet layout. Divide the work area as shown. Dimensions do not have to be put on your finished drawing.

AIRCRAFT INSIGNIA OF THE WORLD

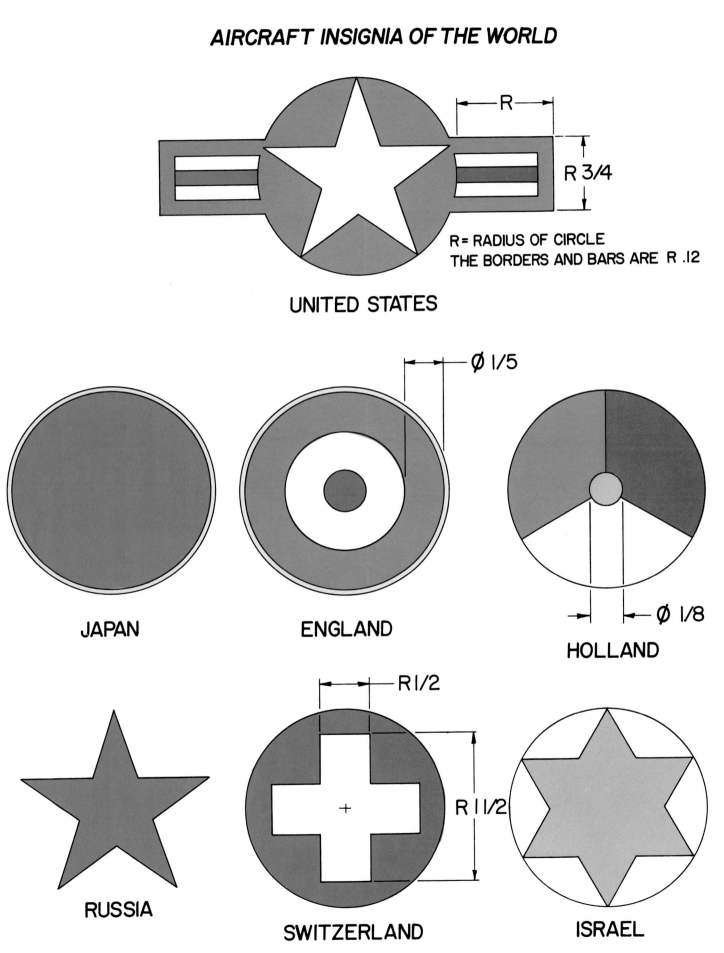

R
R 3/4
R= RADIUS OF CIRCLE
THE BORDERS AND BARS ARE R .12

UNITED STATES

JAPAN

ENGLAND

Ø 1/5

Ø 1/8

HOLLAND

RUSSIA

R 1/2
R 1 1/2

SWITZERLAND

ISRAEL

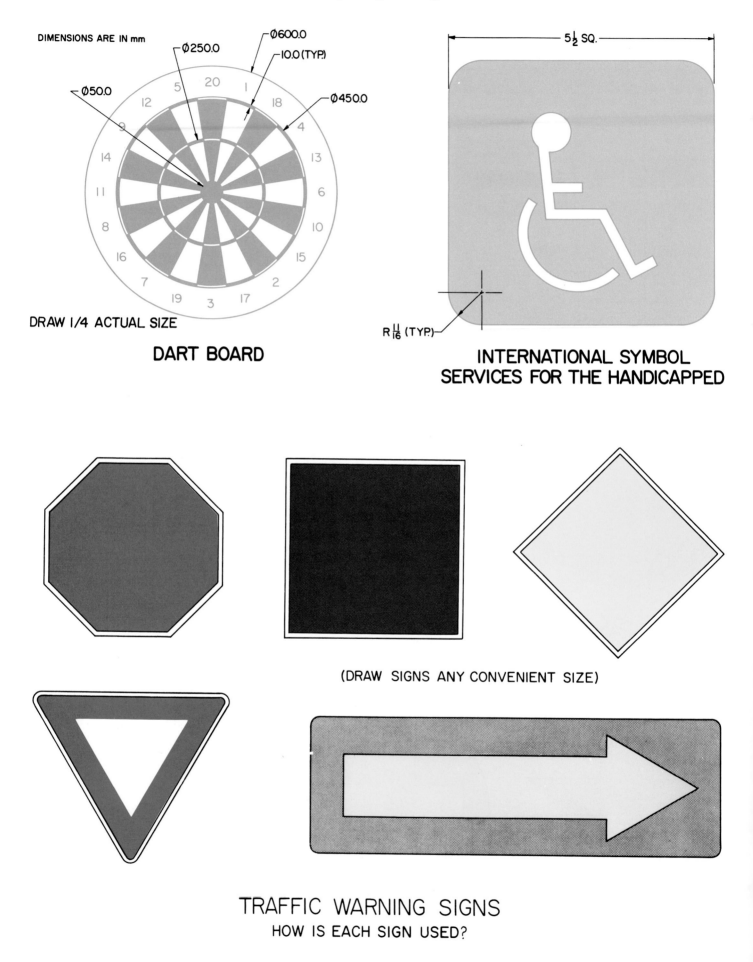

DIMENSIONS ARE IN mm

Ø600.0

Ø250.0

10.0 (TYP.)

Ø50.0

Ø450.0

DRAW 1/4 ACTUAL SIZE

DART BOARD

5½ SQ.

R 11/16 (TYP.)

INTERNATIONAL SYMBOL
SERVICES FOR THE HANDICAPPED

(DRAW SIGNS ANY CONVENIENT SIZE)

TRAFFIC WARNING SIGNS
HOW IS EACH SIGN USED?

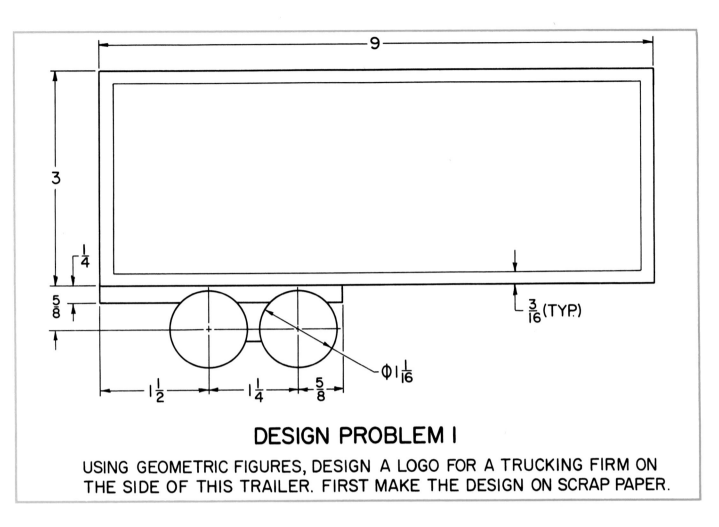

DESIGN PROBLEM I

USING GEOMETRIC FIGURES, DESIGN A LOGO FOR A TRUCKING FIRM ON THE SIDE OF THIS TRAILER. FIRST MAKE THE DESIGN ON SCRAP PAPER.

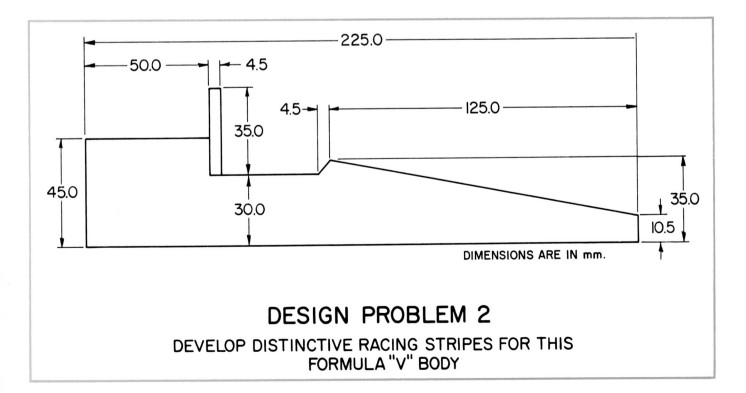

DESIGN PROBLEM 2

DEVELOP DISTINCTIVE RACING STRIPES FOR THIS
FORMULA "V" BODY

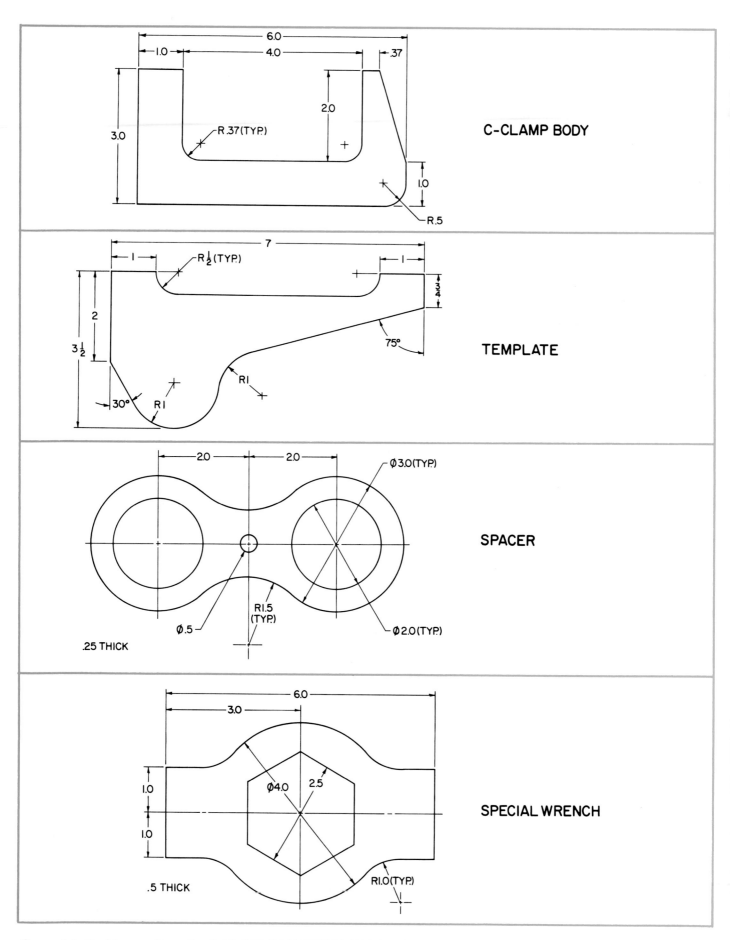

C-CLAMP BODY

TEMPLATE

SPACER

SPECIAL WRENCH

Geometric Problems.   Draw the **C-CLAMP BODY, TEMPLATE, SPACER** and **SPECIAL WRENCH** full size. Center each problem on the drawing sheet. It will not be necessary to draw top and/or side views.

Unit 7
Lettering

After studying this unit and completing the assigned problems, you should be able to:

◆ Produce neat lettering.

◆ Demonstrate the difference between vertical and inclined lettering.

◆ Identify several lettering aids and devices.

◆ Use preprinted lettering.

Lettering is used on drawings to give dimensions and other pertinent information needed to fully describe the item. The lettering must be neat and legible if it is to be easily read and understood.

A drawing will be improved by good lettering. A good drawing will look sloppy and unprofessional if the lettering is poorly done.

Single-Stroke Gothic Alphabet

The American National Standards Institute (ANSI) recommends that the *Single-Stroke Gothic Alphabet* be the accepted lettering standard. It can be drawn rapidly and is highly legible, Figure 7-1. It is called single-stroke lettering not because each letter is made with a single stroke of the pencil (most letters require several strokes to complete), but because each line is only as wide as the point of the pencil or pen.

Single-stroke lettering may be vertical or inclined. There is no definite rule stating that it should be one way or the other. However, mixing styles on a drawing should be avoided, Figure 7-2.

You can do first-class lettering if you learn the basic shapes of the letters, the proper stroke sequence for making them, and the recommended spacing between letters and words. You must also practice regularly.

Lettering with a Pencil

An H or 2H pencil is used by most drafters for lettering. Sharpen the pencil to a sharp conical point. Rotate it as you letter, Figure 7-3, to keep the point sharp and the letters uniform in weight and line width. Resharpen the pencil when the lines become wide and "fuzzy."

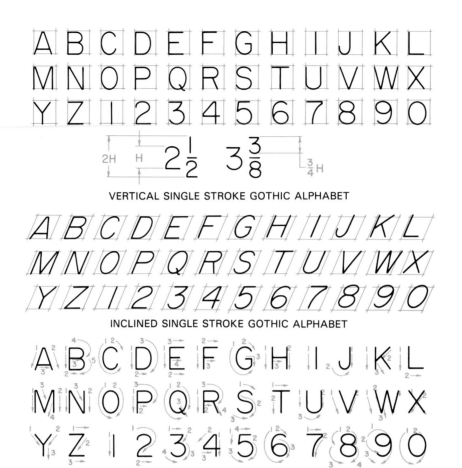

VERTICAL SINGLE STROKE GOTHIC ALPHABET

INCLINED SINGLE STROKE GOTHIC ALPHABET

RECOMMENDED SEQUENCE FOR MAKING SINGLE STROKE GOTHIC ALPHABET

Figure 7-1. The Single-Stroke Gothic Alphabet.

ONLY ONE FORM OF LETTERING SHOULD APPEAR ON A DRAWING.

AVOID COMbINING *SEVERAL* foRMS Of LETTERING.

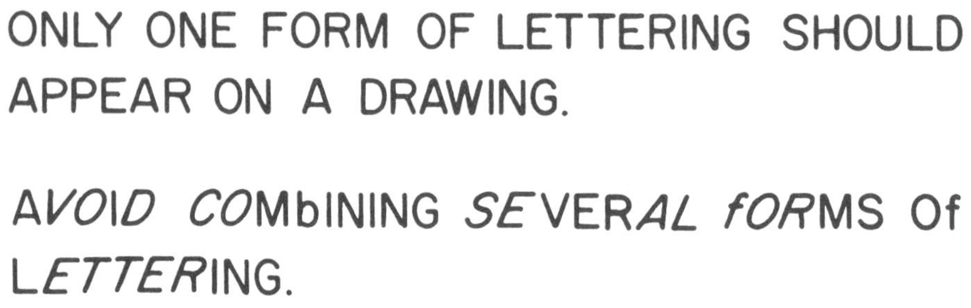

Figure 7-2. Avoid mixing several forms of lettering on a drawing.

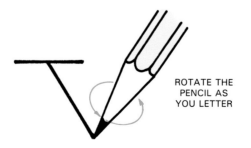

ROTATE THE PENCIL AS YOU LETTER

Figure 7-3. Rotate the pencil as you letter to keep the point sharp and lettering uniform in weight.

Guide Lines

Good lettering requires the use of **guide lines**, Figure 7-4. Guide lines are very fine lines made with a "needle sharp" 4H or 6H pencil. Guide lines should be drawn so lightly they will not show up on a print made from the drawing. **Vertical guide lines**, Figure 7-5, may be used to assure that the letters will be vertical. Use **inclined guide lines**, drawn at 67 1/2 degrees to the horizontal line, where inclined lettering is to be used, Figure 7-6.

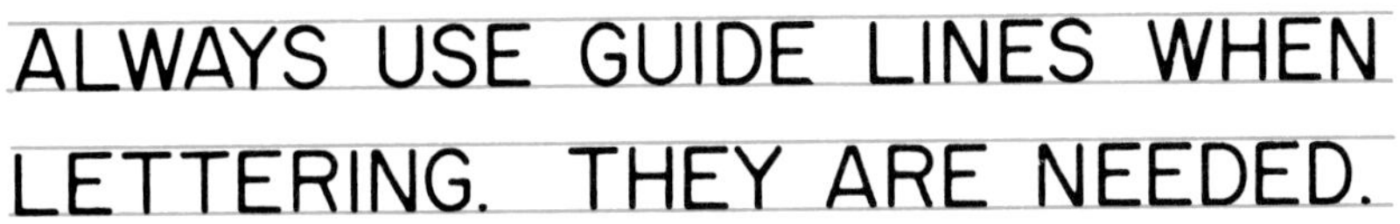

ALWAYS USE GUIDE LINES WHEN
LETTERING. THEY ARE NEEDED.

Figure 7-4. Guide lines must be used when lettering to keep the letters uniform in height.

VERTICAL GUIDE LINES HELP IN KEEPING
LETTERS UNIFORMLY VERTICAL.

Figure 7-5. Vertical guide lines.

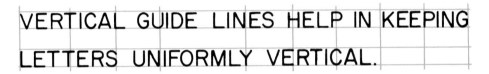

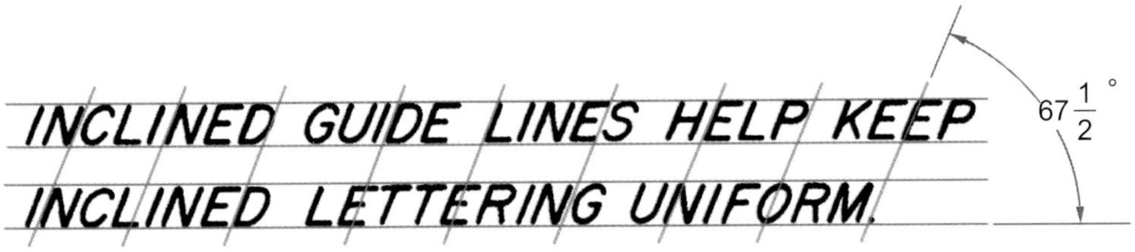

INCLINED GUIDE LINES HELP KEEP
INCLINED LETTERING UNIFORM.

$67\frac{1}{2}°$

Figure 7-6. Inclined guide lines.

Spacing

In lettering, proper spacing of the letters is important. There is no hard and fast rule which indicates how far apart the letters should be spaced. The letters should be placed so spaces between the letters appear to be about the same. Adjacent letters with straight lines require more space than curved letters. Letter spacing is judged by eye rather than by measuring, Figure 7-7.

DEVALUATION

SPACED BY MEASURING

DEVALUATION

SPACED VISUALLY

Figure 7-7. Letter spacing is judged by eye rather than by measuring.

Spacing between words and between sentences is another matter. The spacing between words and sentences should be clearly separated, Figure 7-8.

WORDS AND LETTERS MUST BE CLEARLY SEPARATED. SPACING BETWEEN WORDS AND SENTENCES IS EQUAL TO THE HEIGHT OF THE LETTER USED.

Figure 7-8.   Spacing between words and sentences.

Vertical spacing between lines of lettering should be no more than the height of the letters, and no less than half the height of the letters, Figure 7-9.

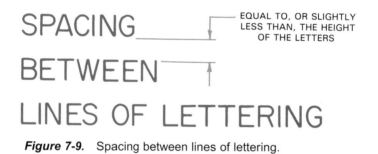

Figure 7-9.   Spacing between lines of lettering.

Letter Height

On most drawings, 1/8 inch letters will be satisfactory. Titles are usually 3/16 to 1/4 inch in height. To make the information easier to read, these sizes may be increased on large drawings.

Lettering Aids and Devices

The **Braddock lettering triangle**, shown in Figure 7-10, and the **Ames lettering instrument**, Figure 7-11, are devices that may be used as aids for drawing guide lines.

The numbers engraved below each series of holes in the Braddock lettering triangle indicate the height of the letters in thirty-seconds. The series marked 3 indicates that the guide lines will be 3/32 inch apart; 4 indicates that they will be 4/32 or 1/8 inch apart.

The numbers on the disk of the Ames lettering instrument are rotated until they are even with a line engraved on the base of the tool. The numbers also indicate the spacing of the guide lines in thirty-seconds.

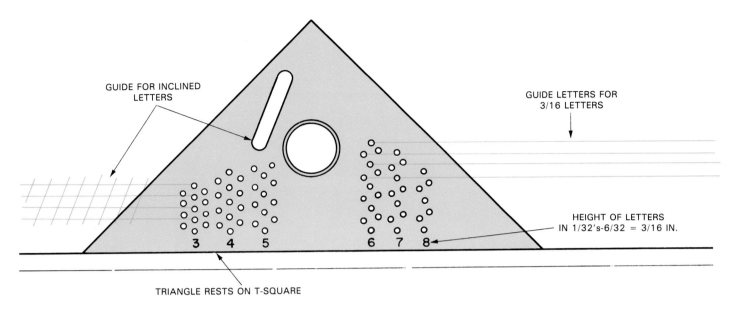

Figure 7-10. The Braddock lettering triangle.

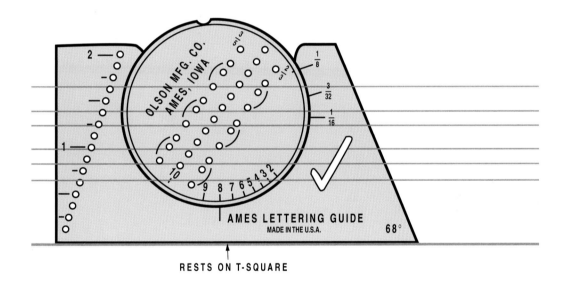

Figure 7-11. The Ames lettering instrument.

Mechanical Lettering Devices

Hand lettering is expensive because it requires so much time. For this reason, industry utilizes many mechanical devices to save time and improve the legibility of the letters and figures.

With the use of microfilm, where the drawings are reduced and enlarged photographically, it is necessary for lettering to be highly legible. Good lettering can be done easily and rapidly with mechanical lettering devices, such as the one shown in Figure 7-12. A *stylus* (pin) following a design cut into a metal or plastic *matrix* (pattern), guides a pen to draw individual letters. India ink is used in the pen. Many different letter and symbol patterns are available.

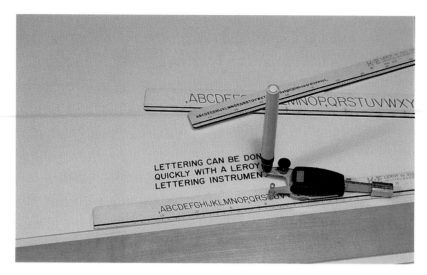

Figure 7-12. Mechanical-type lettering device. Many styles and sizes of letters, figures, and symbols are available.

Preprinted Lettering

A wide selection of alphabets and symbols are available in preprinted lettering, Figure 7-13.

Figure 7-13. A few of the hundreds of styles and sizes of preprinted transfer lettering that are readily obtainable. (Formatt)

Rub-on materials are also known as **_dry transfer materials_**, Figure 7-14. The lettering is transferred from the backing sheet by lightly rubbing over the design with a burnisher. This is done after the letter has been positioned on the drawing sheet.

CAD Lettering

This unit is concerned with helping you to develop the skills and dexterity to letter clearly and accurately and, eventually, rapidly on your drawings.

Many types of lettering can be used on drawings produced by computer-aided drafting (CAD). Some of the styles are similar to the Single-Stroke Gothic lettering you are now learning. Others are quite different. In any case, they are first generated on the screen by keyboarding in the appropriate text and numerals. The shape and size of the letters and numerals produced on the drawing printout is determined by the CAD program used.

More information on CAD lettering can be found in Unit 24, Computer-Aided Drafting and Design.

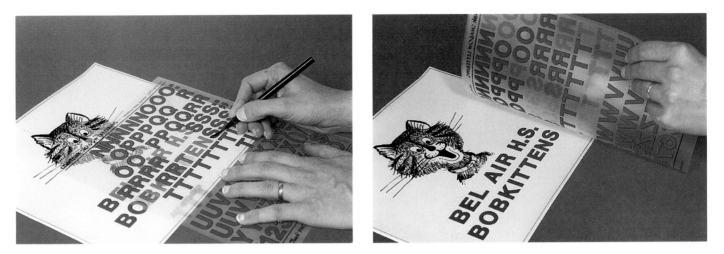

Figure 7-14.   Left. Transfer letters are applied by placing the letter in position and burnishing (rubbing) the back of the carrier sheet with a burnisher. Right. The lettering sheet is carefully removed after the letter is applied. Some transfer sheets have guide lines which are erased after an entire line of letters have been applied. Drafters are sometimes called upon to produce work other than production drawings. These illustrations show a drafter completing an advertising sheet.

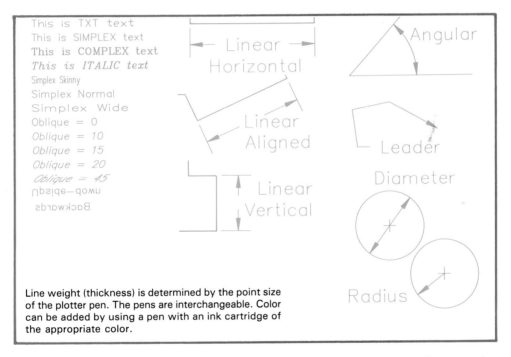

Figure 7-15.   When adding dimensions and text to a CAD drawing, the drafter can select text and numeral size, style, and orientation (position and/or location) on the graphics. The material shown here was prepared using AutoCAD® computer-aided design and drafting software. (Autodesk, Inc.)

Drafting Vocabulary

Ames lettering instrument
Braddock lettering triangle
Burnisher
Device
Dry transfer materials
Engraved

Gothic
Legible
Lettering
Matrix
Microfilm

Pressure-sensitive
  materials
Single-Stroke Gothic
  Alphabet
Stencil
Stylus

Test Your Knowledge—Unit 7

Please do not write in the book. Place your answers on another sheet of paper.

1. Lettering is used on drawings to give _____ and other pertinent _____ needed to fully describe the item.
2. Why must lettering on a drawing be neat and legible?
3. The Single-Stroke Gothic letter is recommended because it can be drawn _____ and is highly _____.
4. The single-stroke lettering on a drawing sheet may be _____ or _____ because there is no definite rule stating that it has to be one or the other.
5. A(n) _____ or _____ pencil is usually used for lettering.
6. Why should guide lines be used when lettering?
7. Name three types of lettering aids and mechanical lettering devices. Briefly describe how each works.
8. Why does industry use mechanical lettering devices?

Outside Activities

1. Using signs and posters from your school's bulletin board, make a display of different lettering examples. Point out good spacing and poor spacing. Check to see if lettering styles are mixed in each example.
2. Research the variety of lettering styles, alphabets, and symbols available with mechanical lettering devices.
3. Letter the sentence: "Good lettering technique requires practice and concentration." by hand using vertical Single-Stroke Gothic. Letter the same sentence using a mechanical lettering device such as the Leroy lettering equipment. Time yourself and report to the class which is faster and which is easier to read.
4. Demonstrate the use of transfer lettering by producing a safety poster for another classroom in your school.

Problems

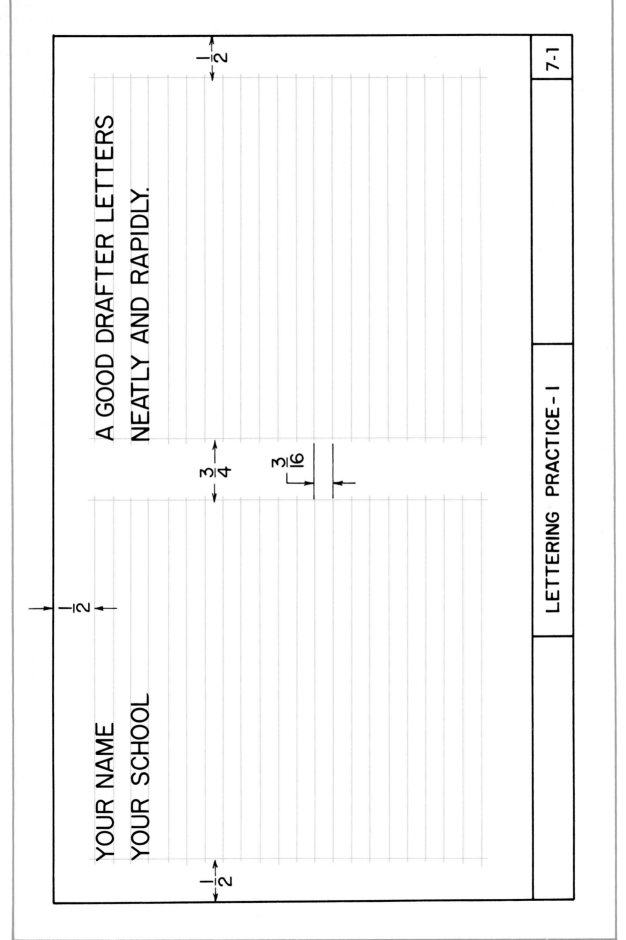

A GOOD DRAFTER LETTERS
NEATLY AND RAPIDLY.

YOUR NAME
YOUR SCHOOL

LETTERING PRACTICE - I

7-1

Problem Sheet 7-1. LETTERING PRACTICE.

THE QUICK RED FOX JUMPED
OVER THE LAZY BROWN DOG.

1 2 3 4 5 6 7 8 9 0 $\frac{1}{2}$ $\frac{1}{16}$ $\frac{3}{8}$ $\frac{5}{32}$

LETTERING PRACTICE-2

7-2

Problem Sheet 7-2. LETTERING PRACTICE.

7-3

"THE BEST THING ABOUT THE

FUTURE IS THAT IT COMES ONE

DAY AT A TIME."

MAKE THE MOST OF YOURSELF

FOR THAT IS ALL THERE IS OF

YOU.

LETTERING PRACTICE-3

Problem Sheet 7-3.   LETTERING PRACTICE.

1 2 3 4 5 6 7 8 9 0 $\frac{1}{2}$ $\frac{3}{4}$ $\frac{5}{8}$ $\frac{7}{9}$

PACK EACH BOX WITH SEVEN
DOZEN GIANT JUGS.

7-4

LETTERING PRACTICE-4

Problem Sheet 7-4.   LETTERING PRACTICE.

SELECT A FAVORITE SAYING
OR QUOTATION AND LETTER
IT IN $\frac{1}{8}$, $\frac{3}{16}$ AND $\frac{1}{4}$ VERTICAL
OR INCLINED LETTERS.

LETTERING PRACTICE–5

7-5

Problem Sheet 7-5.   LETTERING PRACTICE.

Multiview drawings are used to produce most products or structures today. Typically, the complexity of the product or structure determines the number of drawings needed to fully describe its construction.

Unit 8
Multiview Drawings

After studying this unit and completing the assigned problems, you should be able to:

◆ Explain what a multiview drawing is.

◆ Define orthographic projection.

◆ Use orthographic projection to develop multiview drawings.

◆ Identify the views necessary to make a multiview drawing.

◆ Locate multiview drawings on the drawing sheet.

◆ Transfer points between views when making multiview drawings.

When a drawing is made with the aid of instruments, it is called a *mechanical drawing*. Straight lines are made with a T-square and triangle or a drafting machine. Circles, arcs, and curved lines are drawn with a compass, French curve, or suitable template.

Drawings can also be generated on a computer using computer-aided design and drafting software,

Figure 8-1. They are converted into hard copy (paper drawings) on a high-speed plotter. Many large firms use CAD almost exclusively.

Regardless of the technique, whether traditional or computer-aided, the principles of drafting remain the same. The drafter must be familiar with the standards and procedures necessary to develop the graphics that will accurately describe the part. It is

Figure 8-1. Computer-aided design, or CAD, permits design changes to be made rapidly. The engineer, designer, or drafter can see the modifications immediately by calling them onto the computer screen. Some programs will also produce performance changes, if any, that will result from the design changes. (Beech Aircraft Corp.)

possible to use advanced computer technology to transfer the CAD-generated information directly to a computer-controlled machine tool that will manufacture the part. No drawings are necessary. This process is called ***CAD/CAM—Computer-aided design/Computer-aided manufacturing***.

A large number of the drawings used by industry are in the form of ***multiview drawings***. That is, more than one view is required to give an accurate shape and size description of the object being drawn or generated. In developing the needed views, the object is normally viewed from six directions, as shown in Figure 8-2.

The various directions of sight will give the ***front, top, right side, left side, rear,*** and ***bottom views***, Figure 8-3. To obtain the views, think of the object as being enclosed in a hinged glass box, Figure 8-4, with the views projected into the side of the box.

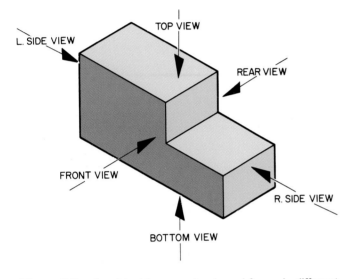

Figure 8-2. An object is normally viewed from six different directions.

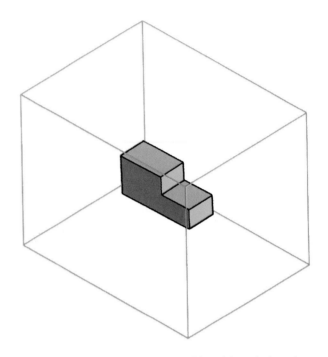

Figure 8-4. The object is enclosed in a hinged glass box.

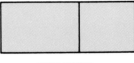

TOP VIEW

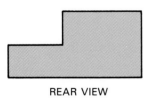

REAR VIEW

L. SIDE VIEW

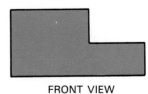

FRONT VIEW

R. SIDE VIEW

BOTTOM VIEW

Figure 8-3. The six directions of sight give these views.

The method for developing multiview drawings is called ***orthographic projection***. It permits three-dimensional objects to be shown on a flat surface having only two dimensions. It reveals the width, depth, and height of the object.

Orthographic projection forms the basis for engineering drawing. Two methods of projection are used, Figure 8-5. ***Third angle projection*** is preferred in the United States and several other nations. ***First angle projection*** is typically used in Europe.

With third angle projection, the object is drawn as viewed in the glass box, Figure 8-6. That is, the views are projected to the six sides of the box. The projected views are drawn as shown when the box is opened out.

With first angle projection, the object is drawn *as if the object were placed on each side of the glass box.* That is, as if the object is viewed as projected onto the drawing surface. See Figure 8-7.

The ISO (International Standards Organization) symbols shown in Figure 8-8 are employed to indicate the projected system of a given drawing.

Selecting Views to be Used

As can be seen on Figures 8-6 and 8-7, at least six views of an object can be drawn. This does not mean that all six of the views must be used, or are needed. Only those views required to give a shape description of the object should be drawn. Any view that repeats the same shape description as another view can be eliminated, Figure 8-9.

In many instances, two or three views are sufficient to show the shape of an object.

Those views of a drawing showing a large number of hidden lines are used only when absolutely necessary. The use of too many hidden lines on a drawing tends to make the drawing confusing to the person reading the drawing or fabricating the part. Use another view, without as many

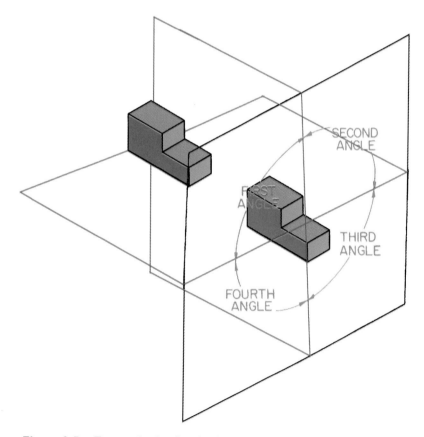

Figure 8-5.   Two methods of projection are generally employed: first angle projection in Europe; third angle projection in the United States and many other countries.

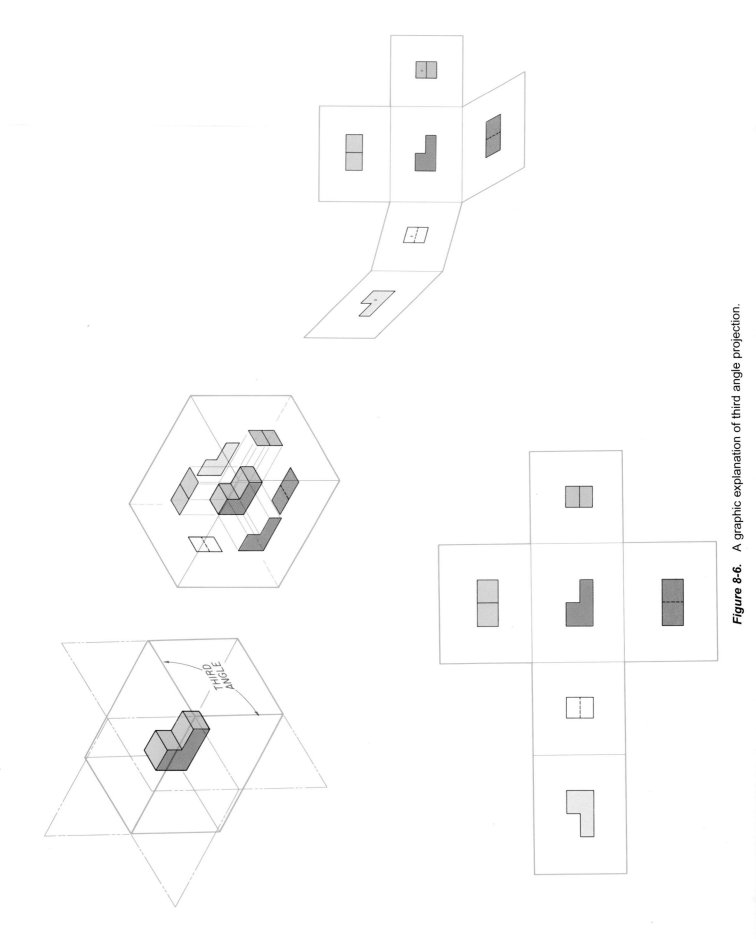

Figure 8-6. A graphic explanation of third angle projection.

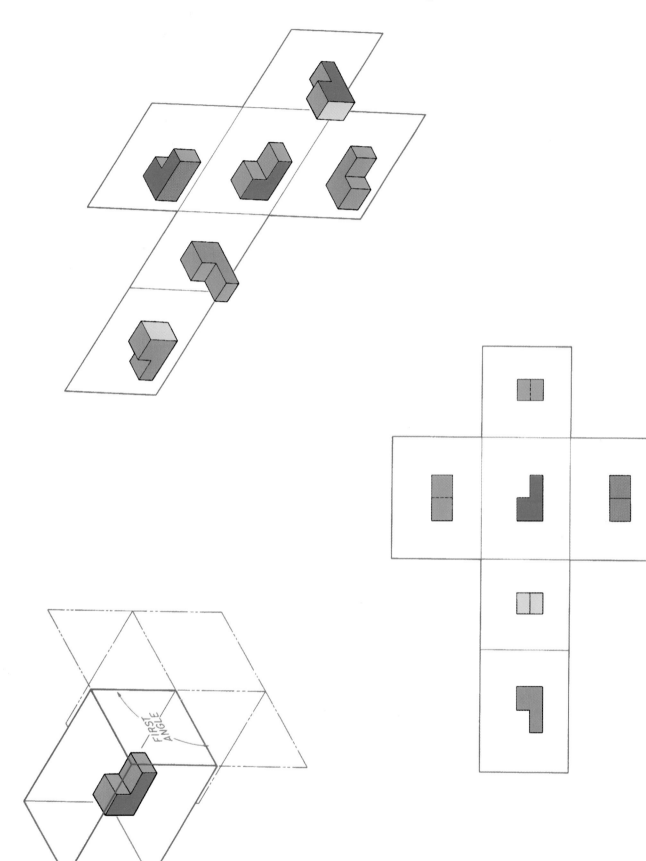

Figure 8-7.   A graphic explanation of first angle projection.

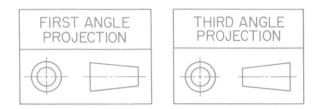

Figure 8-8. The appropriate ISO symbol is placed on a drawing to show which method of projection was used on that sheet.

hidden lines, as in Figure 8-10. Remember, the goal is to communicate clearly.

Transferring Points

Each view will show a minimum of two dimensions. Any two views of an object will have at least one dimension in common. Time can be saved if a dimension from one view is projected to the other view instead of measuring the dimension a second

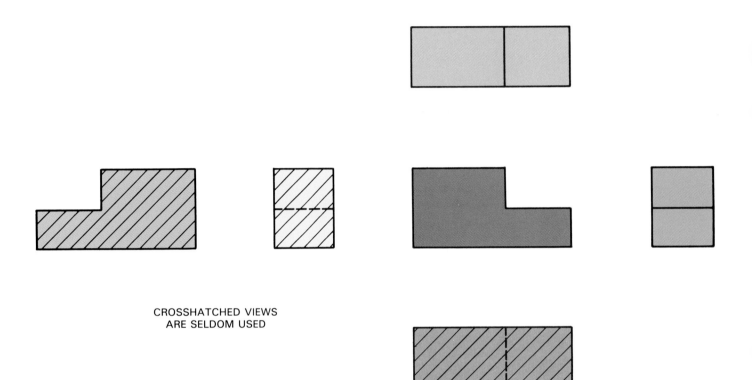

CROSSHATCHED VIEWS
ARE SELDOM USED

Figure 8-9. Not all views are needed. Eliminate any view that repeats the same shape description as another view.

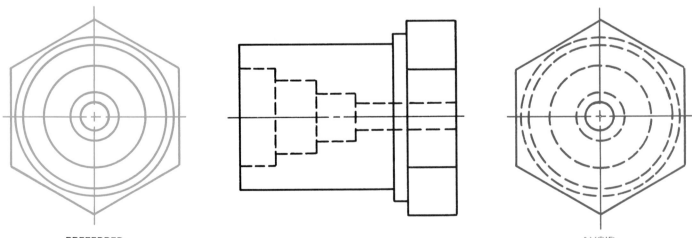

PREFERRED

AVOID

Figure 8-10. Views showing a large number of hidden lines are used only if absolutely necessary. Too many hidden lines tend to make the drawing confusing.

time, Figure 8-11. Transfer the points with construction lines.

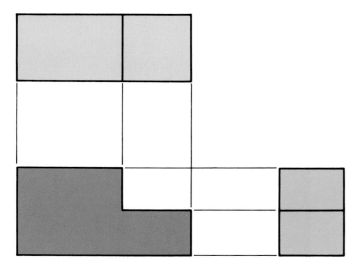

Figure 8-11. Transferring points from view to view.

Additional time can be saved in transferring the depth of the top view to the side view. Two methods of projection are shown in Figure 8-12. Projection provides for greater accuracy in the alignment of the views. It is faster than measuring each view separately with a scale or dividers.

How to Center a Drawing on the Sheet

A drawing looks more professional if the views are evenly spaced and centered on the drawing sheet. Centering the views on a sheet is not difficult if the following procedure is used:

1. Examine the object to be drawn. Observe its dimensions—width, depth, and height, Figure 8-13. Determine the position in which the object will be drawn.

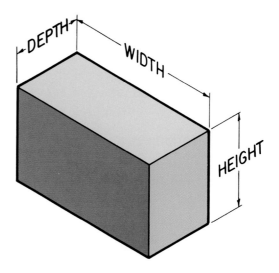

Figure 8-13. How an object is described.

2. Measure the working area of the sheet *after* the border and title block have been drawn. It should measure 7 in. by 10 in., if you used an 8 1/2 in. by 11 in. drawing sheet. Refer to how to prepare a drawing sheet on page 77.

3. Allow one inch between views.

4. To locate the front view, add the width of the front view, one inch spacing between views, and the depth of the right side view.

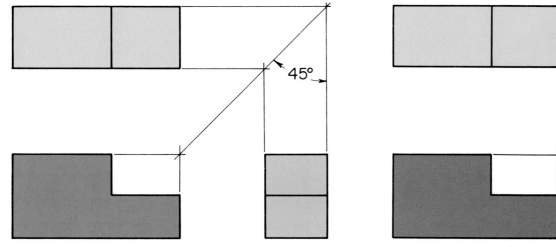
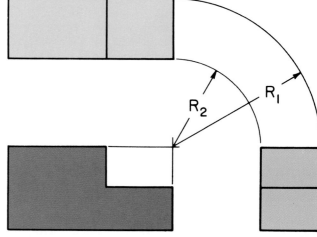

Figure 8-12. Two accepted methods employed to transfer the depth of the top view to the side view.

Subtract this total from the horizontal width of the working surface (10 in.). Divide this answer by 2. This will be the starting point for laying out the sheet horizontally.

Using the object shown in Figure 8-14, for example, it would be as follows:

> Width of front view = 5 in.
> Space between views = 1 in.
> Depth of right side view = 1 1/2 in.
> Total = 7 1/2 in.

> Width of working area = 10 in.
> Total dimensions
>   of views and spacing = –7 1/2 in.
>   Result = 2 1/2 in.

Divide 2 1/2 inches by 2 = 1 1/4 in.

This is the distance in from left border line to locate the starting point for drawing.

5. Measure in 1 1/4 in. from the left border line. Draw a vertical construction line through this point.

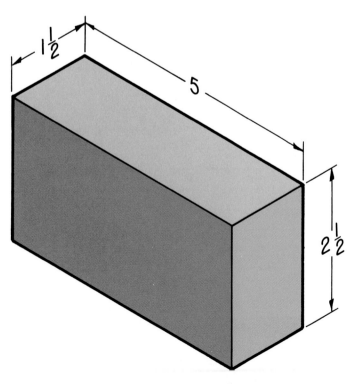

Figure 8-14. Object used as an example for centering views on a drawing sheet.

6. From this line, measure over a distance equal to the width of the front view. Draw another vertical construction line. See Figure 8-15.

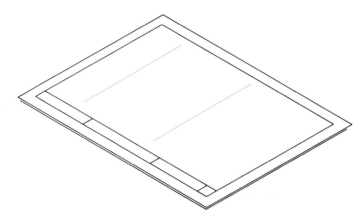

Figure 8-15. The first step in locating the front and top views on the drawing sheet.

7. The same procedure is followed to center the views vertically. The height of the front view and the depth of the top view are used. A one-inch space will separate the views. Add these distances together. Subtract the sum from the vertical working space (7 in.). Divide the answer by 2.

For example:

> Height of front view = 2 1/2 in.
> Space between views = 1 in.
> Depth of top view = 1 1/2 in.
> Total = 5 in.

> Height of working area = 7 in.
> Total dimensions
>   of views and spacing = –5 in.
>   Result = 2 in.

Divide 2 in. by 2 = 1 inch

This is the distance up from lower border line to locate starting point for drawing.

8. Measure up 1 in. from the lower border line. Draw a horizontal construction line through this point.

9. From this line, measure up the height of the front view and mark a point. Mark a point at

the one-inch spacing that separates the views. Mark one more point for the depth of the top view. Draw construction lines through these points. See Figure 8-16.

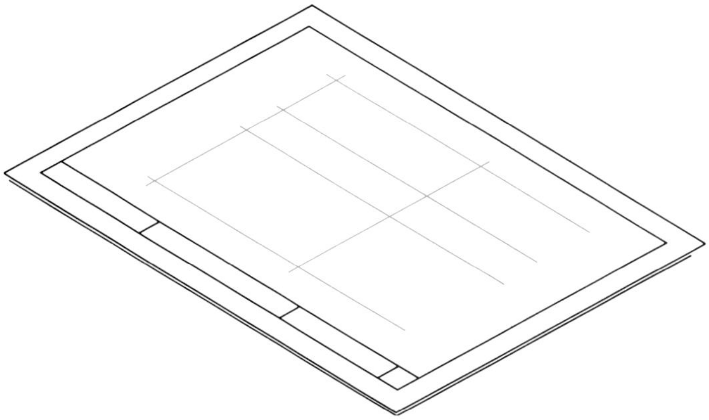

type of line (object line, hidden line, center-line, etc.), Figure 8-18. Use the erasing shield when erasing the remaining construction lines.

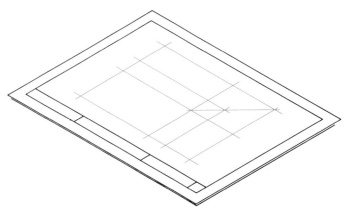

Figure 8-17. Projecting information from the top view to block in the right side view.

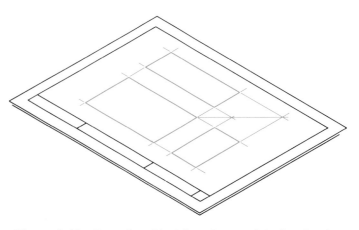

Figure 8-16. The front and top views blocked in with construction lines.

10. Use either the 45 degree angle method or the radius method to transfer the depth of the top view to the right side of the object, Figure 8-17.

11. Draw in the right side view. Use construction lines.

12. Complete the drawing by going over the construction lines. Use the correct weight for the

Figure 8-18. Draw the object lines to complete the drawing. Construction lines may be erased.

Drafting Vocabulary

Computer-aided design/Computer-aided manufacturing (CAD/CAM)

Direction of sight
First angle projection
International Standards Organization (ISO)

Mechanical drawing
Multiview drawing
Orthographic projection
Third angle projection

Test Your Knowledge—Unit 8

Please do not write in the book. Place your answers on another sheet of paper.

1. A drawing is said to be a MECHANICAL DRAWING when it is drawn using _____.

2. What does the term CADD mean?

3. A drawing that uses two or more views to describe an object is known as a(n) _____.

4. In developing the needed views, the object being drawn is normally viewed from six directions. What are the six views that will be seen?

5. The method employed to develop these six views is called _____ _____.

6. Each of the views will show a minimum of _____ dimensions. Any two of the views will have at least _____ dimension in common.

7. Orthographic projection permits _____ dimensional objects to be shown on a flat surface having only _____ dimensions.

Outside Activities

1. Collect props for the class to draw using instruments. One prop should require only a two-view drawing; another prop should require a three-view drawing. Find other props which require more than three views to give a complete shape description.

2. Build a hinged box out of clear plastic which can be used to demonstrate the unfolding of an object into its multiview parts; the front, top, bottom, and sides. Place a prop inside the box, trace the profile of the object on the side of the plastic with chalk, then unfold the box to show the multiview projections.

3. Make a large poster for your drafting room showing the step-by-step procedure to follow in centering a drawing on a sheet.

Problems

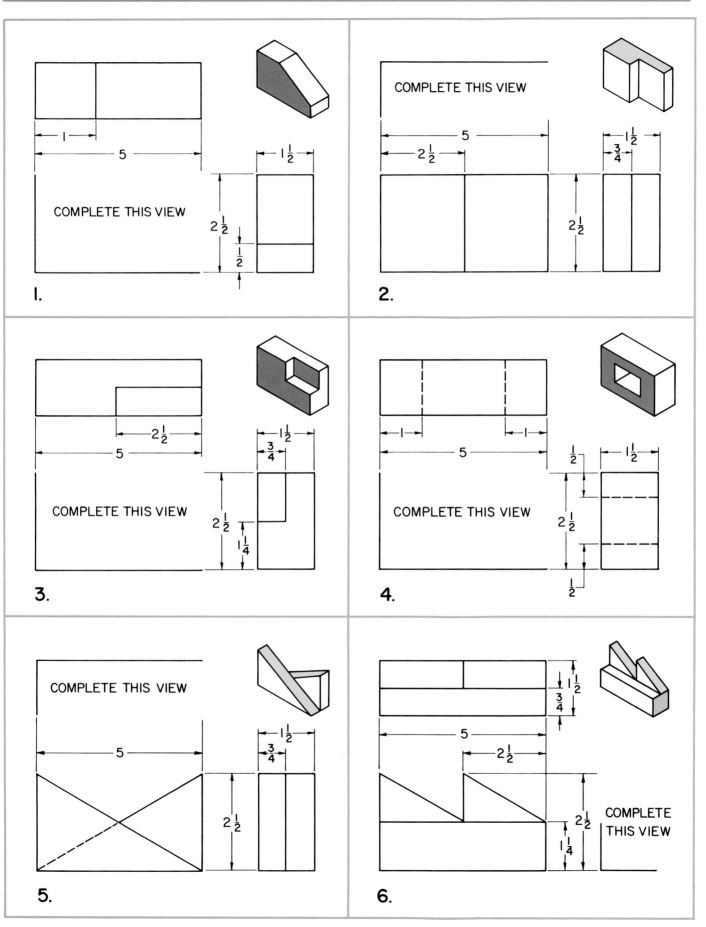

1. COMPLETE THIS VIEW

2. COMPLETE THIS VIEW

3. COMPLETE THIS VIEW

4. COMPLETE THIS VIEW

5. COMPLETE THIS VIEW

6. COMPLETE THIS VIEW

Problem Sheet 8-1. MULTIVIEW DRAWINGS. Draw each problem on a separate sheet and complete as indicated.

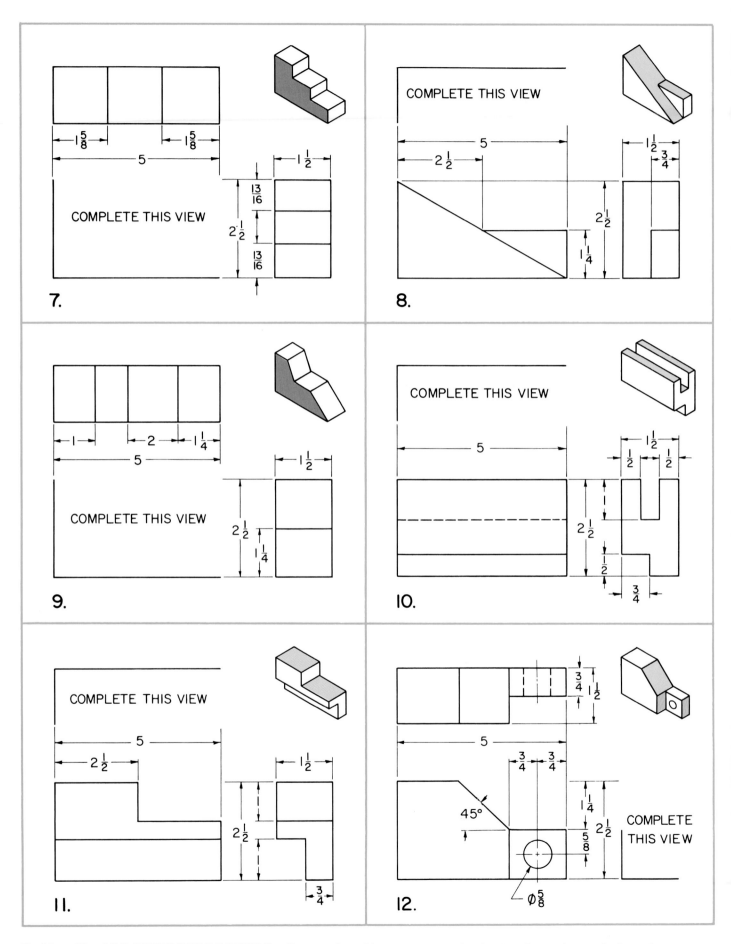

Problem Sheet 8-2. MULTIVIEW DRAWINGS.   Draw each problem on a separate sheet and complete as indicated.

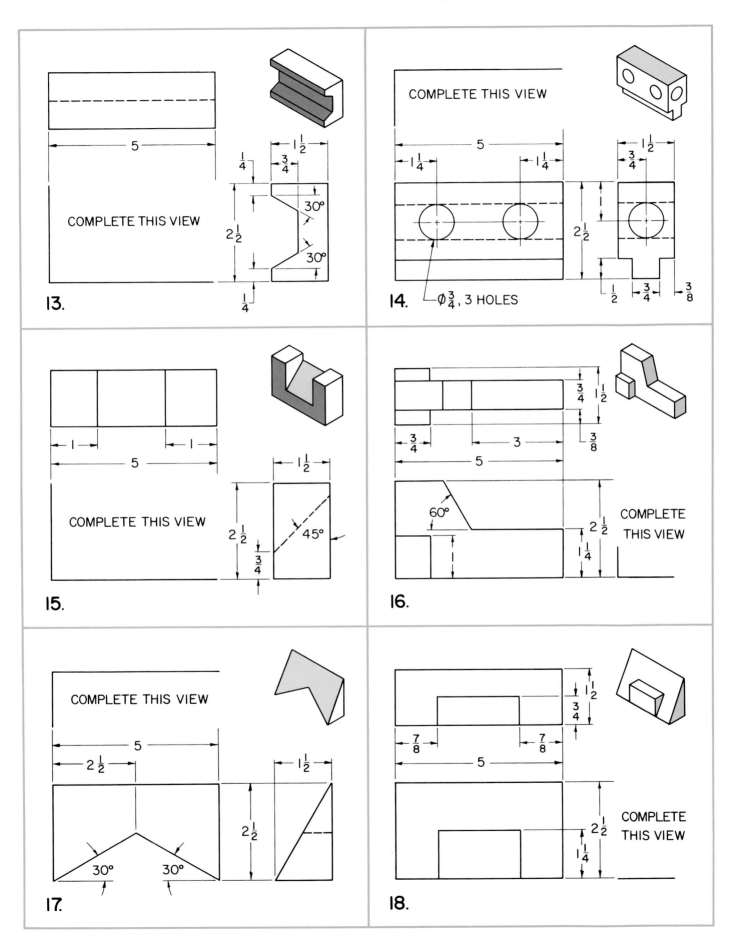

13. COMPLETE THIS VIEW
5
1½
¼
¾
30°
2½
30°
¼

14. COMPLETE THIS VIEW
5
1¼
1¼
1½
¾
2½
½
¾
⅜
Ø ¾, 3 HOLES

15. COMPLETE THIS VIEW
1
1
5
1½
2½
¾
45°

16. COMPLETE THIS VIEW
¾
3
⅜
5
3¾ 1½
60°
2½
1¼

17. COMPLETE THIS VIEW
5
2½
1½
2½
30°   30°

18. COMPLETE THIS VIEW
1½
¾
⅞   ⅞
5
2½
1¼

Problem Sheet 8-3. MULTIVIEW DRAWINGS. Draw each problem on a separate sheet. Draw as many views as necessary to fully describe each problem.

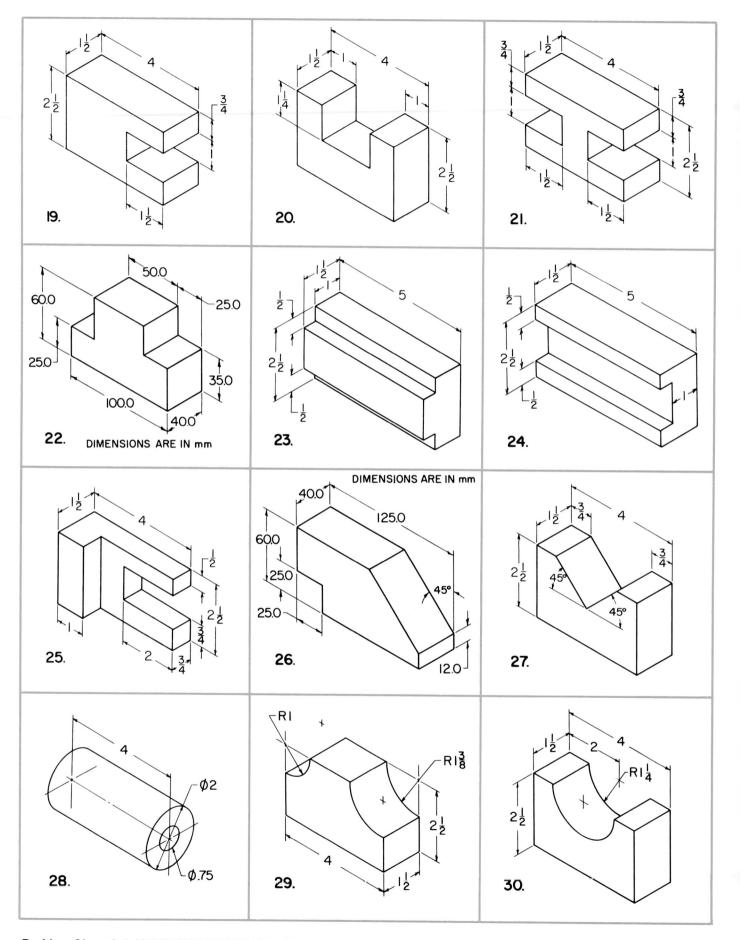

Problem Sheet 8-4. MULTIVIEW DRAWINGS. Draw each problem on a separate sheet. Draw as many views as necessary to fully describe each problem.

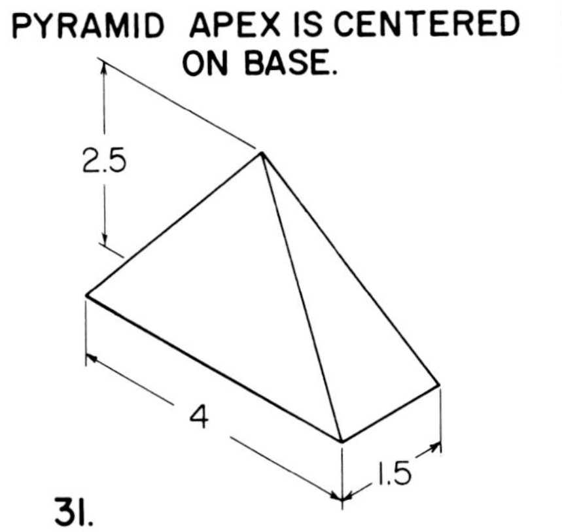
PYRAMID APEX IS CENTERED ON BASE.
2.5
4
1.5

31.

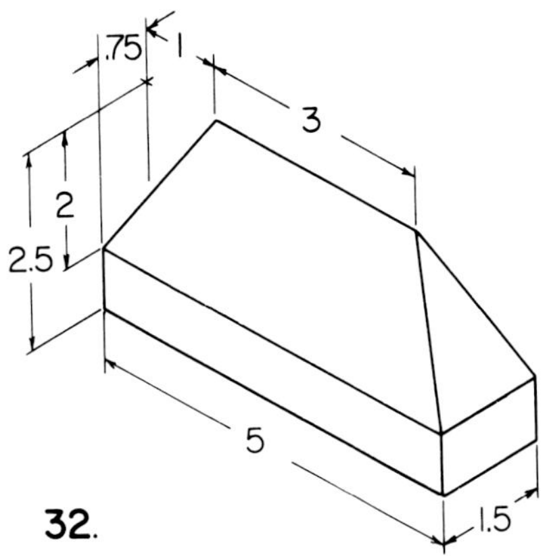
.75
1
3
2
2.5
5
1.5

32.

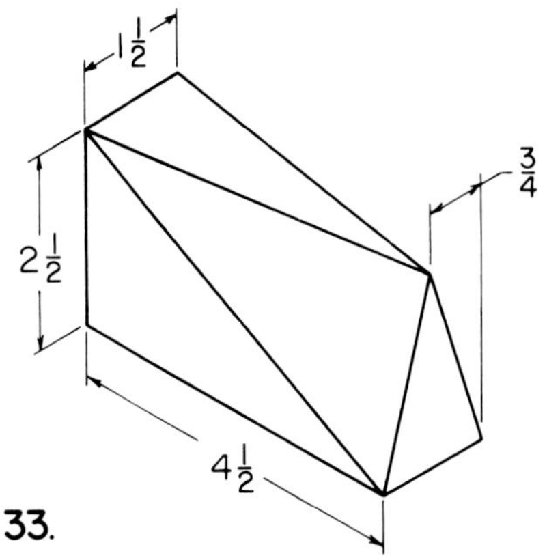
1 1/2
3/4
2 1/2
4 1/2

33.

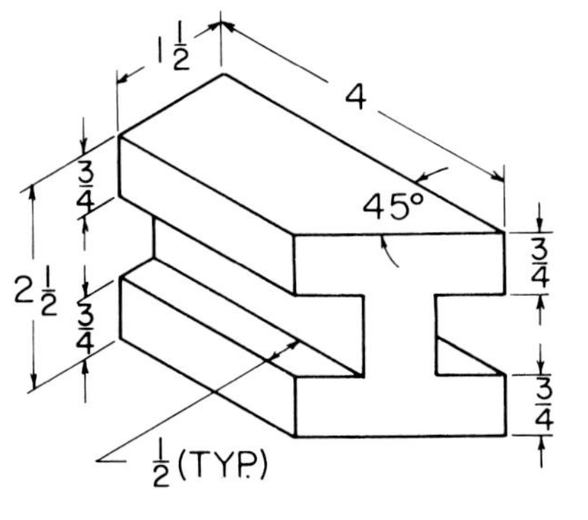
1 1/2
4
45°
3/4
2 1/2
3/4
3/4
3/4
1/2 (TYP.)

34.

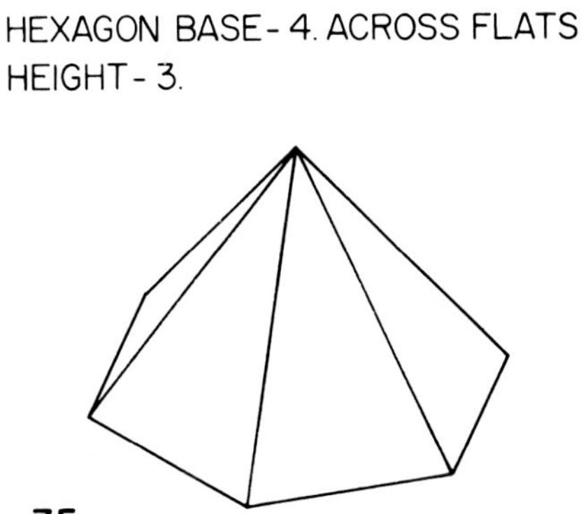
HEXAGON BASE - 4. ACROSS FLATS
HEIGHT - 3.

35.

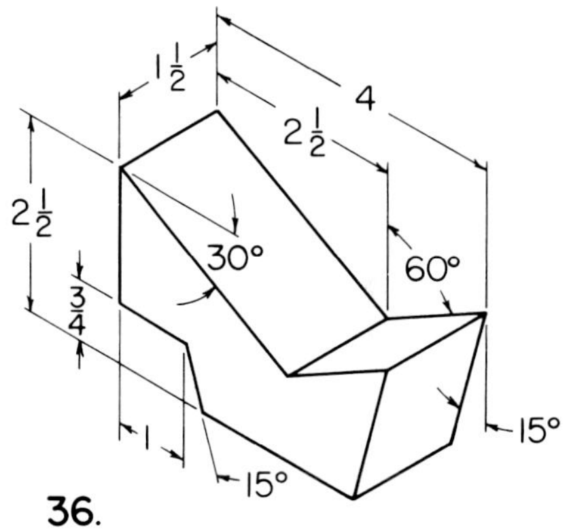
1 1/2
4
2 1/2
2 1/2
30°
60°
3/4
1
15°
15°

36.

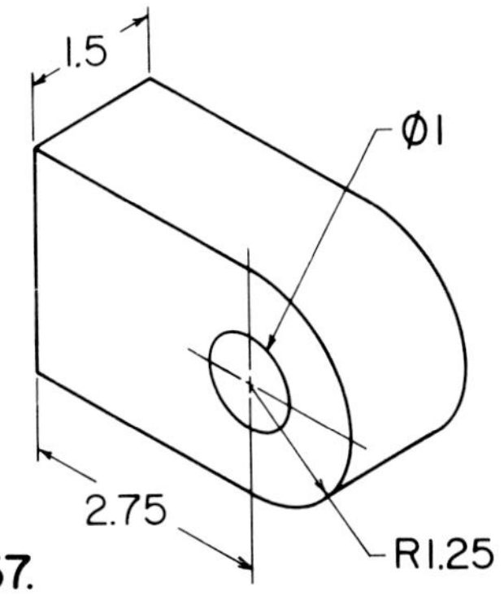
1.5
Ø1
2.75
R1.25

37.

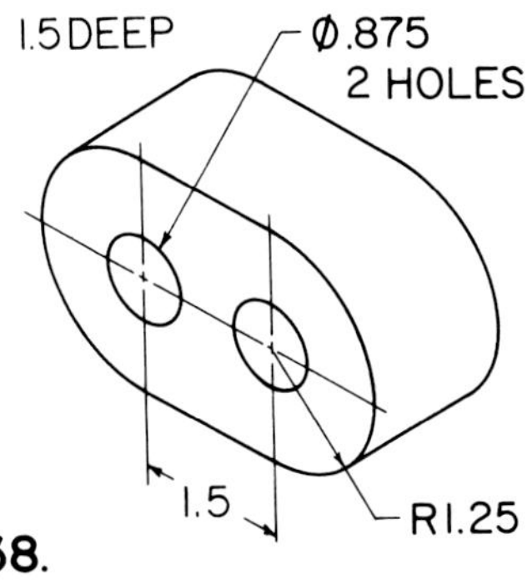
1.5 DEEP
Ø.875
2 HOLES
1.5
R1.25

38.

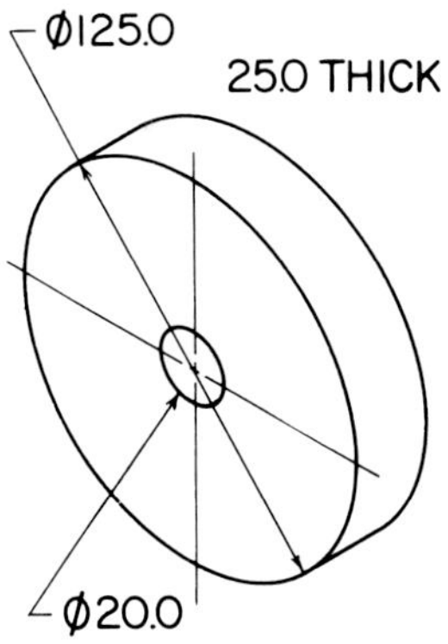
Ø125.0
25.0 THICK
Ø20.0

39.

DIMENSIONS ARE IN mm

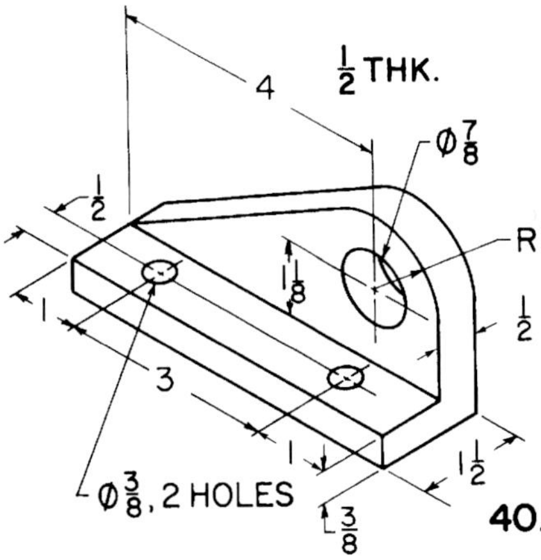
4
1/2 THK.
Ø 7/8
1/2
R1
1 1/8
1
1/2
3
1
1 1/2
Ø 3/8, 2 HOLES
3/8

40.

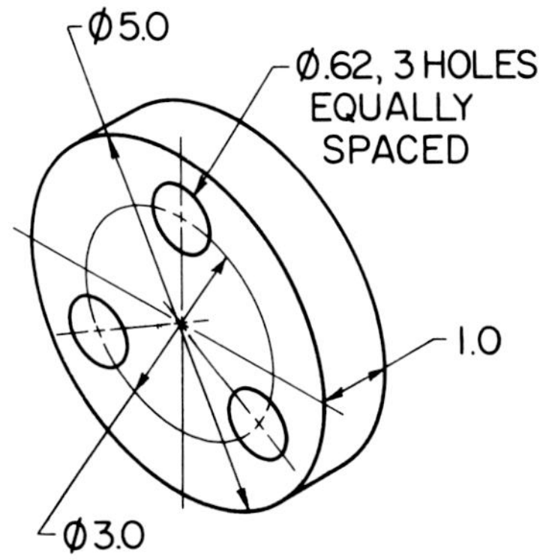
Ø5.0
Ø.62, 3 HOLES
EQUALLY
SPACED
1.0
Ø3.0

41.

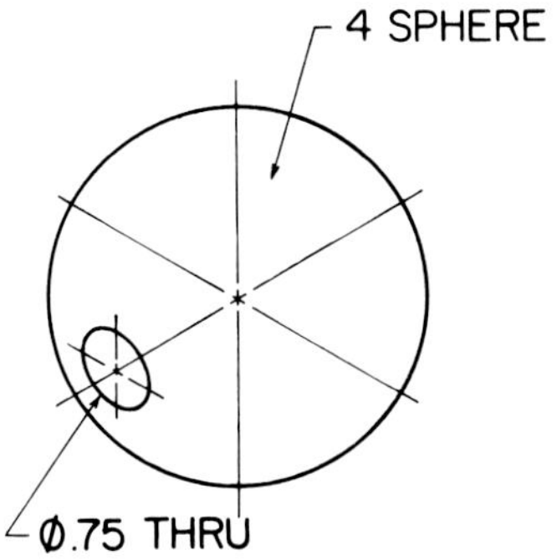
4 SPHERE
Ø.75 THRU

42.

Problem Sheet 8-5. MULTIVIEW DRAWINGS. Draw each problem on a separate sheet. Draw as many views as necessary to fully describe each problem.

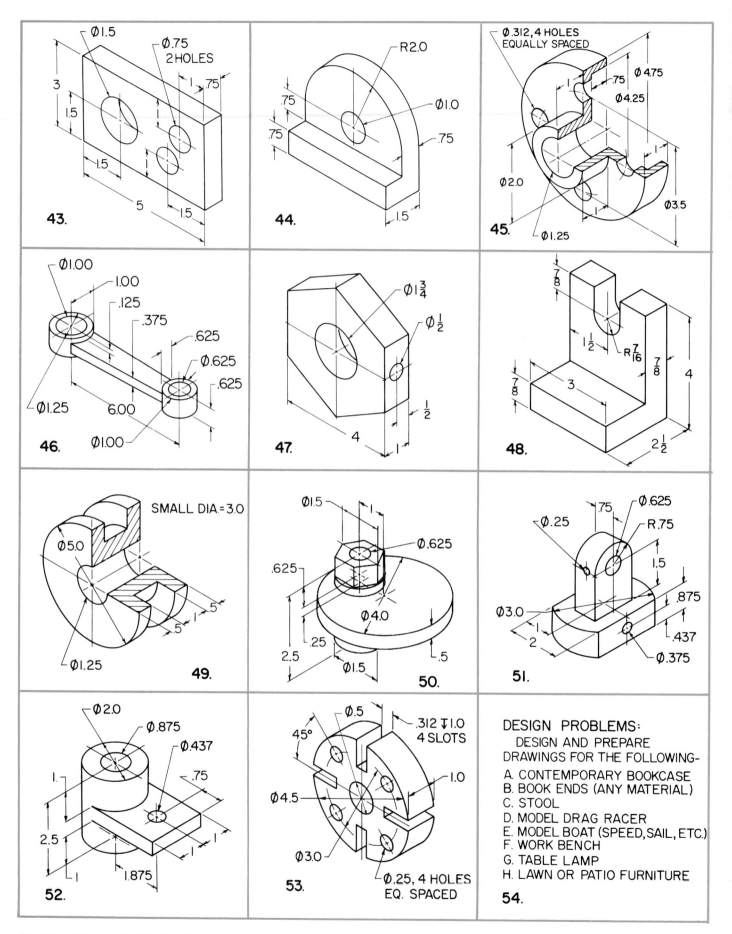

43. Ø1.5, Ø.75 2 HOLES, 3, 1.5, 1.5, .75, .75, 5, 1.5

44. R2.0, Ø1.0, .75, .75, .75, 1.5

45. Ø.312, 4 HOLES EQUALLY SPACED, Ø4.75, .75, Ø4.25, Ø2.0, Ø3.5, Ø1.25

46. Ø1.00, 1.00, .125, .375, .625, Ø0.625, .625, Ø1.25, 6.00, Ø1.00

47. Ø1¾, Ø½, ½, 4, 1

48. ⅞, 1½, R⁷⁄₁₆, ⅞, 4, ⅞, 3, 2½

49. SMALL DIA.= 3.0, Ø5.0, Ø1.25, .5, .5

50. Ø1.5, Ø0.625, .625, Ø4.0, .25, 2.5, .5, Ø1.5

51. .75, Ø0.625, Ø.25, R.75, 1.5, Ø3.0, .875, 2, .437, Ø.375

52. Ø2.0, Ø0.875, Ø0.437, 1., .75, 2.5, 1, 1, 1.875

53. Ø.5, 45°, .312 ▼1.0 4 SLOTS, 1.0, Ø4.5, Ø3.0, Ø.25, 4 HOLES EQ. SPACED

54. DESIGN PROBLEMS:

DESIGN AND PREPARE DRAWINGS FOR THE FOLLOWING—

A. CONTEMPORARY BOOKCASE
B. BOOK ENDS (ANY MATERIAL)
C. STOOL
D. MODEL DRAG RACER
E. MODEL BOAT (SPEED, SAIL, ETC.)
F. WORK BENCH
G. TABLE LAMP
H. LAWN OR PATIO FURNITURE

Problem Sheet 8-6. MULTIVIEW DRAWINGS. Draw each problem on a separate sheet. Draw as many views as necessary to fully describe each problem.

Dimensioning and Notes

After studying this unit and completing the assigned problems, you should be able to:

◆ Explain why dimensions and notes are needed on drawings.

◆ Describe the difference between unidirectional and aligned dimensioning.

◆ Apply the general rules for dimensioning inch and/or metric drawings.

◆ Dimension circles, holes, and arcs.

◆ Explain the five methods used in the changeover from conventional inch measurements to metric measurements.

If an object is to be manufactured according to the designer's specifications, the person crafting the product usually needs more information than that furnished by a scale drawing of its shape. **Dimensions** and **shop notes** are needed, Figure 9-1.

Dimensions define the sizes of the geometrical features of an object. **Notes** provide additional information not found in the dimensions. Dimensions and notes are added by hand or by using the keyboard in CAD applications. They are presented in appropriate units of measure—inches, millimeters, etc. The latest drafting standards recommend decimal inch dimensioning. Decimals are easier to add, subtract, multiply, and divide. However, fractional inch dimensioning is still widely employed. Both types will be found in this text.

Metric drawings are usually dimensioned in millimeters (mm).

Reading Direction for Dimensions

Dimensions are placed on the drawing in either a **unidirectional** or an **aligned** manner, Figure 9-2. Unidirectional dimensioning is preferred.

Unidirectional dimensions are placed to read from the bottom of the drawing. Aligned dimensions are placed parallel to the dimension line. The numerals are read from the bottom and from the right side of the drawing.

Regardless of the method used, dimensions shown with leaders and all notes are lettered parallel to the bottom of the drawing.

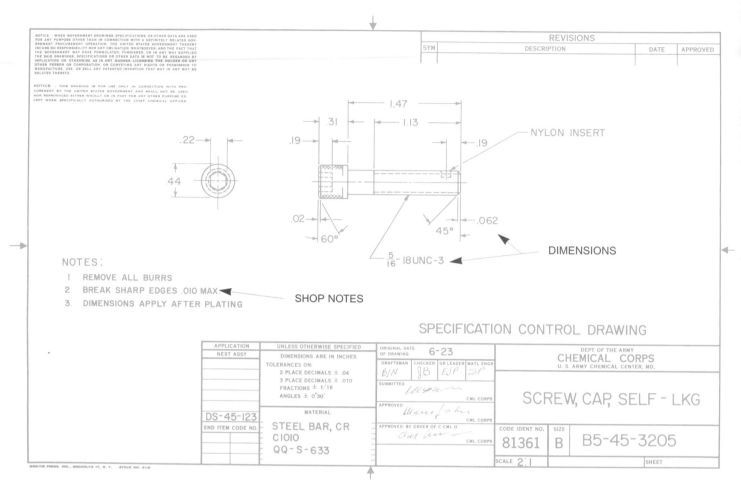

Figure 9-1. Industry drawing showing dimensions and shop notes.

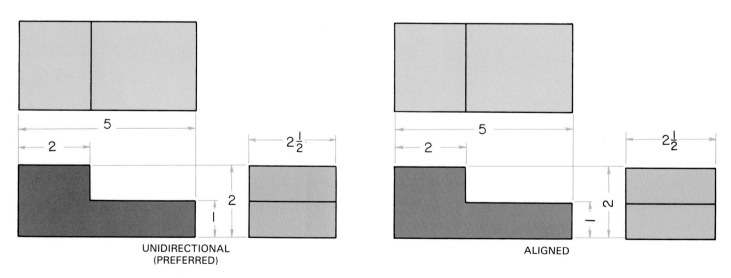

Figure 9-2. The two accepted methods for dimensioning drawings. The unidirectional method is preferred.

Dimensioning a Drawing

From your study of the ***alphabet of lines***, you will remember that special lines are used for dimensioning, Figure 9-3. The ***dimension line*** is a fine solid line used to indicate distance and location. It should be fine enough to contrast with the object lines. It is broken for the insertion of the dimension. The line is capped with arrowheads.

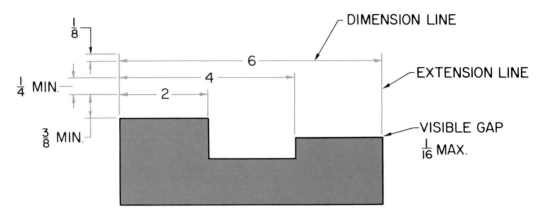

Figure 9-3.   The lines used for dimensioning. Note how they contrast with the object lines.

The line that extends out from the drawing is called an ***extension line.*** It projects about 1/8 in. (3.0 mm) beyond the last dimension line. It should not touch the drawing. The smaller or detail dimensions are nearest the view, while the larger or overall dimensions are farthest from the view.

Arrowheads are drawn freehand and should be carefully made, Figure 9-4. The solid arrowhead is generally preferred. It is made narrower and slightly longer than the open arrowhead. For most applications, 1/8 in. (3.0 mm) long arrowheads are satisfactory. Rub-on arrowheads are sometimes used by the drafter to save time.

Figure 9-4.   Arrowheads are drawn freehand. The solid arrowhead is generally preferred.

General Rules for Dimensioning

To be easily understood, dimensions should conform to the following general rules:

1. Place dimensions on the views that show the true shape of the object, Figure 9-5.
2. Unless absolutely necessary, dimensions should not be placed within the views.

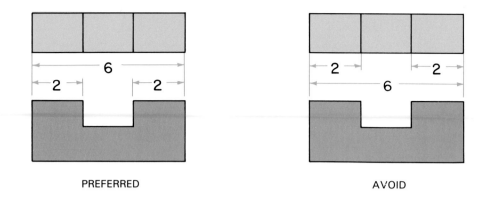

PREFERRED   AVOID

Figure 9-5. Keep the dimensions on the view that shows the true shape of the object.

3. If possible, dimensions should be grouped together rather than scattered about the drawing, Figure 9-6.

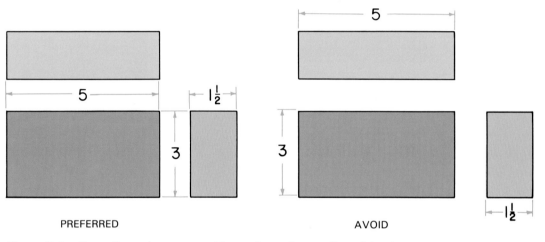

PREFERRED   AVOID

Figure 9-6. Keep dimensions grouped for easier understanding of the drawing.

4. Dimensions must be complete, so no scaling of the drawing is required. It should be possible to determine sizes and shapes without assuming any measurements.

5. Draw dimension lines parallel to the direction of measurement. If there are several parallel dimension lines, the numerals should be staggered to make them easier to read, Figure 9-7.

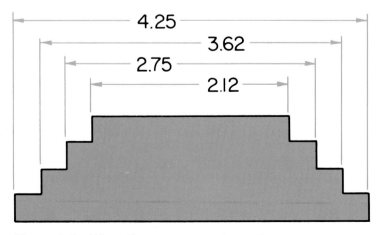

Figure 9-7. When there are several parallel dimension lines, numerals should be staggered to make them easier to read.

6. Dimensions should not be duplicated unless they are absolutely necessary to the understanding of the drawing. Omit all unnecessary dimensions, Figure 9-8.

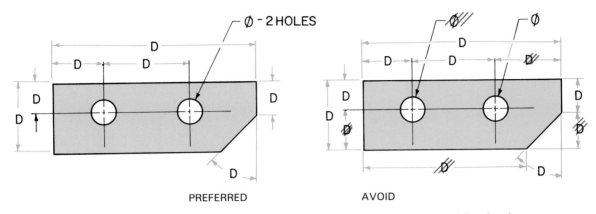

Figure 9-8. Avoid duplicating dimensions unless they are necessary to understand the drawing.

7. Plan your work carefully so the dimension lines do not cross extension lines, Figure 9-9.

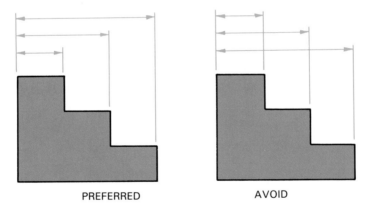

Figure 9-9. Line crossing can be kept to a minimum if the shortest dimension lines are placed next to the object outline. Longer dimension lines are placed farther from the drawing.

8. When all dimensions on a drawing are in inches, the inch symbol (") should not be used. Dimensions on metric drawings are in millimeters unless otherwise noted.
9. Numerals and fractions must be drawn in proper relation to one another. See Figure 9-10.

Dimensioning Circles, Holes, and Arcs

Many products are manufactured from parts that contain circles, round holes, or arcs in their design, Figure 9-11. They are often produced by reaming, boring, turning, spotfacing, drilling, counterboring, and countersinking, Figure 9-12.

PREFERRED

AVOID

Figure 9-10. Numerals and fractions must be drawn in proper size relation to one another.

Figure 9-11. These jet engine castings are typical of manufactured parts that contain circles, round holes, and arcs in their design. (Precision Castparts Corp.)

Figure 9-12.   Machining operations which produce circles or holes. A—Reaming produces a very accurately sized hole. The hole is first drilled slightly undersize before reaming. B—Boring is an internal machining operation. C—Turning work on a lathe. D—Spotfacing machines a surface that permits a bolt head or nut to bear uniformly over its entire surface. E—Drilling is an operation often performed on a drill press. F—Counterboring prepares a hole to receive a fillister or socket head screw. (Clausing)

The American National Standards Institute (ANSI) has set standards on how to call out a diameter or a radius. The Greek letter φ (phi, pronounced "fi") indicates that the dimension is a diameter. The symbol is placed *before* the dimension. Circles and round holes are dimensioned as shown in Figure 9-13.

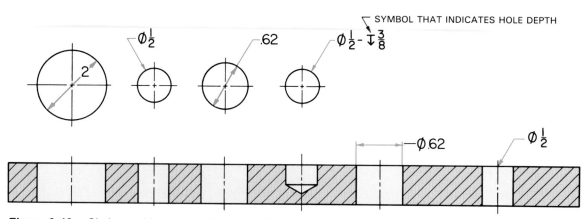

Figure 9-13. Circles and holes are dimensioned by giving the diameter.

Where it is not clear that a hole goes through the part, the abbreviation THRU follows the dimension. Symbols are recommended to show hole depth, and whether the hole is to be spotfaced, counterbored, or countersunk. See Figure 9-14. Also refer to page 381 in Useful Information.

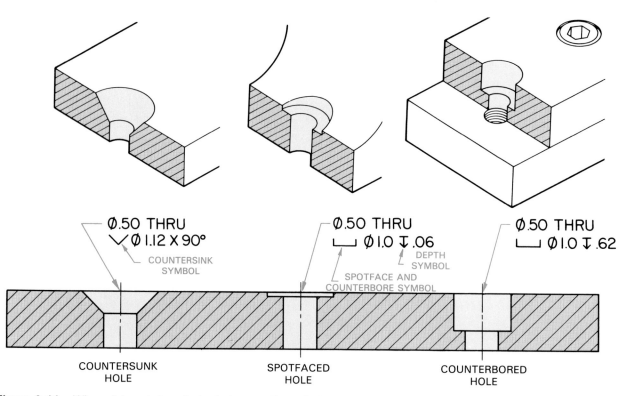

Figure 9-14. Where it is not clear that a hole goes through the part, the abbreviation THRU follows the dimension. Symbols indicate hole depth, and whether the hole is spotfaced, counterbored, or countersunk.

If the diameters of several concentric circles must be dimensioned on a drawing, it may be more convenient to show them on the front view, Figure 9-15.

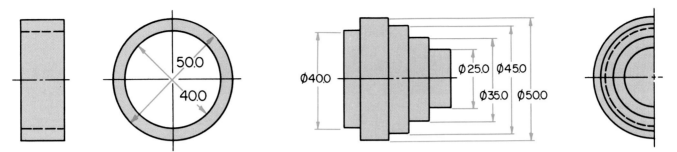

Figure 9-15. Recommended ways to dimension concentric circles. A diameter symbol (φ) is not necessary if the dimension(s) is on the circular view itself. Note that when the method on the right is used, only half a right side view is needed.

The correct way to use a leader to indicate a diameter is shown in Figure 9-16. The leader *always* points to the center of the diameter.

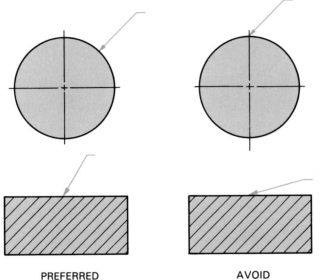

PREFERRED   AVOID

Figure 9-16. A leader is employed to direct attention to a note or to indicate sizes of arcs and circles. When used with arcs and circles, the leader should radiate from their centers.

Arcs are dimensioned by giving the radius, Figure 9-17. The capital letter R indicates that the dimension is a radius. It is placed *before* the dimension.

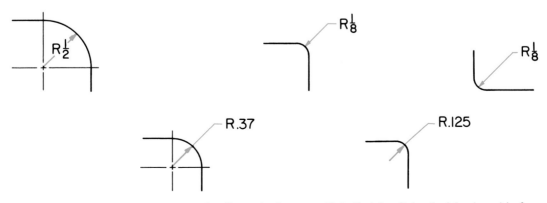

Figure 9-17. Recommended ways for dimensioning arcs. Note that the R (radius) is placed before the dimension.

Round holes and cylindrical parts are dimensioned from centers, *never* from the edges, Figure 9-18.

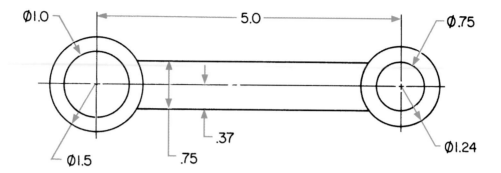

Figure 9-18. Where feasible, round holes and cylindrical parts are dimensioned from centers.

When it is necessary to dimension a series of holes around a circle, use a note to designate the number of holes, their size, and the diameter of the circle on which they are located, Figures 9-19 and 9-20.

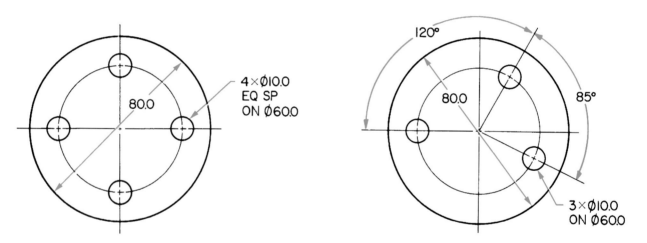

Figure 9-19. Left. The correct way to dimension equally spaced holes. Right. Holes that are not equally spaced are dimensioned this way.

Dimensioning Angles

Angular dimensions are expressed in degrees (°), minutes ('), and seconds ("). Angles are dimensioned as shown in Figure 9-21.

Dimensioning Small Portions of an Object

When the space between extension lines is too small for both the numerals and the arrowheads, dimensions are indicated as shown in Figure 9-22.

Conventional Measurement vs. Metric Measurement

In many companies, metric measurement is used extensively. Some companies have started changing to the metric system to compete successfully in the international market.

It is important that industries working together on projects use the same measurement system. One problem encountered in the transition (changeover) from the conventional system to the metric system has

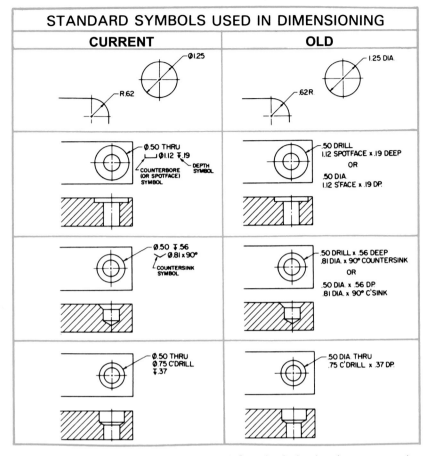

STANDARD SYMBOLS USED IN DIMENSIONING

CURRENT	OLD
Ø1.25 / R.62	1.25 DIA. / .62R.
Ø.50 THRU / ⌴Ø1.12 ⩗.19 COUNTERBORE (OR SPOTFACE) SYMBOL / DEPTH SYMBOL	.50 DRILL 1.12 SPOTFACE x .19 DEEP OR .50 DIA. 1.12 S'FACE x .19 DP.
Ø.50 ⩗.56 ⩗Ø.81 x 90° COUNTERSINK SYMBOL	.50 DRILL x .56 DEEP .81 DIA. x 90° COUNTERSINK OR .50 DIA. x .56 DP. .81 DIA. x 90° C'SINK
Ø.50 THRU Ø.75 C'DRILL ⩗.37	.50 DIA. THRU .75 C'DRILL x .37 DP.

Figure 9-20. ANSI (American National Standards Institute) recommendations for specifying dimensions to circles and holes. Study and compare the examples shown here.

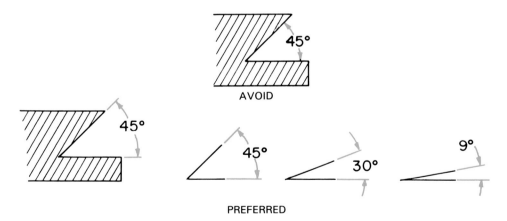

AVOID

PREFERRED

Figure 9-21. Dimensioning angles.

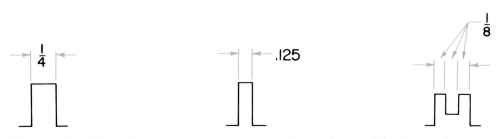

Figure 9-22. When the space between extension lines is too small to place both arrowheads and numerals, dimensions may be indicated as shown.

been at the *interface* (surfaces where parts come together) of a part. It is difficult to interface a part designed to metric standards to fit an existing part that was designed to inch standards.

Five methods have been devised for drafters to provide the information necessary to make metric parts that must interface with existing parts dimensioned using conventional measurement.

1. Dual dimensioning.
2. Dimensioning with letters and tabular chart.
3. Metric dimensioning with readout chart.
4. Dimensioning with metric units only.
5. Undimensioned master drawings.

Dual Dimensioning

The use of *dual dimensioning* was the first method devised to dimension engineering drawings with both inch and metric units, Figure 9-23.

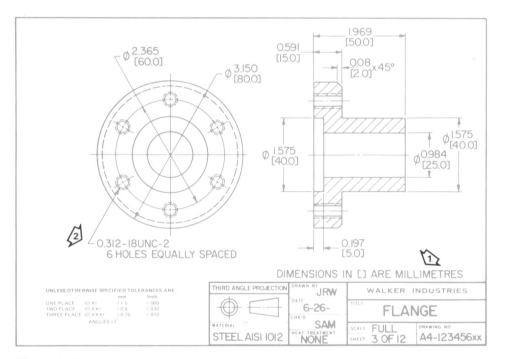

Figure 9-23. Dual-dimensioned drawing. Dual dimensioning has limited use. It was the first method devised to dimension engineering drawings with both inch and metric units. Notice the following important points on the drawing indicated by the numbered arrows 1. Note indicating how metric dimensions are identified. 2. Thread size is not given in metric units because there is no metric thread this size.

The dual dimensions are presented by the *position method* or the *bracket method*, Figure 9-24. The inch dimension was placed first on drawings of products to be made in the United States. The metric dimension was placed first on drawings to be produced where metrics was the basic form of measurement.

Dual dimensioning is the most complicated dimensioning system. It is seldom used. However, large numbers of dual-dimensioned drawings are still in use. It is presented in this text so you will be aware of its existence.

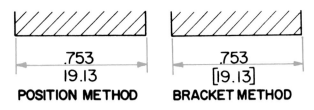

Figure 9-24. Methods of indicating inches and millimeters on a dual-dimensioned drawing.

Dimensioning with Letters and Tabular Chart

Dimensioning with letters, Figure 9-25, is a technique that is sometimes used. Letters (A, B, C, etc.) are used in place of either inch or millimeter dimensions. A *tabular chart* is added to the drawing. The chart shows the metric and inch equivalents of each letter.

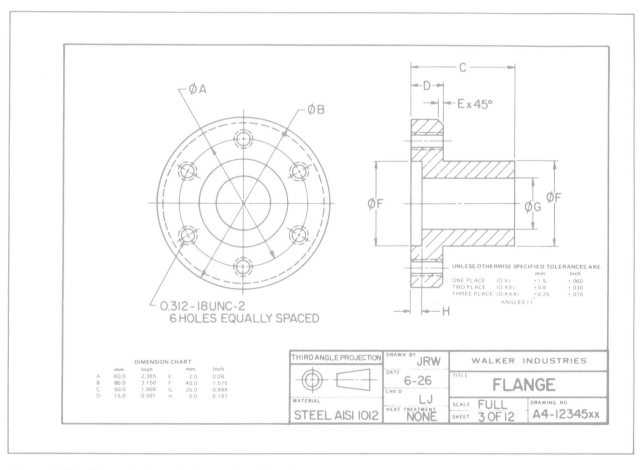

Figure 9-25. Dimensioning with letters. A tabular chart at the lower-left corner of the drawing shows the inch and millimeter equivalents for each letter.

Metric Dimensioning with Readout Chart

In doing metric dimensioning with a *readout chart*, the part is designed to metric standards. Only metric dimensions are placed on the drawing, Figure 9-26. A readout chart is added to the drawing. It shows metric dimensions in the left column and the inch equivalents in the right column. This technique permits a comparison of values.

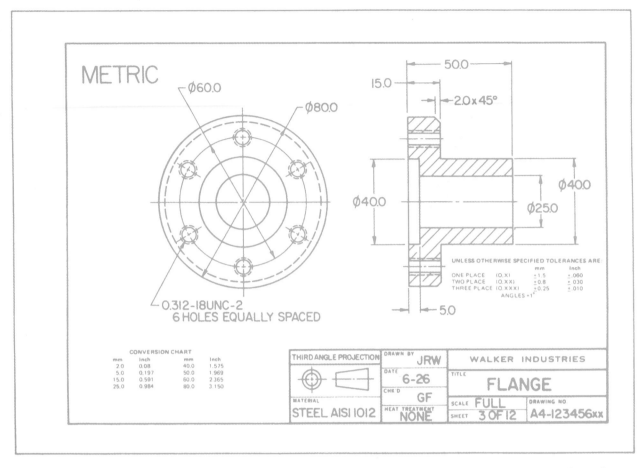

Figure 9-26. Metric dimensioned drawing with readout chart that shows inch equivalents to the metric dimensions.

Dimensions with Metric Units Only

Dimensioning with only metric units is the quickest way to get engineers, designers, drafters, and craftworkers to "think metric." Only metric dimensions are used on the drawing. See Figure 9-27.

Undimensioned Master Drawings

A *master drawing* is first made without dimensions, Figure 9-28. Next, prints are made. Metric dimensions may be added to one print, inch dimensions to another print. Notes and details can be added in whatever language is needed to produce the part—German, French, English, Japanese, etc.

Whatever dimensioning technique is used, drafting personnel must have both scales and templates on hand. They must also have a thorough knowledge of the metric system.

Drafters, engineers, or designers will find they cannot specify a 9/16 bolt by merely listing the metric equivalent of 9/16 in. (15.29 mm). There is no metric bolt that corresponds to this diameter. The same problem will occur if a 1.0 inch diameter is specified as a 25.4 mm diameter. There is no metric-sized shaft 1.0 inch in diameter. The closest size would be 25.0 mm diameter. To get the 25.4 mm diameter shaft, an expensive machining operation would be necessary to turn a larger shaft to the required 25.4 mm diameter.

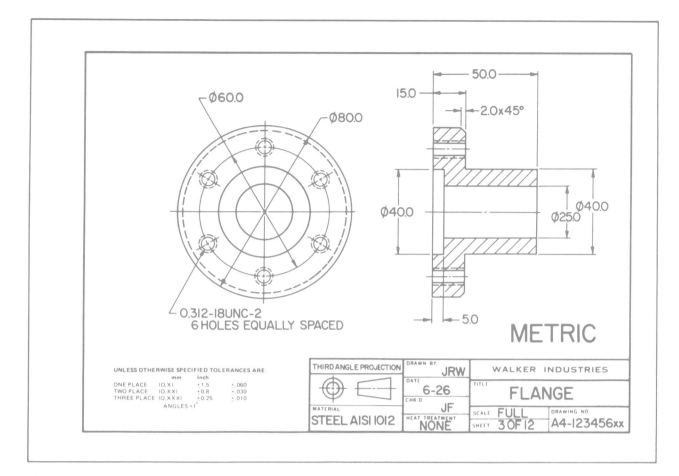

Figure 9-27. Drawings dimensioned only in metric units. (Since there is *no* metric equivalent for the thread indicated on the drawing, thread size is given in inch units.)

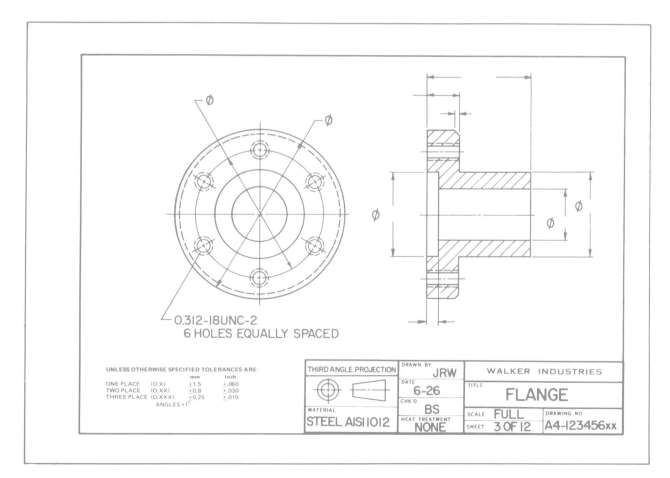

Figure 9-28. No dimensions are shown on the master drawing (except for thread size). After the print is made, dimensions in either inch or metric units are added.

General Rules for Metric Dimensioning

Metric dimensioning should conform to the following rules:

1. The millimeter is the standard metric unit for dimensioning engineering drawings.
2. Millimeter dimensioning on engineering drawings is based on the use of one-place decimals. All metric dimensioning should have one digit to the right of the decimal point. Two and three digits are used where critical tolerances are required. A zero will be shown to the right of the decimal point when full millimeter dimensions are shown.

125.0, not 125

3. When a dimension is less than one millimeter, a zero will be shown to the left of the decimal point.

0.5, not .5

4. All metric drawings should be clearly identified as such. The symbol for millimeter (mm) does not have to be added to every dimension if one of the notes shown in Figure 9-29 is placed on the drawing.

METRIC

ALL DIMENSIONS ARE IN MILLIMETERS.

Figure 9-29. Metric drawings should be clearly identified by use of either of these notes.

5. When the millimeter symbol is used on a drawing, a space is placed between the dimension and the symbol.

125.0 mm, not 125.0mm

Geometric Dimensioning and Tolerancing (GD&T)

The design of parts for many complex applications requires a highly precise system of specifying dimensions and tolerances on drawings. Machining processes have improved to the point where highly accurate definitions of tolerances for *form* (shape and size) and *position* (location) are needed to design products.

Geometric dimensioning and tolerancing (GD&T) is a standard system devised to control interpretation of the form, profile, orientation, location, and runout of features on drawings. This type of tolerancing provides the necessary precision for the most economical manufacture of parts and *interchangeable* parts. Because components for specialized products are typically manufactured in a number of locations (and often in different countries), different dimensioning standards exist and a common drawing *language* must be applied. The GD&T system provides an international language that standardizes the dimensioning and tolerancing process.

GD&T uses *geometric characteristic symbols* to specify and explain form and positional tolerances, Figure 9-29. These symbols relate to such variables as the form of an object, the profile or *outline* of an object, the orientation of features, the location of features, and the *runout* of surfaces or relationship of features to an axis.

Geometric tolerances are often used to specify the location or relationship of features that originate from a *datum*. A **datum** is an exact point, axis, plane, or surface from which features of a part are located. Datums are identified by letters enclosed in a box attached to a triangle, Figure 9-30. Datum feature symbols are positioned next to the datum point or surface, on an extension line, or under a dimension along with a note.

Characteristic symbols, datums, and other geometric tolerance specifications are often indicated in a *feature control frame*. A **feature control frame** is a rectangular compartment that contains a divided series of symbols identifying geometric tolerance. See Figure 9-31. The characteristic symbol is typically given first and is followed by the allowable tolerance and a datum reference letter. (Datums are also identified by squares containing a reference letter.) Feature control frames may be shown along with the dimension or attached to a dimension line or extension line.

Type of Tolerance	Geometric Characteristic	Symbol
Form	Straightness	—
	Flatness	▱
	Circularity	○
	Cylindricity	⌭
Profile	Profile of a line	⌒
	Profile of a surface	⌓
Orientation	Angularity	∠
	Perpendicularity	⊥
	Parallelism	//
Location	Position	⊕
	Concentricity	◎
	Symmetry	⌯
Runout	Circular runout	↗ or ↗
	Total runout	⌰ or ⌰

Geometric Tolerance Specification	Symbol
At maximum material condition	Ⓜ
At least material condition	Ⓛ
Diameter	⌀
Radius	R
Basic dimension	30
Reference dimension	(30)
Counterbore/spotface	⌴
Countersink	⌵
Depth/deep	⌲
Square (shape)	□
Dimension not to scale	25
Number of times/places	6X
Datum feature	A◄
Between	↔

Figure 9-29. The drafter uses various symbols to identify geometric characteristics in the application of geometric dimensioning and tolerancing. The table on the left shows the five standard types of geometric characteristic symbols applied to communicate form, profile, orientation, location, and runout. The table on the right shows additional symbols used in the GD&T system, including maximum material condition, least material condition, datum feature, and other practical items. The application of GD&T symbols on drawings supports precision manufacturing and the global economy.

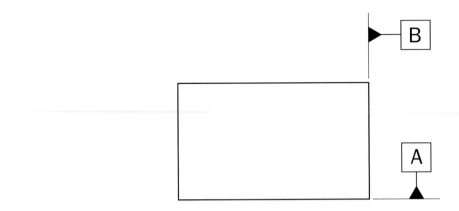

Figure 9-30. A datumis identified by a letter inside a box attached to a solid triangle. A datum is used to indicate an exact point, axis, plane, or surface from which features are located.

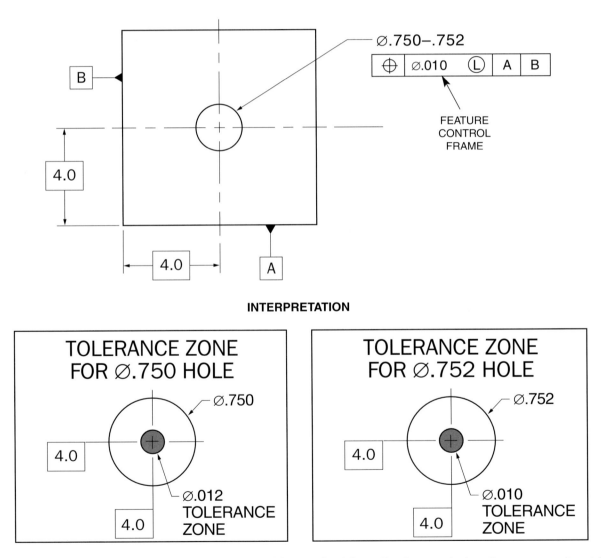

Figure 9-31. The feature control frame on this drawing provides precise information to manufacture the component part. This feature control frame specifies the tolerance for the position of the hole at least material condition. The interpretation of the information in the feature control frame is shown below the drawing. By learning the GD&T system, you will be able to interpret the exact location and form of features.

Drafting Vocabulary

Aligned
Alphabet of lines
American National
   Standards Institute
   (ANSI)
Arrowhead
Bracket method
Concentric
Counterboring
Countersinking
Datum
Degrees
Diameter
Dimension line
Dimensions

Dual dimensioning
Extension line
Feature control frame
Geometric dimensioning
   and tolerancing
   (GD&T)
Geometric characteristic
   symbols
Interface
Leader
Master drawing
Minutes
Notes
Parallel
Position method

Radius
Readout chart
Scale drawing
Scaling
Seconds
Series
Shop notes
Specifications
Spotfacing
Staggered
Symbol
Tabular chart
Unidirectional

Test Your Knowledge—Unit 9

Please do not write in the book. Place your answers on another sheet of paper.

1. Dimensions and shop notes are needed if _____.

2. Dimensions define _____.

3. Notes provide _____.

4. Unidirectional dimensions are read from _____.

5. Aligned dimensions are read from _____.

6. The dimension line is a _____.

   A. heavy solid line

   B. fine solid line

   C. fine dotted line

   D. None of the above.

7. The dimension line is capped with _____.

8. Dimensions are placed on views that show _____.

9. Dimensions should be _____ _____ rather than _____ the drawing.

10. Dimensions must be complete so that _____.

   A. no scaling of the drawing is necessary

   B. sizes and shapes can be determined without assuming any measurements

   C. All of the above.

   D. None of the above.

11. When all dimensions are in inches, the _____ _____ is not used.

12. Circle diameter is indicated by the symbol _____. It is placed _____ the dimension.

13. Sketch the symbols used to indicate the following. Label each symbol on your answer sheet.
    A. Hole depth.
    B. Countersunk hole.
    C. Counterbored or spotfaced hole.
14. When a leader is used to indicate a diameter, it always points to the _____ of the diameter.
15. The letter _____ indicates the radius of an arc. It is placed _____ the dimension.
16. Define "dual dimensioning."
17. Some drawings are dimensioned with letters. Where does the person using the drawing get the inch/millimeter sizes needed to make the part?
18. What is a master drawing and how is it used?
19. A highly precise system of specifying dimensions and tolerances on drawings is called _____.
20. A(n) _____ is an exact point, axis, plane, or surface from which features of a part are located.

Outside Activities

1. Contact several industrial drafting departments. If possible, secure copies of the standards they use for dimensioning. Compare how similar or dissimilar they are. Do they vary by industry?
2. Secure copies of industrial prints. Study how they are dimensioned and how notes are used. Are there any provisions for foreign components to be used with the items on the prints? Will the parts be exported?

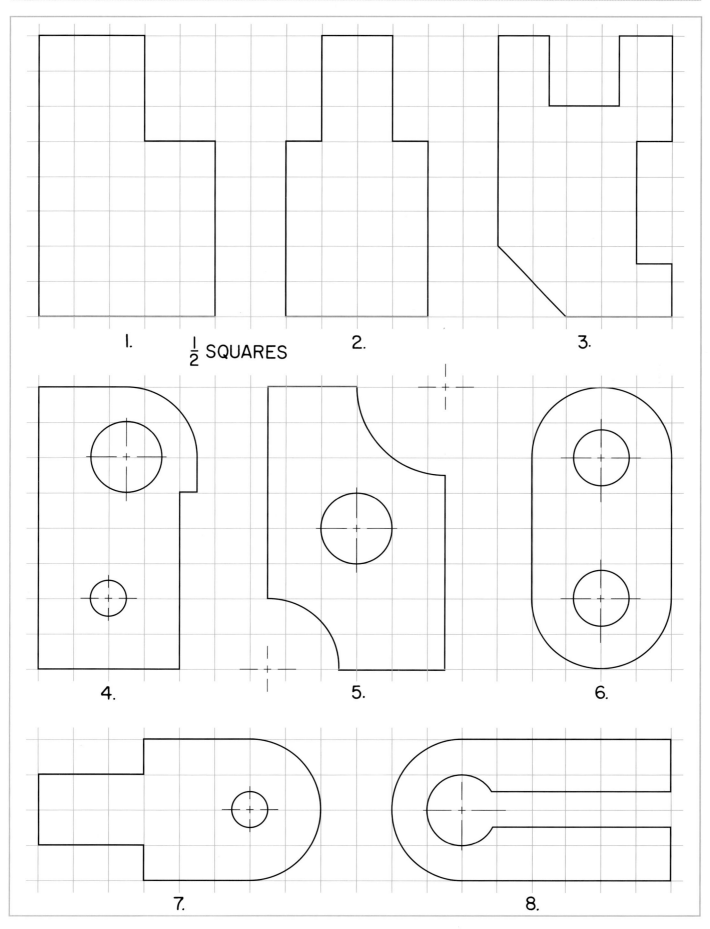

1. $\frac{1}{2}$ SQUARES   2.   3.

4.   5.   6.

7.   8.

Problems 9-1 to 9-8. DIMENSIONING PROBLEMS.   Redraw and dimension. Two of these problems should be drawn on a sheet.

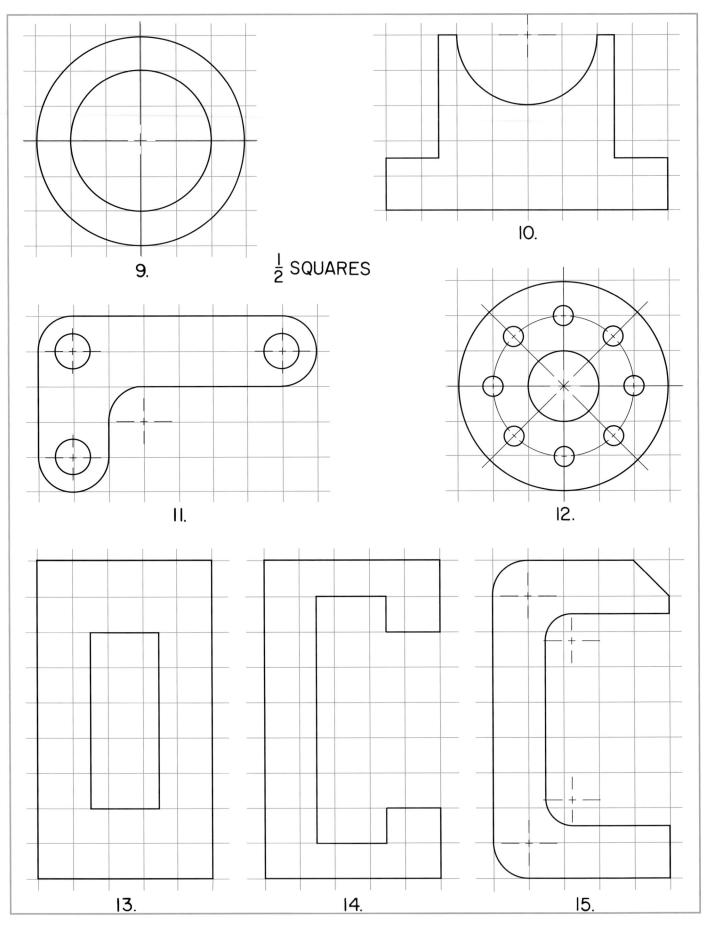

$\frac{1}{2}$ SQUARES

9.

10.

11.

12.

13.

14.

15.

Problems 9-9 to 9-15. DIMENSIONING PROBLEMS. Redraw and dimension problems.

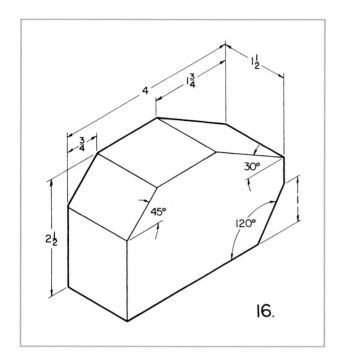

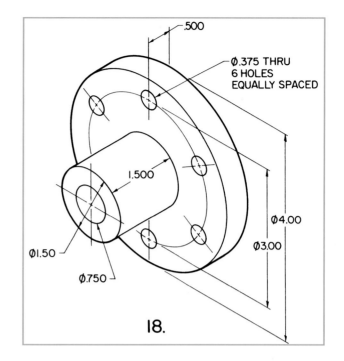

Problem 9-16. SHIM. Prepare the necessary views and correctly dimension them.

Problem 9-18. FACE PLATE. Prepare a two-view drawing and dimension correctly.

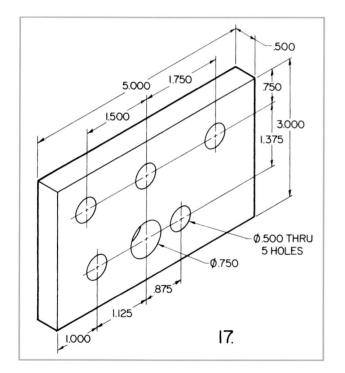

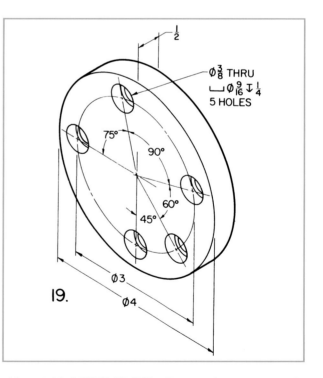

Problem 9-17. ALIGNMENT PLATE. Prepare the necessary views and correctly dimension them.

Problem 9-19. COVER PLATE. Prepare the necessary views and correctly dimension them.

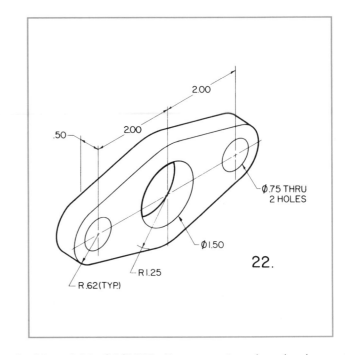

METRIC

100.0
75.0
30.0
10.0
5.0 - 6 FINS
5.0 - 5 SLOTS
15.0
25.0
55.0
12.5
2×Ø6.0

20.

Problem 9-20. HEAT SINK. Draw the necessary views and correctly dimension them.

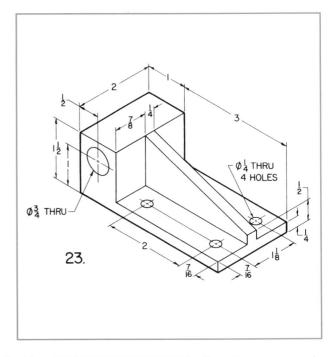

2.00
2.00
.50
Ø.75 THRU
2 HOLES
Ø1.50
R1.25
R.62(TYP.)

22.

Problem 9-22. GASKET. Prepare a two-view drawing and dimension correctly.

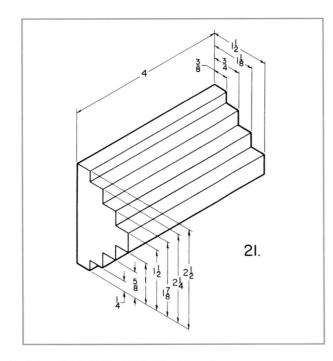

4
1½
3/8
3/4
1⅛
5/8
1½
1⅞
2½
2¼
¼

21.

Problem 9-21. STEP BLOCK. Draw the necessary views and correctly dimension them.

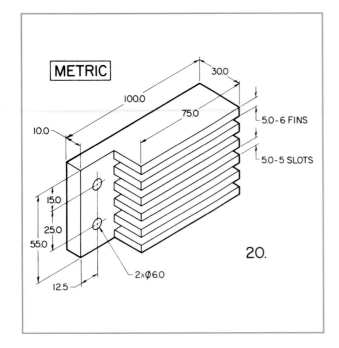

2
1
½
7/8
¼
1½
3
Ø¼ THRU
4 HOLES
½
Ø¾ THRU
2
1¼
2
7/16
7/16
1⅛

23.

Problem 9-23. MOTOR BRACKET. Draw the necessary views and correctly dimension them.

Unit 10
Sectional Views

After studying this unit and completing the assigned problems, you should be able to:

◆ Describe what a sectional view is and why it is used.

◆ Explain when a sectional view is needed.

◆ Draw sectional views when required.

◆ Apply the following types of sectional views to various drawings: full section, half section, revolved section, aligned section, removed section, and broken-out section.

Sectional views permit complicated interior features to be shown on a drawing without a large number of confusing hidden lines. Sectional views are also useful when explaining interior construction or hidden features which cannot be shown clearly by various outside views and hidden lines.

Details on objects that are simple in design can be shown in regular multiview drawings. When an object has some of its design features hidden from view, as in Figure 10-1, it is another matter. It is not possible to show the interior structure without using a "jumble" of hidden lines. Sectional views are employed to make drawings of this type less confusing and easier to understand or visualize, Figure 10-2.

Sectional views show how an object would look if a cut were made through it perpendicular (at an exact right angle) to the direction of sight, Figure 10-3. The section drawing should be placed behind the arrows. Sectional view drawings are necessary for a clear understanding of the shape of complicated parts.

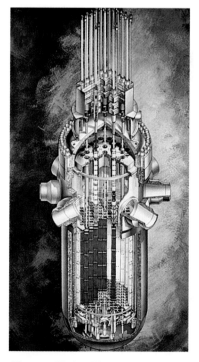

Figure 10-1. Pictorial sectional view showing the interior details of a nuclear reactor vessel. The unit shown is used to make electricity. (Westinghouse)

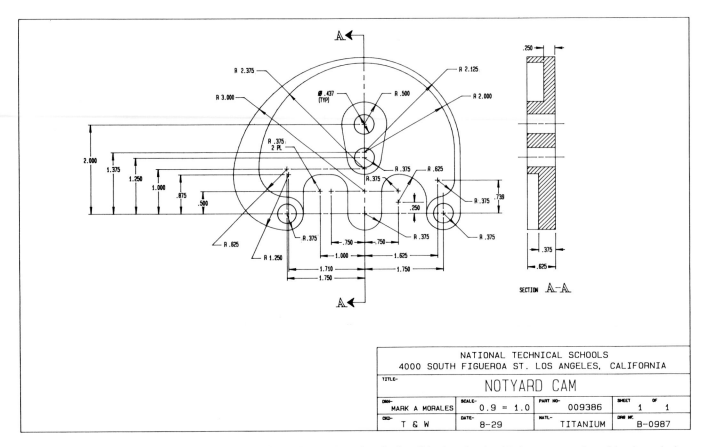

Figure 10-2. Computer-generated sectional view. The student developing this drawing had to have a good working knowledge of basic drafting procedures and techniques.

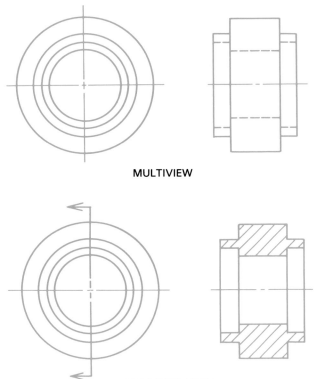

Figure 10-3. Comparison of a conventional multiview drawing and a drawing in section. The sectional view is always placed behind the arrows.

Cutting-Plane Line

The *cutting-plane line* indicates the location of an imaginary cut made through the object to reveal its interior structure, Figure 10-4. The extended lines are capped with arrowheads that show the direction of sight to view the section. Some drafters prefer the use of a simplified cutting-plane line that employs only the ends.

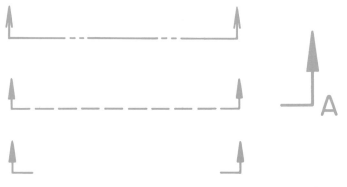

Figure 10-4. Cutting-plane lines. To save time, some drafting departments prefer to use the simplified cutting-plane line that employs only the ends.

Letters A-A, B-B, etc., identify the section if it is moved to another position on the drawing. The same technique is used if several sections are used on a single drawing.

Section Lines

General-purpose section lining is often used in drawings where the material specifications ("specs") are shown elsewhere on the drawing.

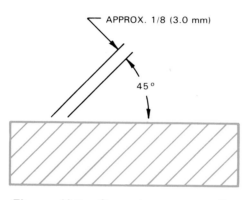

APPROX. 1/8 (3.0 mm)

45°

Figure 10-5. General-purpose section lining is spaced by eye and usually drawn at 45 degrees. Spacing of 1/8 in. (3.0 mm) is commonly used.

Spacing of general-purpose section lining is by eye and usually at 45 degrees, Figure 10-5. Line spacing is somewhat dependent on the drawing size or the area to be sectioned. Spacing of 1/8 in. (3.0 mm) is commonly used. However, larger spac-

ing is recommended when large areas must be sectioned. General-purpose section lining is the same as the symbol used for cast iron. For an expanded list of materials and their section lining symbols, refer to Figure 10-21.

Do not use section lines that are too thick or spaced too closely. Also avoid lines that are not uniformly spaced or are drawn in different directions. See Figure 10-6.

When section lines will be parallel, or nearly parallel with the outline of an object, they should be drawn at some other angle, Figure 10-7.

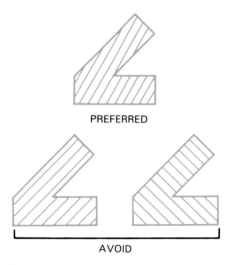

PREFERRED

AVOID

Figure 10-7. The outline shape of the section may require the section lining to be drawn at other than 45 degrees.

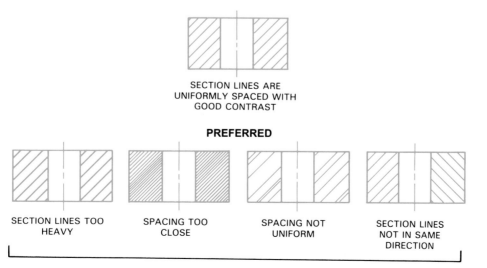

SECTION LINES ARE
UNIFORMLY SPACED WITH
GOOD CONTRAST

PREFERRED

SECTION LINES TOO
HEAVY

SPACING TOO
CLOSE

SPACING NOT
UNIFORM

SECTION LINES
NOT IN SAME
DIRECTION

AVOID

Figure 10-6. Section line spacing.

Should two or more pieces be shown in section, the section lines should be drawn in opposite directions and/or angles to provide contrast, Figure 10-8.

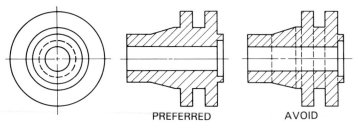

Figure 10-10. Unless absolutely necessary for a complete understanding of a view, hidden lines behind the cutting plane are omitted.

Full Section

A ***full section*** is developed by imagining that the cut has been made through the entire object, Figure 10-11. The part of the object between your eye and the cut is removed to reveal the interior features of the object. The resulting features are drawn as part of the regular multiview projection.

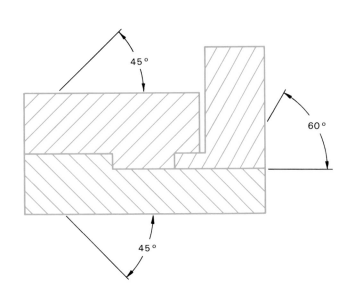

Figure 10-8. Drawing section lines when several parts in the same section are adjacent (next to each other).

Outline sectioning is a permissible and time-saving technique for indicating large sections, Figure 10-9. Section lines are only drawn near the boundary of the sectioned area. The interior portion is left clear.

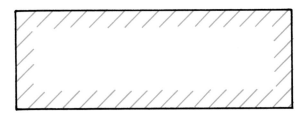

Figure 10-9. Outline sectioning saves time when large areas must be sectioned.

Unless absolutely necessary for a complete understanding of a view, or for dimensioning purposes, hidden lines behind the cutting plane are omitted, Figure 10-10.

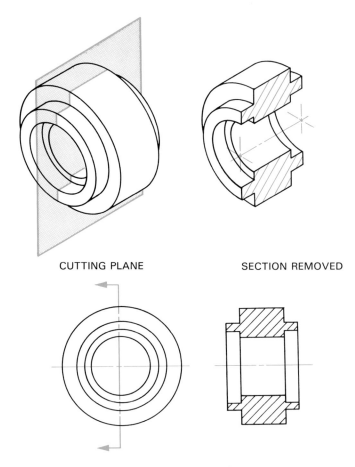

CUTTING PLANE   SECTION REMOVED

THE DRAWING

Figure 10-11. Full section. The cutting plane passes completely through the object; the section behind the cutting plane is exposed.

Half Section

The shape of one-half of the interior features and one-half of the exterior features are shown in a *half section*, Figure 10-12. Half sections are best suited for **symmetrical objects** (those objects having the same size, shape, and relative position on opposite sides of a dividing line or plane). Cutting-plane lines are passed through the piece at right angles to each other. One-quarter of the object is considered removed to show a half section of the interior structure. Unless needed for clarity or dimensioning, hidden lines are not used on half sections.

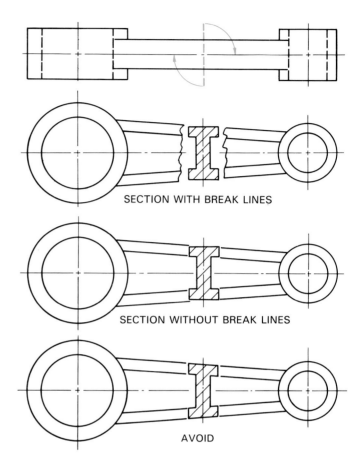

Figure 10-13.   A revolved section may be presented with break lines or without break lines. Note that object lines do not pass through the revolved section when the break lines are omitted.

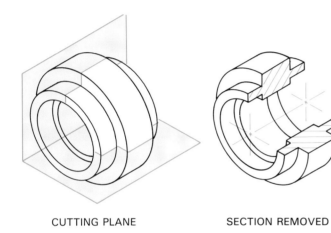

CUTTING PLANE   SECTION REMOVED

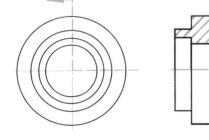

THE DRAWING

Figure 10-12.   Half section. The internal details of a symmetrical object can be fully described by removing one-quarter of the object.

Revolved Section

Revolved sections are primarily utilized to show the shape of such objects as spokes, ribs, and stock metal shapes, Figure 10-13. A revolved section is a drawing within a drawing. To prepare a revolved section, imagine that a section of the part to be shown is cut out and revolved 90 degrees in the same view. Do not draw the lines of the normal view through the revolved section, Figure 10-14.

Aligned Section

It is not considered good drafting practice to make a full section of a symmetrical object that has an odd number of holes, webs, or ribs. An *aligned section* shows two of the holes, webs, or ribs depicted on the view, one of them being revolved into the plane of the other, Figure 10-15. The actual or true projection in a full section may be misleading or confusing.

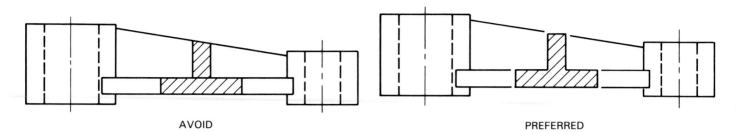

Figure 10-14. A revolved section must show the true shape of the object at the cutting plane. Lines showing the outline of the object in section are never drawn through the revolved section.

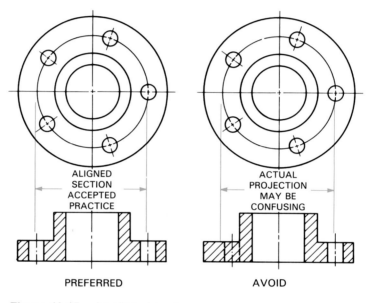

Figure 10-15. An aligned section.

Removed Section

There are times when it is not possible to draw a needed sectional view on one of the regular views. When this occurs, a *removed section* is generally used, Figure 10-16. The section (or sections) are placed elsewhere on the drafting sheet.

A removed section is also employed when the section must be enlarged for better understanding of the drawing.

Offset Section

While the cutting plane is ordinarily taken straight through an object, it may be necessary to show features not located on such a straight cutting plane. An *offset section* uses a cutting plane that is stepped or offset to pass through the required features, Figure 10-17. The features are drawn as if they were in the same plane.

Offsets in the cutting plane are not shown in the sectional view. The section view appears as one complete drawing. Use reference letters A-A, B-B, etc., at the ends of cutting-plane lines.

Broken-Out Section

The *broken-out section* is employed when a small portion of a sectional view will provide the required information. Break lines define the section and are shown on one of the regular views, Figure 10-18.

Conventional Breaks

When an object with a small cross section but of some length must be drawn, the drafter is faced with many choices. If the drawing is made full size, it may be too long to fit on the sheet. If the scale

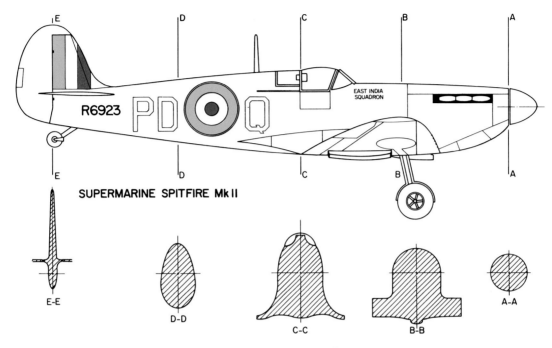

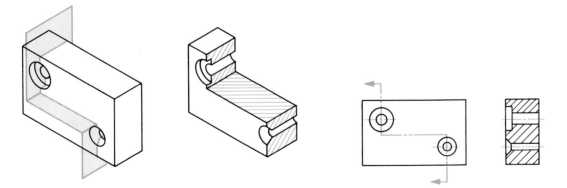

Figure 10-16.   A drawing showing the use of removed sections.

Figure 10-17.   An offset section.

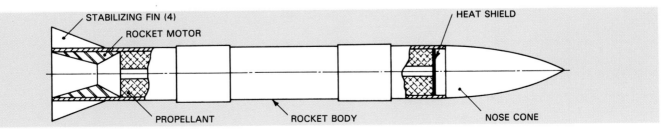

Figure 10-18.   Broken-out section.

used is small enough to fit the part on a sheet, the details may be too small to give the required information or to be dimensioned.

In such cases, the object can be fitted to the sheet by reducing its length by means of a *conventional break*, Figure 10-19. This method permits a portion of the object to be deleted from the drawing. A conventional break can only be used when the cross section of a part is uniform its entire length.

Conventional breaks for round and tubular shapes may be drawn freehand, or with instruments as shown in Figure 10-20.

Symbols to Represent Materials

Figure 10-21 shows symbols which may be employed on sectional views to indicate a number of different materials. The lines should be drawn dark and thin to contrast with the heavier object lines.

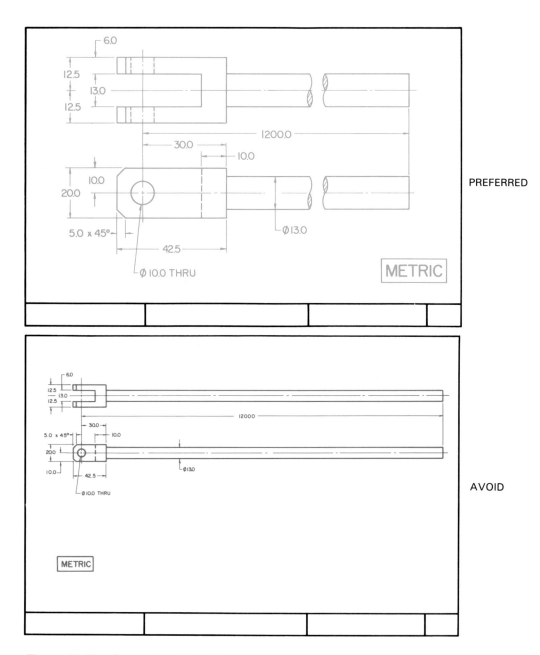

Figure 10-19. Conventional break in use.

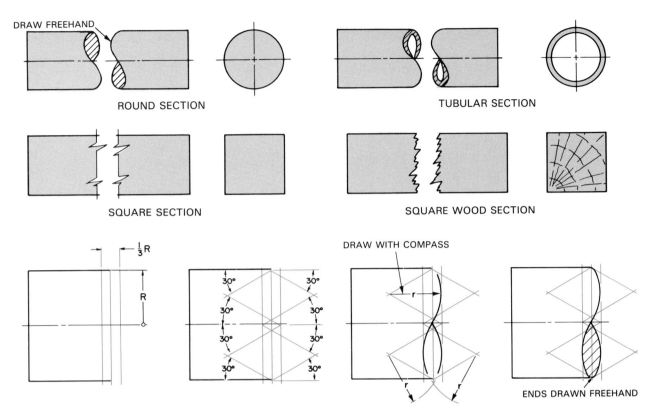

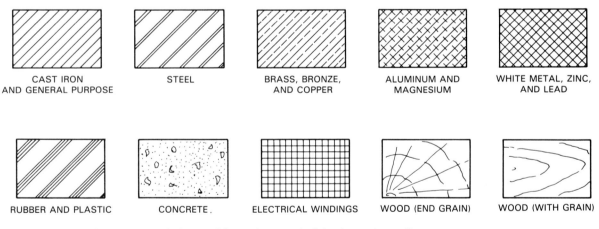

Figure 10-20. Conventional breaks. By using breaks, an object with a small cross section but some length may be drawn full size (or in a larger scale) on a standard-size drawing sheet.

CAST IRON AND GENERAL PURPOSE   STEEL   BRASS, BRONZE, AND COPPER   ALUMINUM AND MAGNESIUM   WHITE METAL, ZINC, AND LEAD

RUBBER AND PLASTIC   CONCRETE   ELECTRICAL WINDINGS   WOOD (END GRAIN)   WOOD (WITH GRAIN)

Figure 10-21. Standard symbols used for various materials shown in section.

Drafting Vocabulary

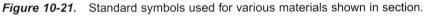

Aligned section
Broken-out section
Conventional break
Cutting-plane line
Details
Full section
Half section

Interior feature
Interior structure
Offset
Offset section
Outline sectioning
Parallel
Perpendicular

Reference
Removed section
Revolved section
Sectional view
Specifications
Symmetrical

Test Your Knowledge—Unit 10

Please do not write in the book. Place your answers on another sheet of paper.

1. What are sectional views?
2. When are section views used?
3. The _____ _____ _____ indicates the location of an imaginary cut made through the object to show its interior details.
4. Sections are identified by the use of _____.
5. The _____ section is used when the cut has been made through the entire object.
6. The _____ section shows half of the interior and half of the exterior of the object.
7. _____ sections are primarily used to show the shape of such things as spokes, ribs, and stock metal shapes.
8. The broken-out section is used when _____.
9. Objects with small cross sections but of considerable length can be fitted on a standard size sheet by making use of the _____.
10. _____ lines may be used to indicate the material that has been cut.

Outside Activities

1. Obtain an industry print using a sectional view. With colored pencils, highlight the different component sections in the plan to demonstrate how the various parts fit together.
2. Using a cutaway object, draft the sectional view showing the internal features. Use both the object and the completed drawing as a classroom teaching aid.
3. Create sectional view drawings of common drafting tools such as a drawing pencil, a T-square head, or a bow compass.

PROBLEMS

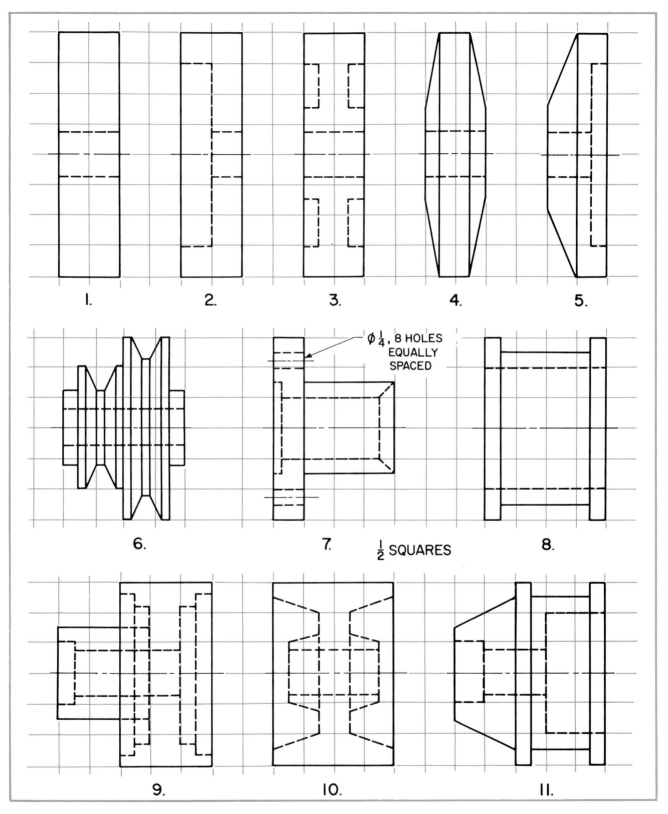

1. 2. 3. 4. 5.

6. 7. $\frac{1}{2}$ SQUARES 8.

$\phi \frac{1}{4}$, 8 HOLES EQUALLY SPACED

9. 10. 11.

Problems 10-1 to 10-5. GRINDING WHEELS. Draw the views necessary to show the shape of each wheel. Make one view a full section or a half section as directed by your instructor. *Problem 10-6. PULLEY.* Draw the views necessary to show the shape of this pulley. Draw one view as a half section. *Problem 10-7. COUPLING.* Draw the views necessary to show the shape of this coupling. Draw one view as a full section. *Problem 10-8. BUSHING.* Draw the views necessary to show the shape of this bushing. Draw one view as a half section. *Problem 10-9. SPACER.* Draw the views necessary to show the shape of this spacer. Draw one view as a half section. *Problem 10-10. FLAT BELT PULLEY.* Draw the views necessary to show the shape of this pulley. Draw one view as a full section. *Problem 10-11. ADAPTER BEARING.* Draw the views necessary to show the shape of this bearing. Draw one view as a half section.

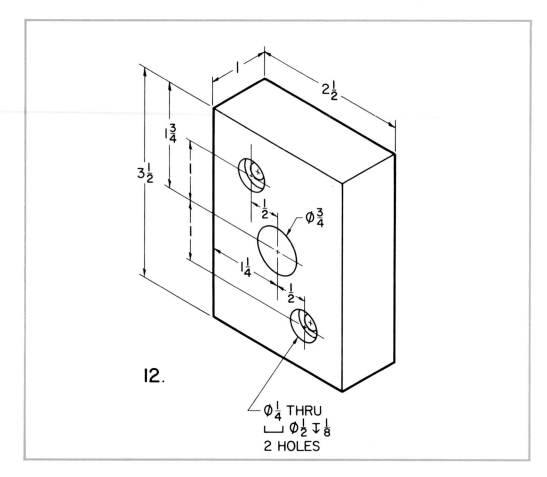

12.

Ø¼ THRU
⌴ Ø½ ⊤⅛
2 HOLES

Problem 10-12. SPACER.   Draw the views necessary to show the shape of the spacer.
Draw one view as an offset section through the three holes.

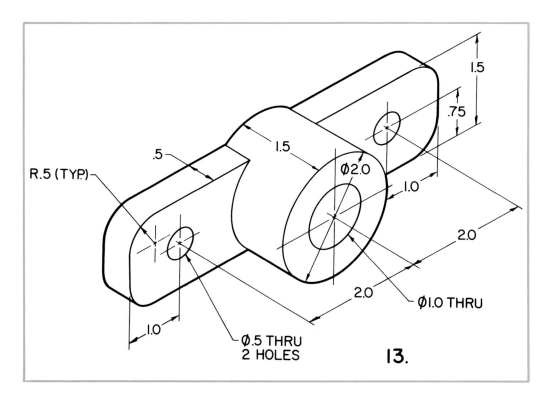

13.

Problem 10-13. SPECIAL BUSHING.   Draw the views necessary to show the shape of the
bushing. Draw one view as a full section along the horizontal plane.

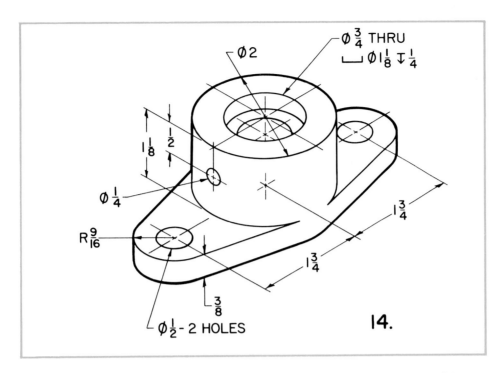

Problem 10-14. BRACKET. Draw the views necessary to show the shape of the bracket. Include a broken-out section through the 1/4 in. diameter hole on one view.

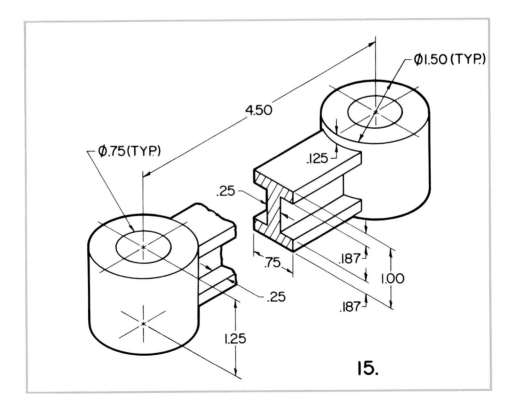

Problem 10-15. CONNECTING ROD. Draw the views necessary to show the shape of the rod. The cross section of the rod may be shown as a revolved section or as a removed section.

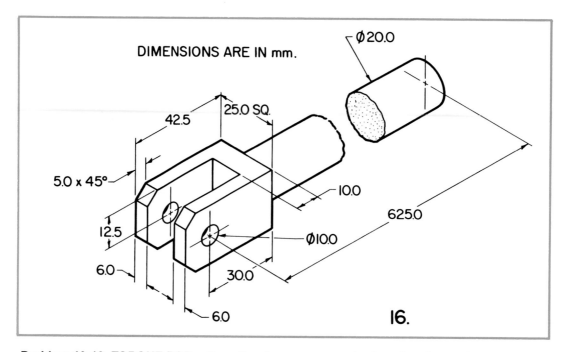

Problem 10-16. TORQUE ROD.   Draw the views necessary to show the shape of the rod. Use the conventional break to fit the object on the drawing sheet.

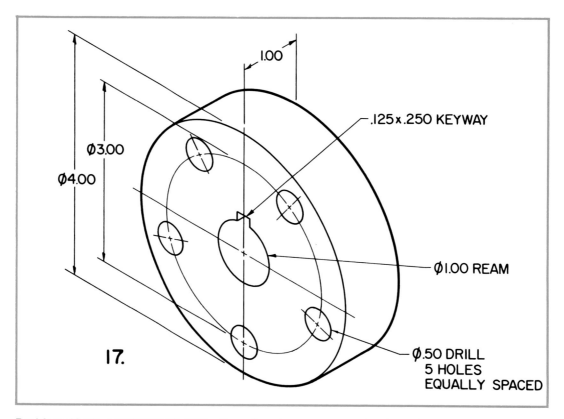

Problem 10-17. ADAPTER PLATE.   Draw the views necessary to show the shape of the plate. Draw one view as a full section.

Auxiliary Views

After studying this unit and completing the assigned problems, you should be able to:

◆ Describe what an auxiliary view is.

◆ Determine when an auxiliary view is needed to fully describe an object.

◆ Explain why an auxiliary view is constructed perpendicular to an angular surface.

◆ Develop and draw simple auxiliary views.

The true shape and size of objects having angular or slanted surfaces cannot be drawn using the regular top, front, and right side views. The **latch plate** shown in Figure 11-1 is such an object. The true length of the angular surface is shown on the front view, but this view does not show its

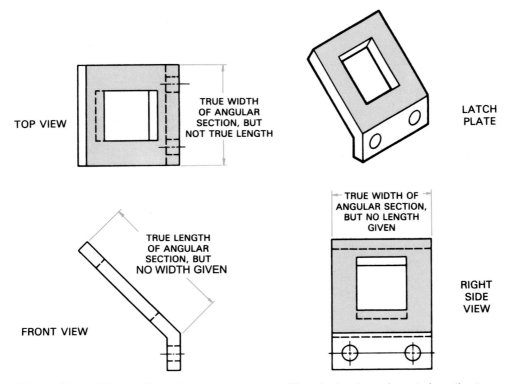

Figure 11-1. Why auxiliary views are necessary. The single views do not show the true shape (length and width) of the angled portion of the object.

width. The true width of the surface is shown on the top and right side views but neither view shows its true length.

An additional or *auxiliary view* is needed to show the true length and true width of the angular surface, Figure 11-2.

When drawing an auxiliary view, remember that the view is *always* projected from the regular view on which the true length of the inclined surface is shown. Also, the construction lines projecting from the angled surface are *always* at right angles to that surface. In other words, the auxiliary view is constructed perpendicular to the inclined line.

When drawing auxiliary views, the usual practice is to show only the angled portion of the view. It is seldom necessary to draw a full projection of the object. See Figure 11-3. It is also often possible to eliminate one of the conventional views when using an auxiliary view, Figure 11-4.

An auxiliary view can be projected from any view that shows the angled surface as a line. It would be called a *front auxiliary* if the view is projected from the front view. A *top auxiliary* is projected from the top view. A *right side auxiliary* is projected from the right side view.

Considerable time can be saved when drawing auxiliary views of symmetrical objects by drawing one-half of the view, Figure 11-5.

Auxiliary views that include rounded surfaces or circular openings may cause minor problems. Figure 11-6 shows how a circular surface would be drawn. Proceed as follows:

1. Draw the needed front, top, or right side views. Divide the circular view into 12 equal parts. This is done with a 30-60 degree triangle. Project the divisions from the circular view to the other view(s). Identify each division with letters as shown. Use A-A for the horizontal centerline.

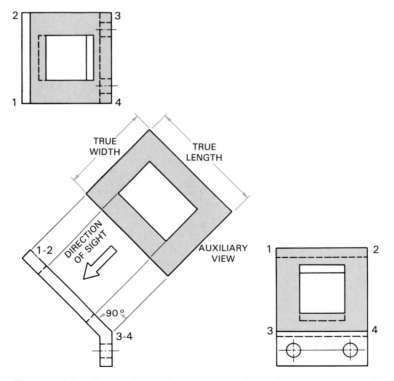

Figure 11-2. The auxiliary view shows the true shape of the angled surface.

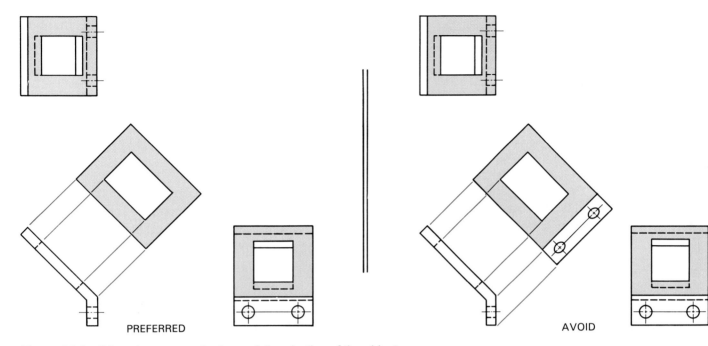

Figure 11-3. It is not necessary to draw a full projection of the object.

PREFERRED

AVOID

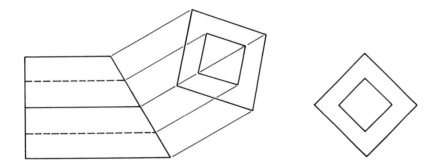

Figure 11-4. One view may often be eliminated when using an auxiliary view.

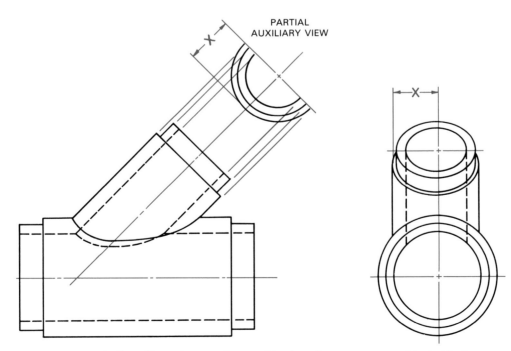

PARTIAL
AUXILIARY VIEW

Figure 11-5. A half auxiliary view can be used to describe some symmetrical objects.

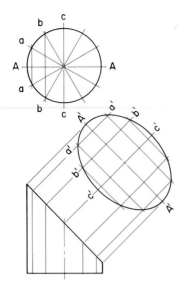

Figure 11-6. Drawing an auxiliary view of a circular object.

2. At any convenient distance from the inclined surface, draw the centerline A'-A' for the auxiliary view. The centerline will be parallel to the inclined face.

3. Project the necessary points from the inclined surface through centerline A'-A'. These lines are at right angles (90 degrees) to the inclined face.

4. Using dividers or a compass, transfer measurements from horizontal centerline A-A to circumference points a, b, c, etc., to the proper projection lines to the right and left of inclined centerline A'-A'. This constructs points a', b', c'.

5. Complete the auxiliary view by connecting the points with a French curve.

Drafting Vocabulary

Angular surface	Project	True length
Auxiliary view	Right angle	True width
Front auxiliary	Right side auxiliary	
Inclined surface	Top auxiliary	

Test Your Knowledge—Unit 11

Please do not write in the book. Place your answers on another sheet of paper.

1. Why are auxiliary views needed?
2. The auxiliary view is always projected from the view that shows the inclined surface as a(n) _____. The construction lines projecting from the inclined surface are always at or _____ to the cut.
3. The auxiliary view when projected from the front view is called a(n) _____ _____.
4. When drawing an auxiliary view of a symmetrical object, much time can be saved by drawing _____ _____ of the view.
5. Connect the points of rounded surfaces or circular openings by using a _____ _____.

Outside Activities

1. Make a sketch of an object that would require an auxiliary view.
2. Make a collection of items from your school that need auxiliary views to show true size and shape. An example would be starting blocks from the track. Draw the views needed to construct such a part.
3. Make a sketch of a home or garage peaked roof as an auxiliary view. Add appropriate dimensions. What is the relationship between the area of roof or the number of shingles used as compared to the steepness of the roof?

Problems

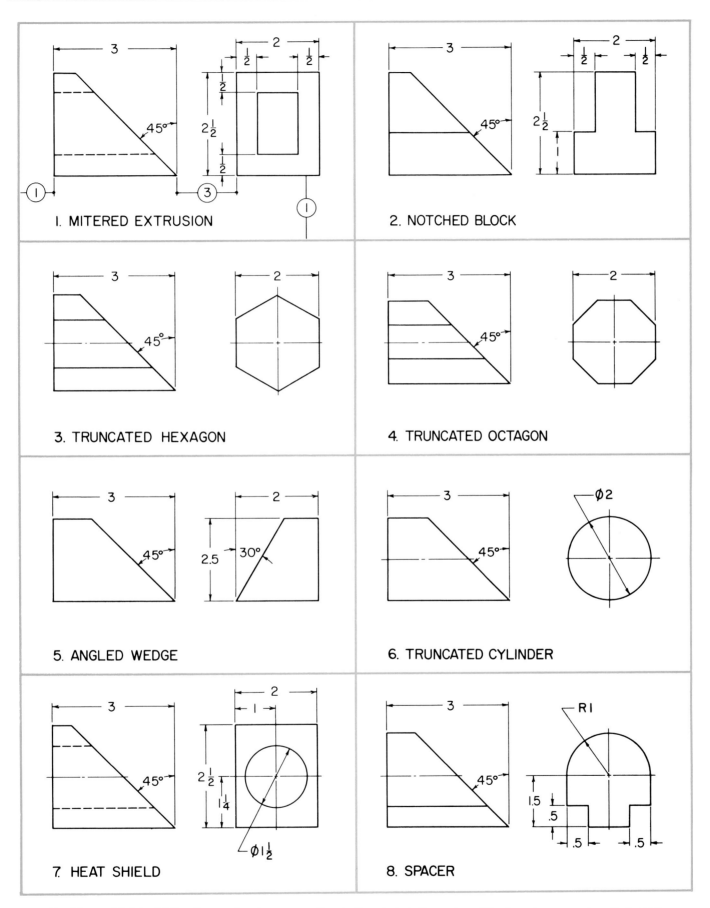

1. MITERED EXTRUSION

2. NOTCHED BLOCK

3. TRUNCATED HEXAGON

4. TRUNCATED OCTAGON

5. ANGLED WEDGE

6. TRUNCATED CYLINDER

7. HEAT SHIELD

8. SPACER

Problems 11-1 to 11-8. Space drawings according to the circled dimensions. The top views may be eliminated.

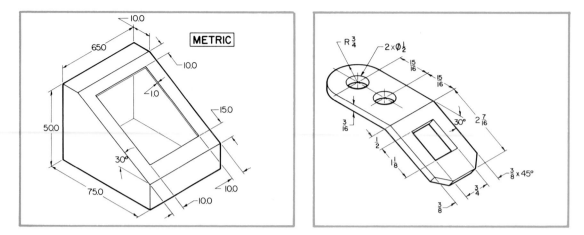

Problem 11-9. Left. INSTRUMENT CASE. 1—Use a vertical drawing sheet format. 2—Allow 4 in. between front and top views. 3—Locate the auxiliary view 1 1/2 in. from the front view. 4—The front view is 1 in. from the left border. **Problem 11-10. Right. BRACKET.** 1—Use a horizontal drawing sheet format. 2—Allow 3 in. between the front and top views. 3—Locate the auxiliary view 1 in. from the front view. 4—The front view is 2 3/4 in. from the left border.

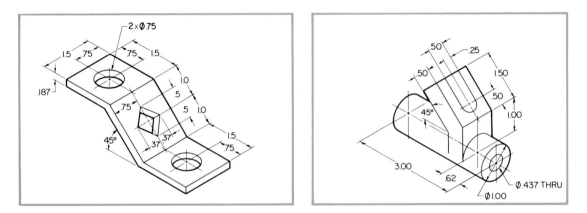

Problem 11-11. Left. HANGER CLAMP. 1—Use a horizontal drawing sheet format. 2—Allow 3 in. between front and top views. 3—Locate the auxiliary view 1 1/2 in. from the front view. 4—Allow 2 1/2 in. between the front and right side view. 5—The front view is 3/4 in. from the left border. **Problem 11-12. Right. SHIFTER BAR.** 1—Use a horizontal drawing sheet format. 2—Allow 2 in. between the front and top views. 3—Locate the auxiliary view 1 in. from the front view. 4—Allow 2 in. between the front and right side view. 5—The front view is 2 in. from the left border.

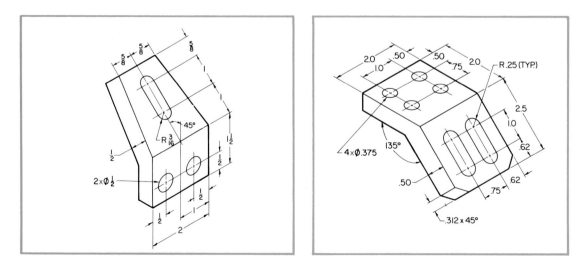

Problem 11-13. Left. SUPPORT. 1—Use a vertical sheet format. 2—Allow 2 1/2 in. between front and top views. 3—Locate the auxiliary view 1 1/2 in. from the front view. 4—The front view is 1 1/4 in. from the left border. **Problem 11-14. Right. ADJUSTABLE BRACKET.** 1—Use a horizontal sheet format. 2—Allow 2 1/8 in. between the front and top views. 3—Locate the auxiliary view 1 in. from the front view. 4—Allow 2 3/4 in. between the front and right side views. 5—The front view is 3/4 in. from the left border.

After studying this unit and completing the assigned problems, you should be able to:

◆ Describe what a pictorial drawing is and why it is used.

◆ Construct from multiview drawings the five types of pictorial views: isometric, cavalier oblique, cabinet oblique, parallel perspective, and angular perspective.

◆ Develop and construct simple exploded assembly drawings and cutaway pictorial drawings.

◆ Center isometric and oblique drawings on a drawing sheet.

A ***pictorial drawing*** shows a likeness (shape) of an object as viewed by the eye. The pictorial of the automobile engine in Figure 12-1 shows many details of its construction.

Figure 12-1. Pictorials have many uses. This one shows the internal details of a V-6 automobile engine. (Buick, Div. of GMC)

Pictorials are also an important part in some computer-aided design projects. They range from basic "wireframe" pictorials in Figure 12-2, to moving, full-color, three-dimensional renditions as in Figure 12-3. The complexity of the computer-generated pictorial is determined by the program level and computer capability.

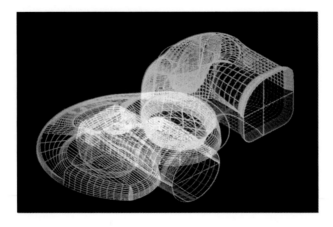

Figure 12-2. Wire frame pictorial of a turbo intercooler being designed for a large diesel engine that will power earth-moving equipment. Engineers will then be able to study airflow turbulence patterns. (American Foundrymen's Society, Inc.)

Figure 12-3. Complex digital computers workstations are being used to design and pre-assemble all of the new Boeing 777 aircraft. The engineer shown here is working on a fuselage section with the various systems shown as three-dimensional images. The engineer can check to see if different systems interfere with one another before sending the design to manufacturing. This improves the accuracy and overall quality of the frame.

If you have worked with radio or electronic kits, you are familiar with pictorial drawings. These drawings show the builder what needs to be done and how to do it to complete the project, Figure 12-4.

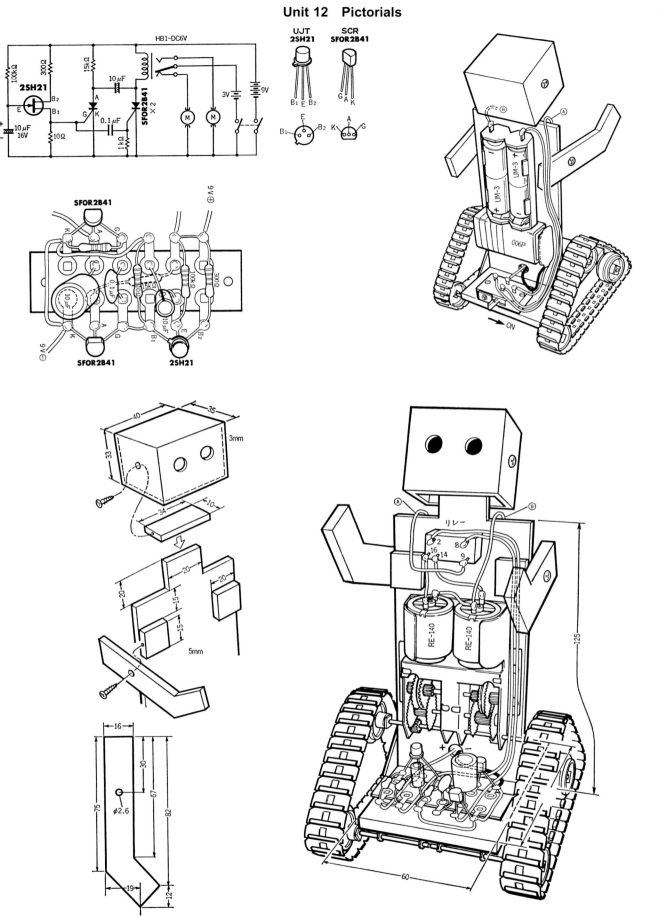

Figure 12-4. Pictorials are used to give instructions in many hobby areas. Shown is a model robot that can be programmed to follow a specified travel pattern.

You should become familiar with several types of pictorial drawings, Figure 12-5. Only the basics of pictorial drawing will be covered in this text.

Isometric Drawings

All *isometric drawing* lines which show the width, length, and depth are drawn full size (or in some proportionate scale). An isometric drawing shows an object as it is. Edges which are upright are shown as vertical lines. The object is assumed to be in position with its corners toward you, and its horizontal edges sloping away at angles of 30 degrees to the right and to the left. Refer to Figure 12-5.

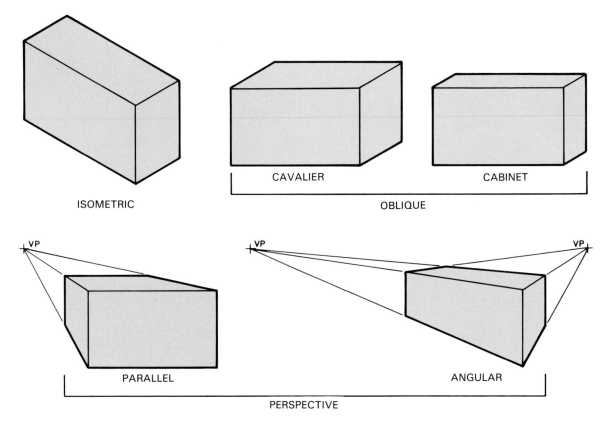

Figure 12-5. Types of pictorial drawings.

An isometric drawing is made entirely with instruments, using three base lines, Figure 12-6. One of the lines is vertical, the other two are drawn at an angle of 30 degrees to the horizontal. The base lines may be reversed if more information can be shown with the object in this position, Figure 12-7.

Lines which are not parallel to the three base lines are called *nonisometric lines*. They are drawn by transferring reference points from a multiview drawing to the isometric drawing, Figure 12-8. The length of nonisometric lines *cannot* be measured directly on the isometric drawing.

Isometric circles are shown in Figure 12-9. Note that isometric circles are circles which are not true ellipses. Isometric circles may be drawn using a compass and a 30-60 degree triangle.

1. Using 30 degree angles, construct an isometric square the same size as the circle. The side of the square is the same length as the circle diameter, as shown to the left in Figure 12-10.

2. Using 60 degree angles, draw lines from the corners of the isometric squares as shown. The intersection points of these lines provide the centers for drawing radius R′. Construct arcs tangent to the sides

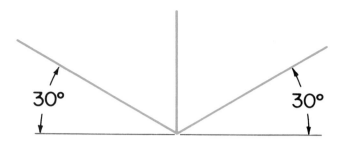

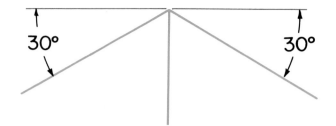

Figure 12-6. Base lines required to make isometric drawings.

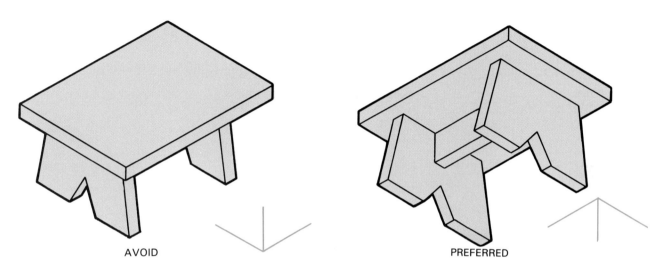

AVOID

PREFERRED

Figure 12-7. Reverse the base lines if more information can be shown with the object in that position. Note the details in the underside view.

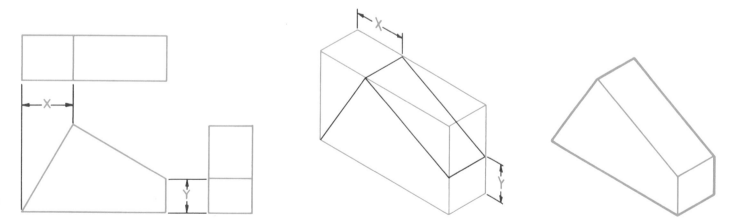

Figure 12-8. Nonisometric lines are made by transferring reference points from a multiview drawing of the object.

Figure 12-9. Isometric circles. Note true ellipse for comparison.

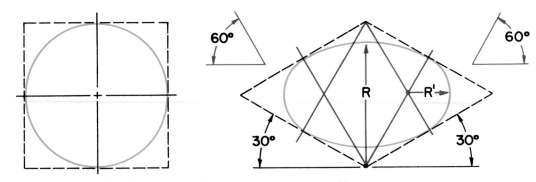

Figure 12-10. Drawing an isometric circle on a flat plane with a compass.

of the isometric square between the 60 degree construction lines as shown in Figures 12-10 and 12-11.

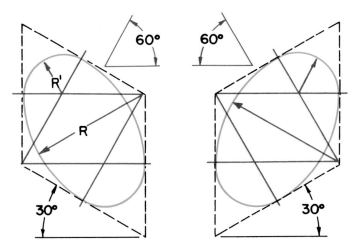

Figure 12-11. Drawing an isometric circle on a vertical plane with a compass.

It is handy to know how to draw isometric circles using drafting tools. An easier way to draw such circles and arcs is to use an ellipse template of appropriate size, Figure 12-12.

Oblique Pictorial Drawings

An *oblique drawing* is a pictorial drawing with the longest dimension or front parallel to the picture plane. Refer to Figure 12-5. The front view is shown in true shape and size. The other views are similar to isometric drawings.

Three types of oblique drawings are shown in Figure 12-13. Each type shows the front of the object as it appears in multiview form. The angle of the depth axis may be any angle, but 15, 30, or 45 degrees are generally used.

Cavalier oblique. An oblique drawing in which the depth axis lines are full scale (full size).

Cabinet oblique. Depth axis lines are drawn one-half scale.

General oblique. Depth axis lines vary from one-half to full scale.

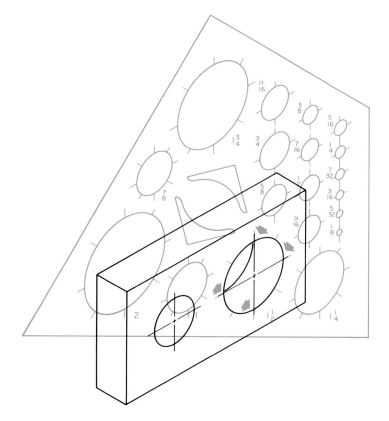

Figure 12-12. A template is a quick way to draw isometric circles.

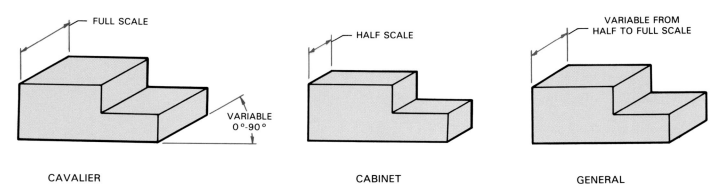

Figure 12-13. Three types of oblique drawings. The cavalier and cabinet drawings are most frequently used.

Dimensioning Pictorial Drawings

To dimension isometric and oblique drawings, draw the dimension lines parallel to the corresponding planes of the object (lines of the drawing). See Figure 12-14 for examples.

It is preferred that dimensions be located outside the outline of the object. They should be placed to read in a horizontal direction, regardless of the direction of the dimension line. Acceptable alternate dimensioning is also shown in Figure 12-14.

Perspective Drawings

A *perspective drawing* is a drawing used by an architect, artist, or drafter to show an object as it would appear to the eye when viewed from a certain position, Figure 12-15. Compare the perspective drawings to the isometric drawing and the oblique drawings shown in Figure 12-13.

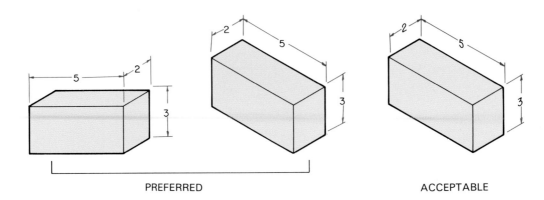

Figure 12-14. Dimensioning pictorials.

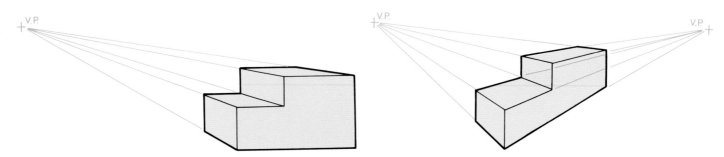

Figure 12-15. Perspective drawings with vanishing points indicated.

A perspective drawing can best be described by imagining that you are landing a jet aircraft. As you approach the runway, you notice that the parallel sides of the runway appear to converge (meet) on the horizon, Figure 12-16. The point where the runway edges meet is called the ***vanishing point (V.P. or VP).*** In making a perspective drawing, lines used to show the depth of the object (receding lines), if continued, will converge to form a vanishing point. Perspective drawings are based on the principle of parallel lines converging in the distance. Refer again to Figure 12-15.

Figure 12-16. Objects appear to become smaller the farther away they are from the eye. Note how the parallel sides of the runway seem to converge or run together in the distance.

To make a simplified ***one-point*** or ***parallel perspective drawing***, the front view is drawn in its true shape in full or scale size, Figure 12-17. Draw the horizon line and select the ***vanishing point*** at random. If at first you do not obtain a pleasing effect, try another vanishing point location. A perspective may be located in any position relative to the vanishing point, Figure 12-18. Project construction lines from each point on the front view to the vanishing point, Figure 12-19.

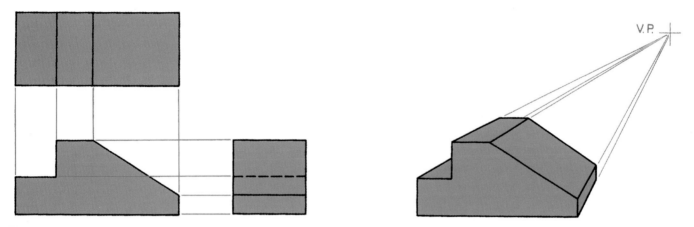

Figure 12-17. On parallel or one-point perspective drawings, the front view is shown in its true shape and in full or scale size.

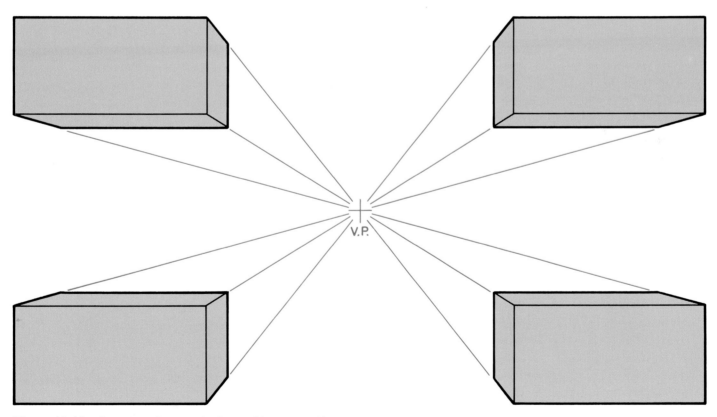

Figure 12-18. A perspective may be located in any position relative to the vanishing point (V.P.).

Simple ***angular*** or ***two-point perspectives*** may be drawn as shown in Figure 12-20. Follow these steps:

1. Locate the position of the horizon and two vanishing points.
2. Locate and draw a vertical line the full-scale height of the object to be drawn.

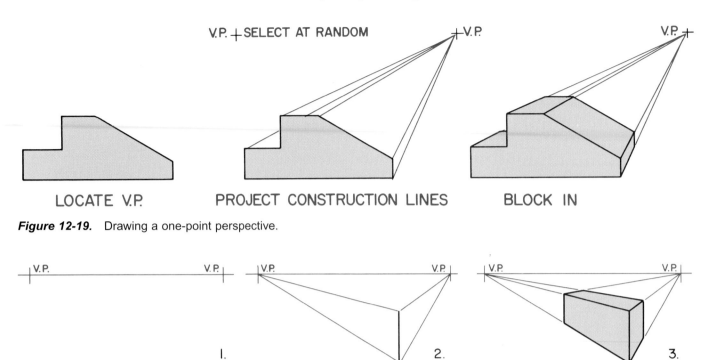

Figure 12-19. Drawing a one-point perspective.

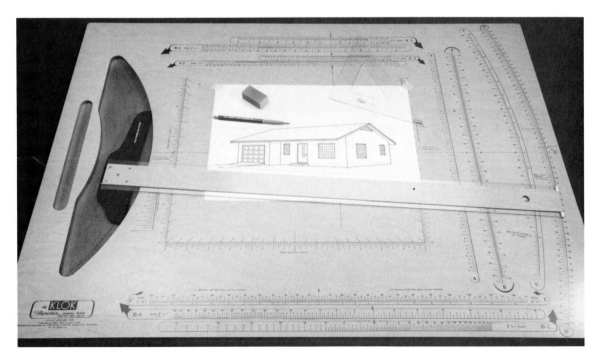

Figure 12-20. Drawing a simple angular or two-point perspective.

3. Draw construction lines from the vertical line to the vanishing points. For simple two-point perspectives like this, it is permissible to locate the length and depth in a position that gives the most pleasing visual effect.

4. Darken the necessary lines with visible object lines.

In drawing perspectives, considerable saving of time plus greater accuracy will result if a perspective drawing board is used, Figure 12-21.

Figure 12-21. Perspective drawing boards help save time when drawing perspectives.

Exploded Assembly Drawings

Exploded assembly drawings are nothing more than a series of pictorial drawings (usually isometrics). They show the parts that make up the object in proper location to one another, Figure 12-22.

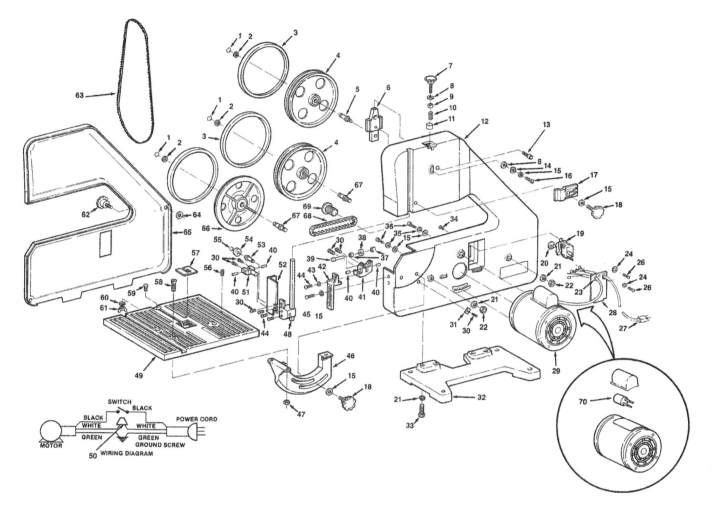

Figure 12-22. An exploded pictorial drawing. Industry uses drawings of this type in areas where the workers have little training in print reading. (Sears-Roebuck)

Exploded assembly drawings are easy to read and can be understood without having an extensive knowledge of print reading. Industry makes extensive use of exploded assembly drawings, especially in areas where semiskilled workers are employed.

Such drawings are also used in many hobby areas. You are probably familiar with them if you build model cars, planes, boats, or rockets.

Cut-Away Pictorial Drawings

Cut-away pictorial drawings have been developed to show the interior details of a product, Figure 12-23. They are usually employed in instructional manuals where the interior details of the product are important in understanding the theory of operation.

Figure 12-23. Cut-away pictorial drawing of a modern automotive engine. (Saturn)

How to Center Pictorial Drawings

There are several ways to center pictorial drawings on a drawing sheet.

One method, which is widely used, is called "eyeballing it." You estimate, by sight, the approximate location of the starting point and develop your drawing from this point.

Another method requires the drawing of the pictorial on a second sheet of paper. When it has been checked for accuracy, a piece of tracing vellum is centered over it and the drawing traced. Or, if regular drawing paper is used, the total width and height of the pictorial is measured. With this information as the starting point, center the drawing on the sheet.

A third method used to center pictorial drawings which works well with most isometric and oblique problems is shown in Figures 12-24 and 12-25.

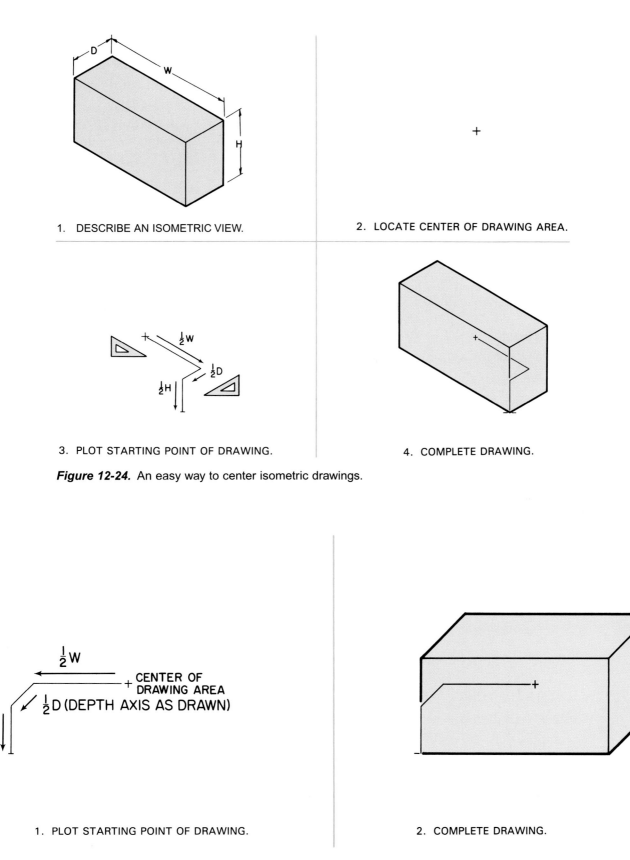

1.   DESCRIBE AN ISOMETRIC VIEW.

2.   LOCATE CENTER OF DRAWING AREA.

3. PLOT STARTING POINT OF DRAWING.

4.   COMPLETE DRAWING.

Figure 12-24. An easy way to center isometric drawings.

½W

CENTER OF
DRAWING AREA

½D (DEPTH AXIS AS DRAWN)

½H

1.   PLOT STARTING POINT OF DRAWING.

2.   COMPLETE DRAWING.

Figure 12-25.   An easy way to center oblique drawings.

Drafting Vocabulary

Angular perspective
Cabinet oblique
Cavalier oblique
Cut-away pictorial
   drawing
Depth axis
Exploded assembly
   drawing

General oblique
Isometric drawing
Nonisometric lines
Oblique drawing
One-point perspective
Parallel perspective
Perspective drawing

Pictorial drawing
Receding lines
Structural details
Template
Two-point perspective
Vanishing point
Wireframe

Test Your Knowledge—Unit 12

Please do not write in the book. Place your answers on another sheet of paper.

1. Pictorial drawing is a method of showing an object as _____.
   A. it would look on the drawing
   B. it would appear to the eye
   C. another form of multiview drawing
   D. All of the above.
   E. None of the above.
2. Why are pictorials often used?
3. List four types of pictorial drawings:
4. Isometric pictorial drawings are drawn about three base lines. Sketch these base lines.
5. The lines that represent angles in an isometric drawing are known as _____ lines.
6. Prepare a sketch that shows the difference between a cavalier and a cabinet pictorial drawing.
7. Oblique drawings may be drawn at any angle, but _____ angles are usually used.
8. What is an "exploded assembly drawing?"

Outside Activities

1. Collect pictorial drawings used to advertise real estate. Use both residential and commercial illustrations. Describe to the class why you think pictorials are used to "sell" real estate.
2. Make a bulletin board display of pictorial drawings such as exploded assembly views of models that you and your classmates have built. Explain why the pictorial views assisted in constructing the models.
3. Search through old magazines (not library copies) and find examples of cut-away pictorial drawings. Make a comparison of drawings using full color as compared to only printed in black and white. Try to make sharp black and white photocopies of the full-color cut-aways. Does color help explain or "sell?"

Problems

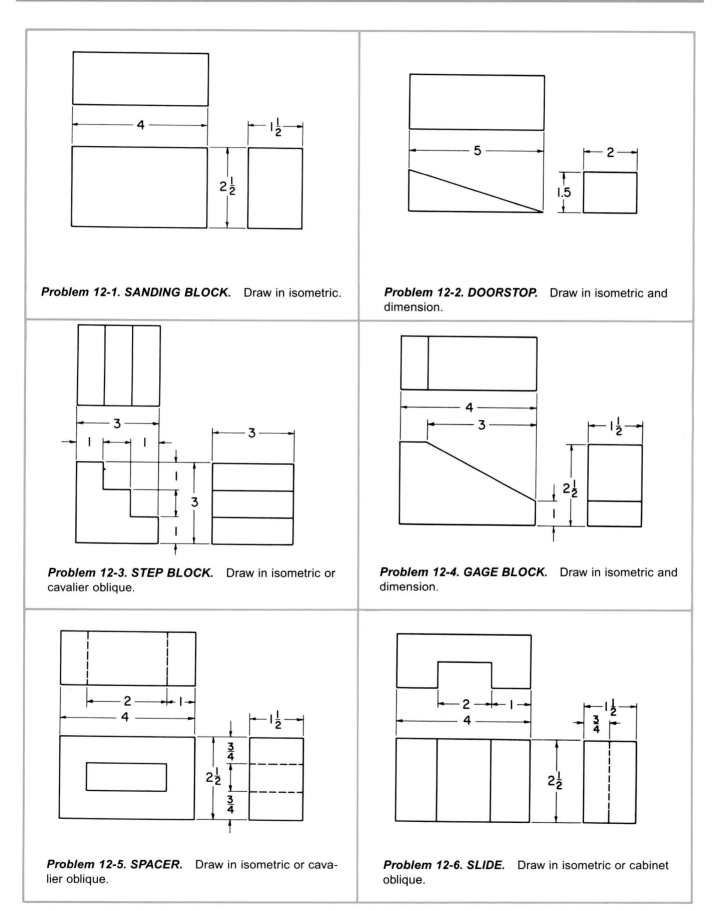

Problem 12-1. SANDING BLOCK.   Draw in isometric.

Problem 12-2. DOORSTOP.   Draw in isometric and dimension.

Problem 12-3. STEP BLOCK.   Draw in isometric or cavalier oblique.

Problem 12-4. GAGE BLOCK.   Draw in isometric and dimension.

Problem 12-5. SPACER.   Draw in isometric or cavalier oblique.

Problem 12-6. SLIDE.   Draw in isometric or cabinet oblique.

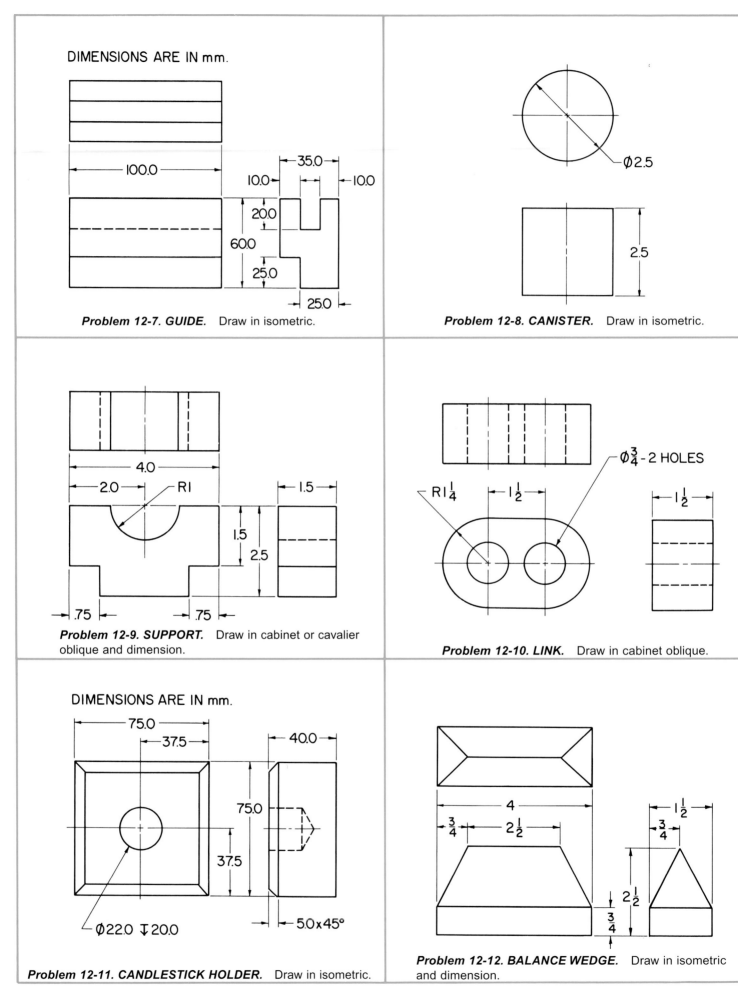

DIMENSIONS ARE IN mm.

100.0

60.0

10.0 35.0 10.0

20.0

25.0

25.0

Problem 12-7. GUIDE. Draw in isometric.

Ø2.5

2.5

Problem 12-8. CANISTER. Draw in isometric.

4.0

2.0 R1

1.5

1.5

2.5

.75 .75

Problem 12-9. SUPPORT. Draw in cabinet or cavalier
oblique and dimension.

R1¼ 1½ Ø¾ - 2 HOLES

1½

Problem 12-10. LINK. Draw in cabinet oblique.

DIMENSIONS ARE IN mm.

75.0

37.5

40.0

75.0

37.5

5.0 x 45°

Ø22.0 ⌄20.0

Problem 12-11. CANDLESTICK HOLDER. Draw in isometric.

4

¾ 2½

1½

¾

2½

¾

Problem 12-12. BALANCE WEDGE. Draw in isometric
and dimension.

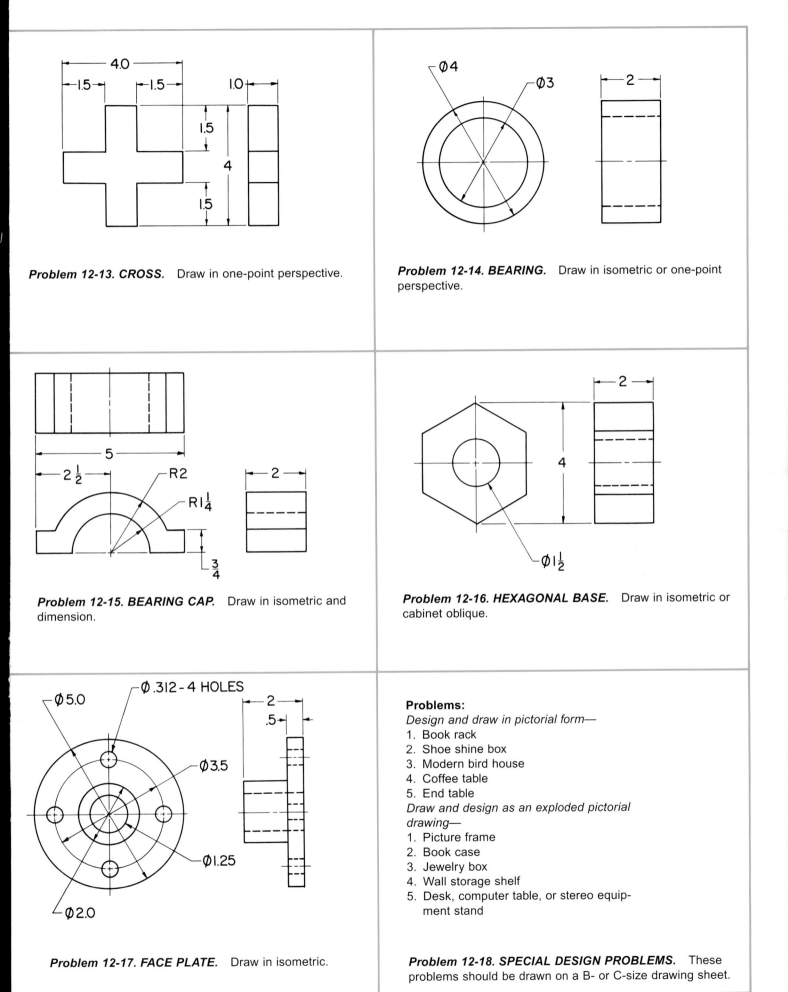

Problem 12-13. CROSS.   Draw in one-point perspective.

Problem 12-14. BEARING.   Draw in isometric or one-point perspective.

Problem 12-15. BEARING CAP.   Draw in isometric and dimension.

Problem 12-16. HEXAGONAL BASE.   Draw in isometric or cabinet oblique.

Problem 12-17. FACE PLATE.   Draw in isometric.

Problems:
Design and draw in pictorial form—
1. Book rack
2. Shoe shine box
3. Modern bird house
4. Coffee table
5. End table
Draw and design as an exploded pictorial drawing—
1. Picture frame
2. Book case
3. Jewelry box
4. Wall storage shelf
5. Desk, computer table, or stereo equipment stand

Problem 12-18. SPECIAL DESIGN PROBLEMS.   These problems should be drawn on a B- or C-size drawing sheet.

Sheet metal patterns for the F-22 Raptor were developed using CAD. Accuracy is extremely important to maintain the stealth capability of the aircraft. (Lockheed-Martin)

Pattern Development

After studying this unit and completing the assigned problems, you should be able to:

◆ Define pattern development.

◆ Prepare simple patterns using parallel line and radial line development methods.

◆ Develop patterns for a cylinder, rectangular prism, truncated prism, truncated cylinder, pyramid, right rectangular pyramid, and cone.

◆ Describe what a wired edge, hem, and seam is and why they are used.

◆ Explain why it is necessary to allow additional material on patterns for hems, edges, and seams.

◆ Use a French curve when developing certain patterns.

A *pattern* is a full-size drawing of the various outside surfaces of an object stretched out on a flat plane, Figure 13-1. A pattern is frequently called a *stretchout*. The pattern can be bent or folded into a three-dimensional shape.

Patterns or stretchouts are produced by utilizing a form of drafting called *pattern development*. The method is also known as *surface development* and *sheet metal drafting*. Pattern development is important to many occupations and hobbies.

Patterns were required to make the clothing and shoes you wear. Wallets and handbags are cut to shape using patterns as guides. Pattern development plays an important part in the fabrication of sheet metal ducts and pipes needed in the installation of heating and air conditioning units. Stoves, refrigerators, and other appliances are fabricated from many sheet metal parts. Accurate patterns had to be developed for the parts before the appliance could be put into production.

Drafters in the aerospace industry must be familiar with pattern development techniques.

Manufacturing space-age materials into components for airplanes, rockets, and hot air balloons requires many patterns, Figure 13-2. The technique used is called *lofting*. The computer-generated patterns are developed directly on materials by a photographic process using lasers.

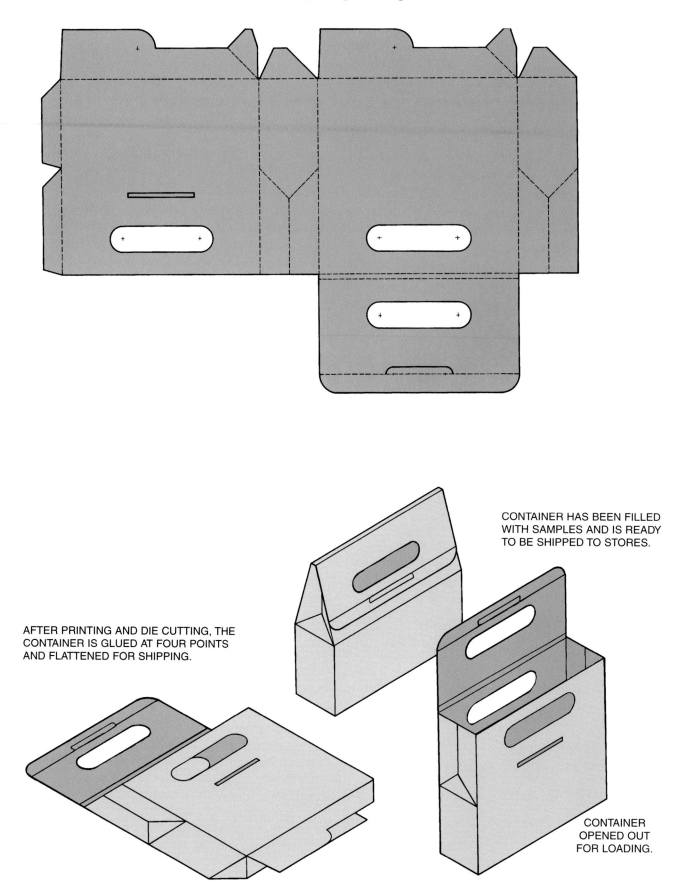

CONTAINER HAS BEEN FILLED
WITH SAMPLES AND IS READY
TO BE SHIPPED TO STORES.

AFTER PRINTING AND DIE CUTTING, THE
CONTAINER IS GLUED AT FOUR POINTS
AND FLATTENED FOR SHIPPING.

CONTAINER
OPENED OUT
FOR LOADING.

Figure 13-1. A computer-generated pattern of a container used to introduce a sample of a new product. The design reduces waste to a minimum, only four points are glued, and it can be folded for easy shipment.

Figure 13-2. Very accurate patterns must be developed, usually by CAD, for hot air balloons. Each panel must be precisely shaped so the stress of the ropes that support the basket is shared equally around the circumference of the vehicle.

Figure 13-3. A flying model airplane depends upon accurate patterns to construct wing ribs, fuselage, formers, and tail sections.

Every plate on a ship's hull and superstructure started with a pattern. Otherwise, they would have required extensive cutting and fitting before they could be used. This would be very costly.

If you have ever built a model boat or flying model airplane, Figure 13-3, you know that many patterns are needed to cut the parts to shape for assembly.

Many pattern developments are prepared by CAD. See Figures 13-1 and 13-2. However, before the drafter can use CAD for this type of work, he or she must have a thorough knowledge of pattern development techniques.

As you can see, patterns play an important part in the manufacture of many products. How many items in the drafting room can you name that use pattern development in their manufacture?

How to Draw Patterns

Regular drafting techniques, as described elsewhere in this text, are used to draw patterns. Basic pattern development falls into the two categories of parallel line development and radial line development. Refer to Figure 13-4.

Parallel Line Development. This technique is employed to make patterns for prisms and cylinders. The lines used to develop the pattern are parallel or at right angles to each other.

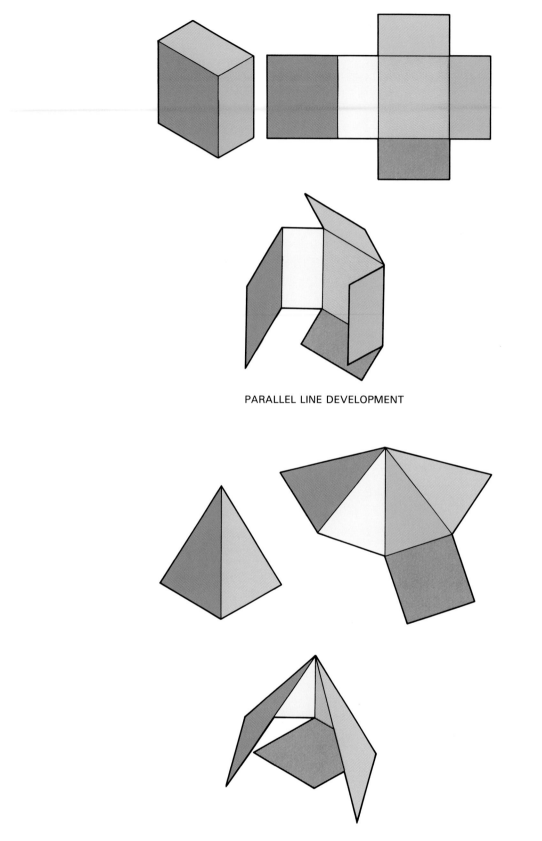

PARALLEL LINE DEVELOPMENT

RADIAL LINE DEVELOPMENT

Figure 13-4. Parallel line development is used to draw patterns for prisms and cylinders. (The lines used are parallel or at right angles to each other.) Patterns for regular tapering, shaped like cones and pyramids, are developed using radial line development (the lines used to radiate out from a single point).

Radial Line Development. Patterns for regular tapering forms such as cones, pyramids, etc., are developed by this technique. The lines used to develop the pattern radiate out from a single point.

Combinations and variations of the basic developments are used to draw patterns for more complex geometric shapes.

While regular drafting techniques are utilized in pattern development, some lines have additional meanings. Sharp folds or bends are indicated on the stretchout by an object line. See **Pattern Development of a Rectangular Prism**, page 210, for an example of this type.

Curved surfaces are shown on the pattern with construction lines or centerline. An example of this is shown in **Pattern Development of a Cylinder**, page 209.

Stretchouts are seldom dimensioned.

When developing a pattern or stretchout for a sheet metal product, it is often necessary to allow additional material for *hems, edges,* and *seams,* Figure 13-5.

Hems strengthen the lips of sheet metal objects. They are made in standard fractional sizes, 3/16 in., 1/4 in., 3/8 in., etc. Standard metric sizes are 4.0 mm, 6.0 mm, 10.0 mm, etc.

A *wired edge* provides extra strength and rigidity to sheet metal edges.

Seams make it possible to join sheet metal sections. They are usually finished by soldering, spot welding, or riveting.

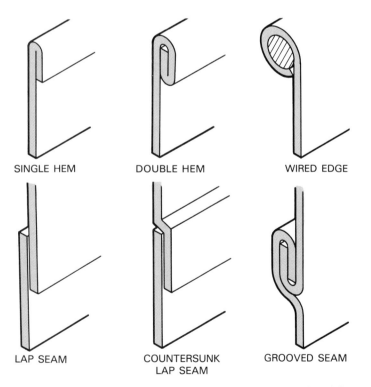

SINGLE HEM   DOUBLE HEM   WIRED EDGE

LAP SEAM   COUNTERSUNK   GROOVED SEAM
                      LAP SEAM

Figure 13-5. Typical hems, edges, and seams used to join and give rigidity to sheet metal. Extra material is required and must be included in the design.

Reference lines and *reference points* are needed when developing the stretchout of a circular object. To provide these, the circle is divided into 12 equal parts. Figure 13-6 shows how to divide a circle to produce the reference points.

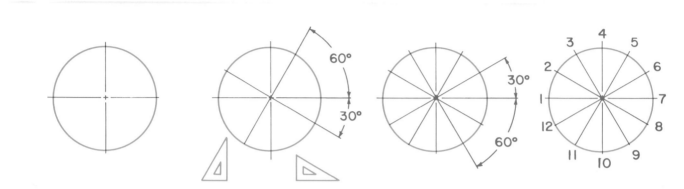

Figure 13-6. How to divide a circle into 12 equal parts.

Irregular curves are drawn using a *French curve.* The points of the irregular curve are first plotted. The points may then be connected with lightly sketched lines. Match the French curve to the sketched line or points, taking care to make the curve of the line flow smoothly. Refer to Figure 13-7 to see how this is done.

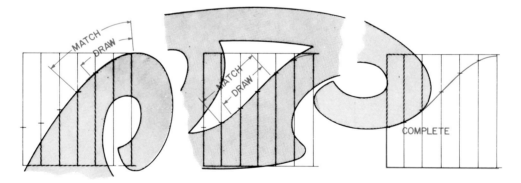

Figure 13-7. Using a French curve to draw an irregular curve. When drawing curved lines take care to make the curve of the line follow smoothly to produce one continuous line.

Pattern Development of a Cylinder

1. Draw front and top views of required cylinder. Divide top view into 12 equal parts and number as shown.

2. The height of the pattern or stretchout is the same as the height of the front view. Project construction lines from the top and bottom of the front view.

3. Allow sufficient space (1 in. is adequate) between the front view and the pattern, and draw a vertical line. This will locate line 1 of the pattern.

4. Calculate the circumference of the circle. Draw a straight line the length of the circumference, and divide it into 12 equal parts (C=πd, where π=3.14). Set your compass or divider to one of the equal parts on the line. Transfer this distance to the extended lines of the pattern to locate reference lines 1, 2, 3 through 12, and 1.

5. Draw the top and bottom tangent to the extended lines.

6. Allow 1/4 in. for seams, and go over all outlines with visible object lines. The lines that represent the curves or circular lines are drawn in color, or, are left as construction lines.

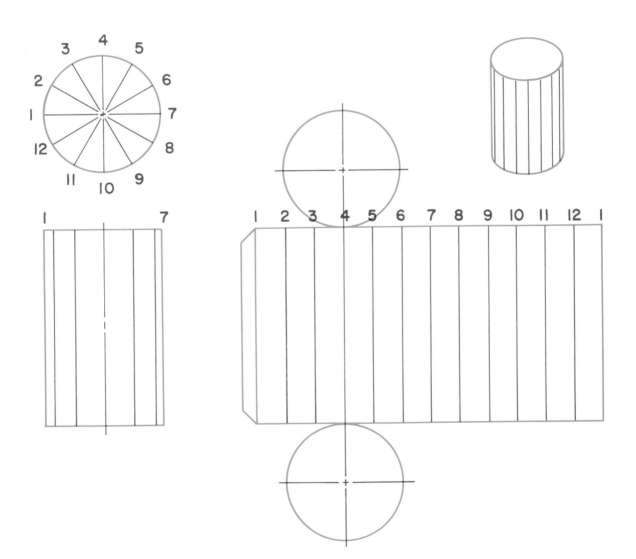

Pattern Development of a Rectangular Prism

1. Draw the front and top views.
2. The height of the pattern is the same as the height of the front view. Project construction lines from the top and bottom of the front view.
3. Measure over 1 in. from the front view and draw a vertical line between the extended lines to locate line 1.
4. Set your compass or divider from 1 to 2 on the top view, and transfer this distance to the extended lines. Locate the other distances in the same manner.
5. Construct the top and bottom as shown.
6. Allow 1/4 in. for seams, and go over all outlines and folds with object lines. The pattern may be cut out, folded to shape, and cemented together using rubber cement.

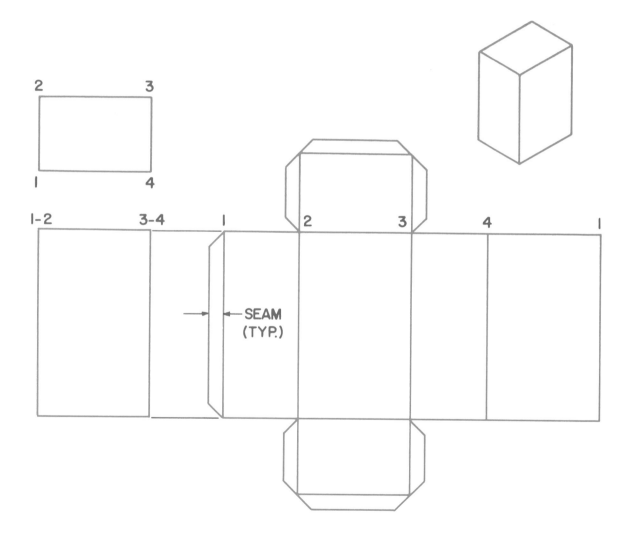

Pattern Development of a Truncated Prism

1. Draw front and top views. Number the points as shown.
2. Proceed as in previous examples of pattern development.
3. Mark off and number the folding points. Project point 1 on the front view to line 1 of the stretchout. Repeat with points 2, 3, and 4 to lines 2, 3, and 4.
4. Connect the points 1 to 2, 2 to 3, 3 to 4, and 4 to 1.
5. Draw the top and bottom in position.
6. Allow material for seams, and go over the outline and fold lines with object lines. The pattern produces a truncated prism.

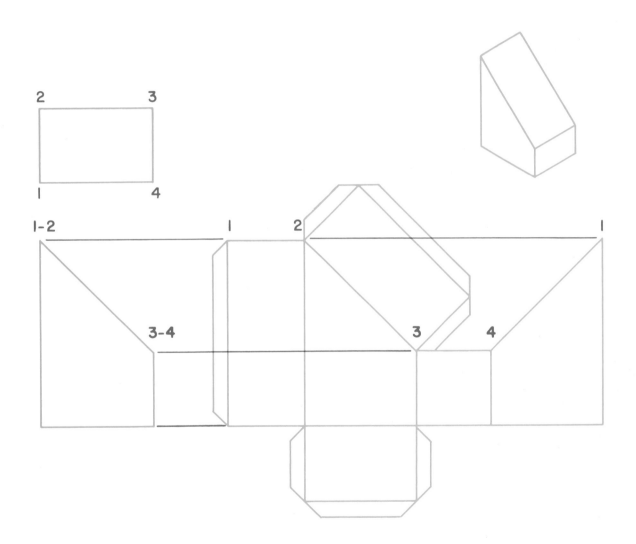

Pattern Development of a Truncated Cylinder

1. Draw the front and top views. Divide and number as shown.
2. Extend lines from the top and bottom of the front view.
3. Allow 1 in. between the front view and the pattern and draw line 1.
4. Set your compass or dividers from 1 to 2 on the top view and step off 12 equal divisions on the extended lines of the pattern. Number them.
5. Draw vertical construction lines at each of the above divisions.
6. The curve of the pattern is developed by projecting lines from the points on the front view. Point 1 is projected over until it intersects line 1 on the pattern; points 2-12 intersect lines 2 and 12; etc. When all of the points are located they are connected with a curved line drawn with a French curve.
7. Complete by adding the top and bottom. The top is developed as an auxiliary view. The pattern produces a truncated prism.

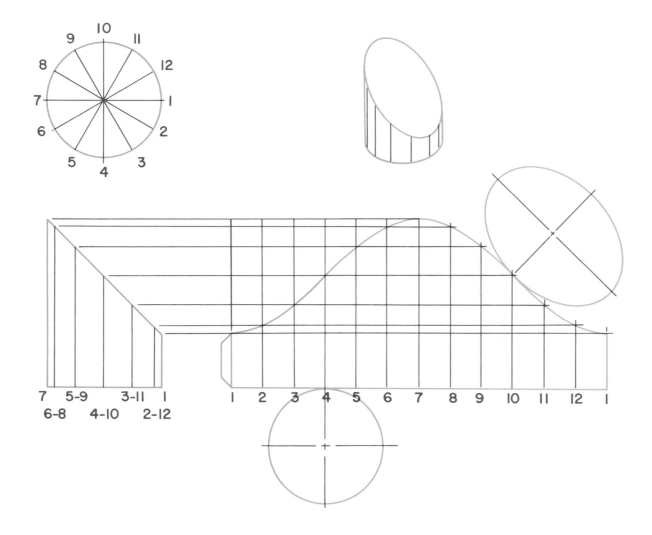

Pattern Development of a Pyramid

1. Draw the front and top views. Number as shown.
2. Locate centerline X of the stretchout.
3. Set your compass to a radius equal to X-1 on the front view and using the above center, draw arc A-B.
4. Draw a vertical line through center X and arc A-B.
5. Set compass from 1 to 2 on top view and at point where vertical line intersects the arc as the starting point, step off two divisions on each side of the line. (Points 1-2-3-4-1 on the stretchout.)
6. Connect the points, draw the bottom in place, and go over the outline and folds with object lines. The completed pattern will produce a pyramid.

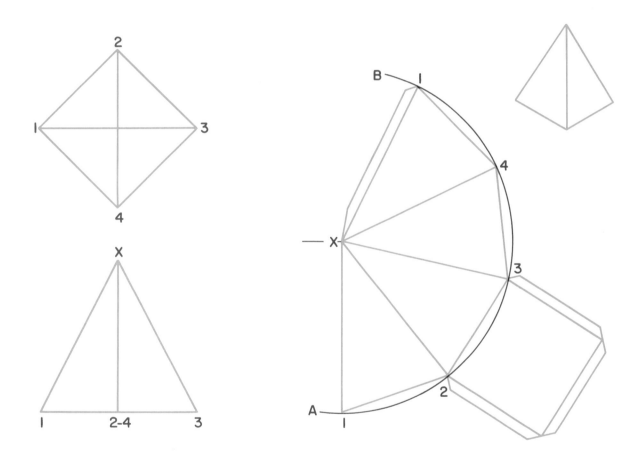

Pattern Development of a Cone

1. Draw the front and top views. Divide the top view into 12 equal parts. Number as shown.
2. Locate centerline X of the stretchout.
3. Set your compass from X to 1 on the front view and with X of the stretchout as the center, draw arc A-B.
4. Draw a vertical construction line through centerline X and the arc.
5. Set your compass from 1 to 2 on the top view, and the point where the vertical line intersects the arc as the starting point, step off six divisions on each side of the line. (Points 1-2-3-4, etc. on the pattern.)
6. Go over the outline carefully with object lines.
7. The lines that represent the curved portion may be drawn in color, or, they may be left as construction lines.

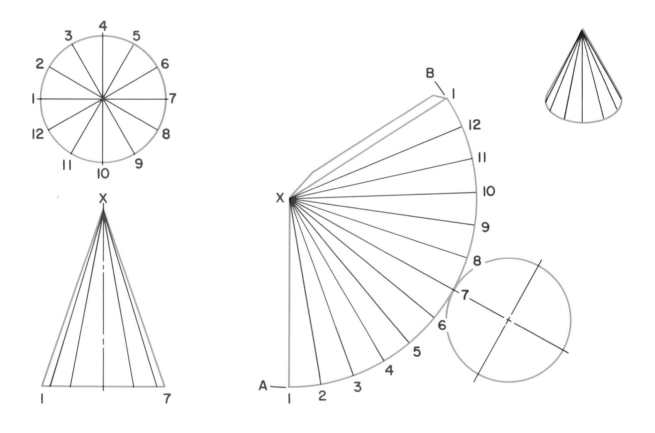

Pattern Development for a Right Rectangular Pyramid

1. Draw the front and top views. Number as shown.
2. Neither view shows the true length of the pyramid edges. To find the true length, rotate one edge as shown and project it to the front view where it appears in true length when the projected point is connected with point X.
3. Locate centerline X of the stretchout.
4. Set compass to true length of the pyramid's edge. With X as the center, draw arc A-B.
5. Draw a vertical line through center X and arc A-B.
6. Set compass from 1-2 on the top view. With the point where the vertical line intersects arc A-B as the starting point, step off pyramid side 1-2.
7. Reset compass to distance 2-3 on the top view and step off side 2-3 on arc A-B.
8. Repeat the sequence and step off sides 3-4 and 4-1 on arc A-B.
9. Connect the points. Draw the bottom in place. Go over the outline and folds with object lines. The pattern produces a right pyramid.

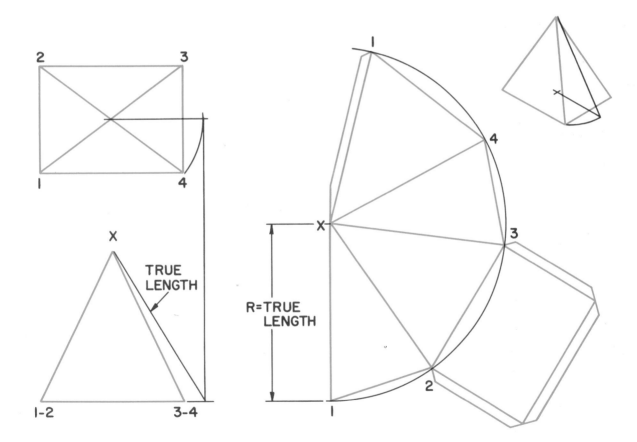

Drafting Vocabulary

Edges	Pattern	Seam
Fabrication	Pattern development	Sheet metal drafting
Fitting	Prism	Stretchout
Hem	Radial line development	Surface development
Lofting	Reference lines	Wired edge
Parallel line development	Reference points	

Test Your Knowledge—Unit 13

Please do not write in the book. Place your answers on another sheet of paper.

1. Pattern development is defined as a form of drafting that _____.
2. Pattern development is also known as _____ _____ _____ and _____ _____ _____.
3. Patterns are also known as _____.
4. List four uses of patterns.
5. In pattern development, a heavy solid line (object line) indicates that _____.
6. Very light lines (construction lines) and centerlines indicate that _____.
7. Irregular curves are drawn with a(n) _____.
8. The lines used to develop patterns for prisms and cylinders use lines that are _____ or at _____ angles to each other.
9. The lines used to develop patterns for cones or pyramids _____ from a single _____.

Outside Activities

1. Secure illustrations of products that require pattern development in their manufacture. They may be cut from discarded magazines and newspapers. Label each product as to how the pattern was developed, by parallel line development or by radial line development.
2. Using a discarded product such as a paper cup, milk carton, or snack food package, carefully unfold the product. Draw a full-size pattern for the product.

Problems

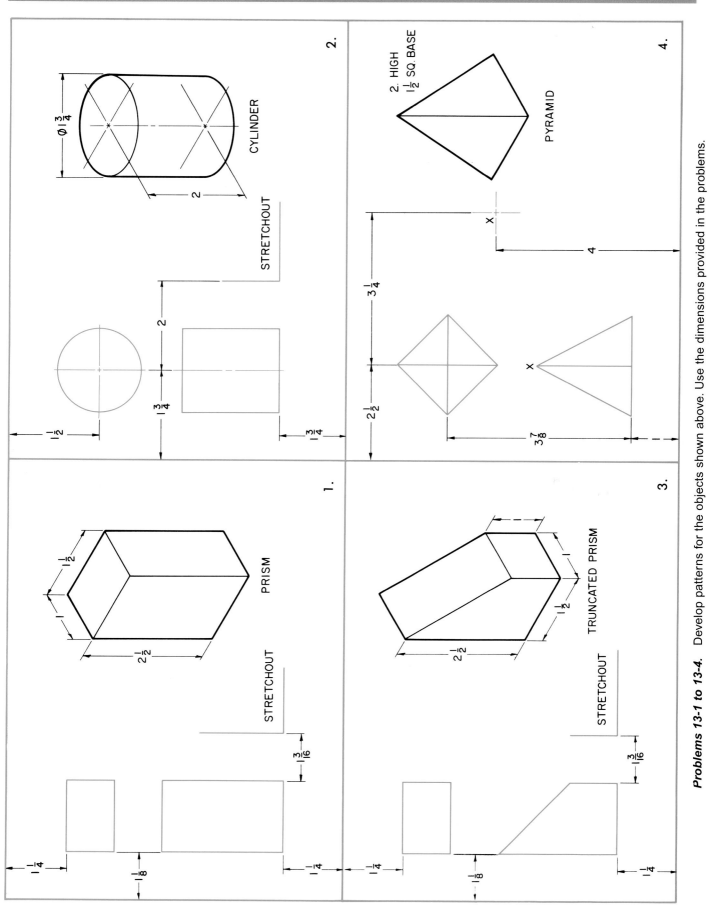

Problems 13-1 to 13-4. Develop patterns for the objects shown above. Use the dimensions provided in the problems.

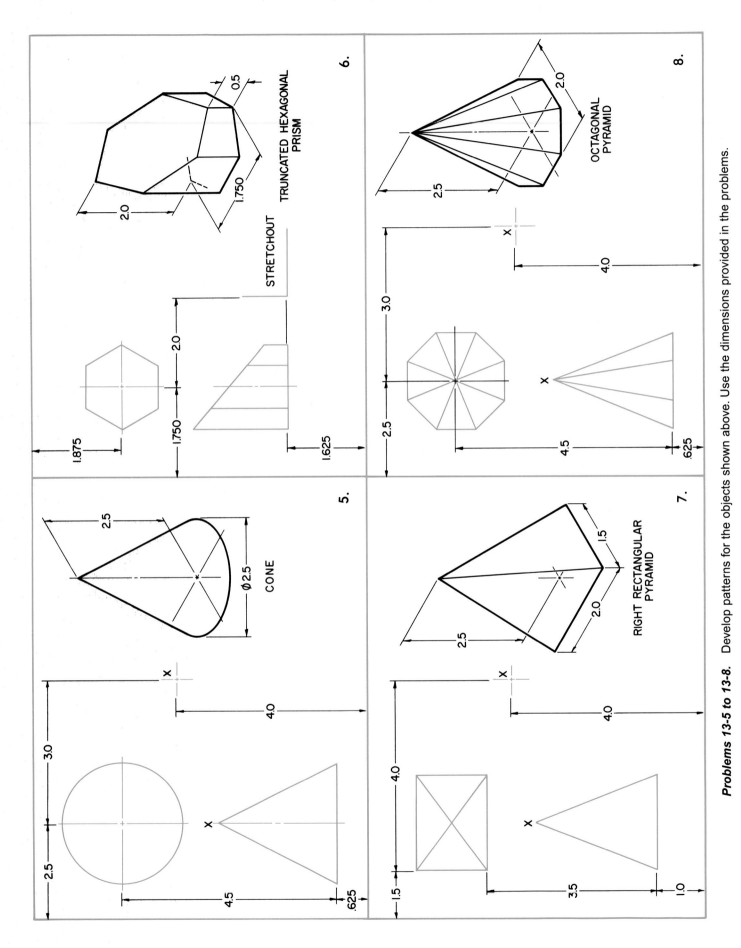

6.

0.5

TRUNCATED HEXAGONAL PRISM

2.0

1.750

STRETCHOUT

1.875

2.0

1.750

1.625

8.

2.0

OCTAGONAL PYRAMID

2.5

X

3.0

4.0

2.5

X

4.5

.625

5.

2.5

Ø 2.5

CONE

X

3.0

4.0

2.5

X

4.5

.625

7.

1.5

RIGHT RECTANGULAR PYRAMID

2.0

2.5

X

4.0

4.0

X

1.5

3.5

1.0

Problems 13-5 to 13-8. Develop patterns for the objects shown above. Use the dimensions provided in the problems.

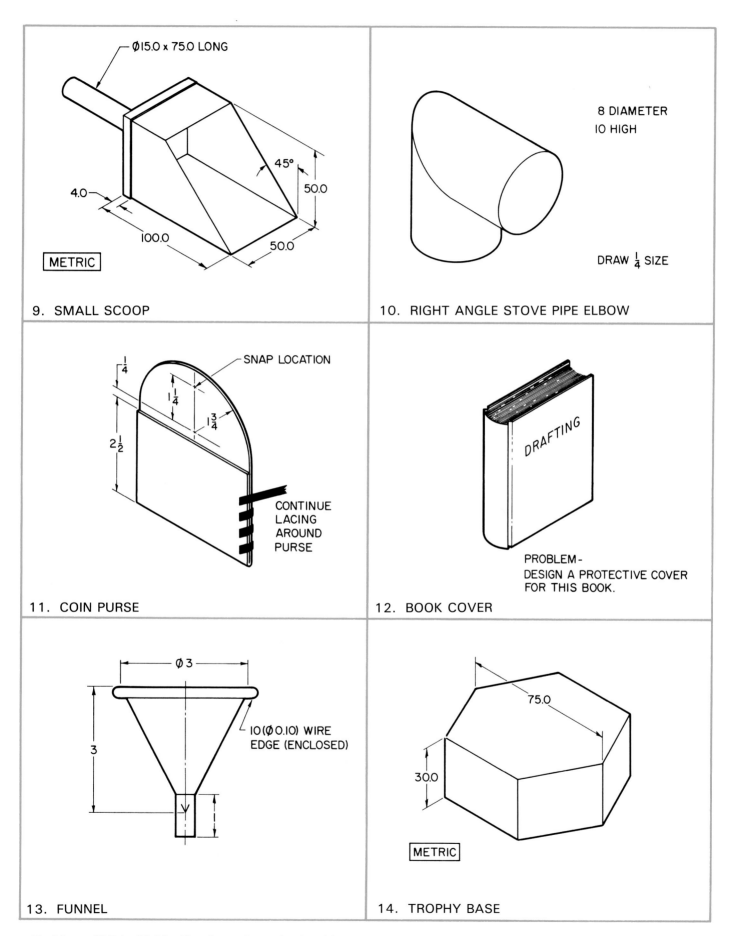

9. SMALL SCOOP

Ø15.0 x 75.0 LONG

45°

50.0

4.0

100.0

50.0

METRIC

10. RIGHT ANGLE STOVE PIPE ELBOW

8 DIAMETER
10 HIGH

DRAW ¼ SIZE

11. COIN PURSE

¼

1¼

1¾

2½

SNAP LOCATION

CONTINUE
LACING
AROUND
PURSE

12. BOOK COVER

DRAFTING

PROBLEM-
DESIGN A PROTECTIVE COVER
FOR THIS BOOK.

13. FUNNEL

Ø 3

3

10 (Ø 0.10) WIRE
EDGE (ENCLOSED)

14. TROPHY BASE

75.0

30.0

METRIC

Problems 13-9 to 13-14.   Develop patterns for the objects shown above. Use the dimensions provided in the problems.

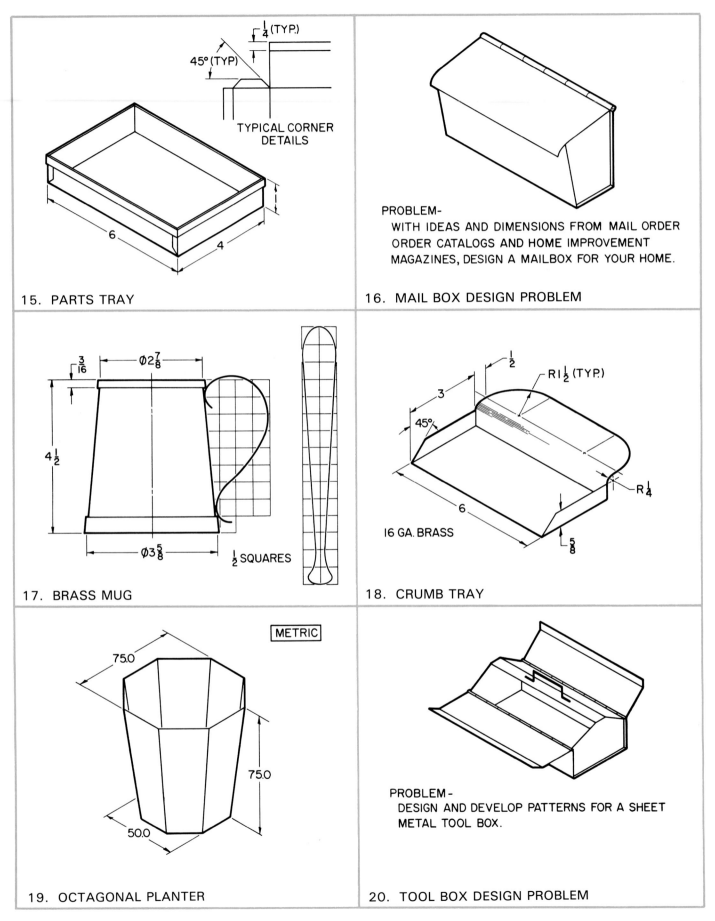

¼ (TYP.)

45° (TYP)

TYPICAL CORNER
DETAILS

6

4

15. PARTS TRAY

PROBLEM-
WITH IDEAS AND DIMENSIONS FROM MAIL ORDER
ORDER CATALOGS AND HOME IMPROVEMENT
MAGAZINES, DESIGN A MAILBOX FOR YOUR HOME.

16. MAIL BOX DESIGN PROBLEM

3/16

Ø2⅞

4½

Ø3⅝

½ SQUARES

17. BRASS MUG

½

R1½ (TYP.)

3

45°

6

R¼

16 GA. BRASS

⅝

18. CRUMB TRAY

METRIC

75.0

75.0

50.0

19. OCTAGONAL PLANTER

PROBLEM-
DESIGN AND DEVELOP PATTERNS FOR A SHEET
METAL TOOL BOX.

20. TOOL BOX DESIGN PROBLEM

Problems 13-15 to 13-20. Develop patterns for the objects shown above. Use the dimensions provided in the problems.

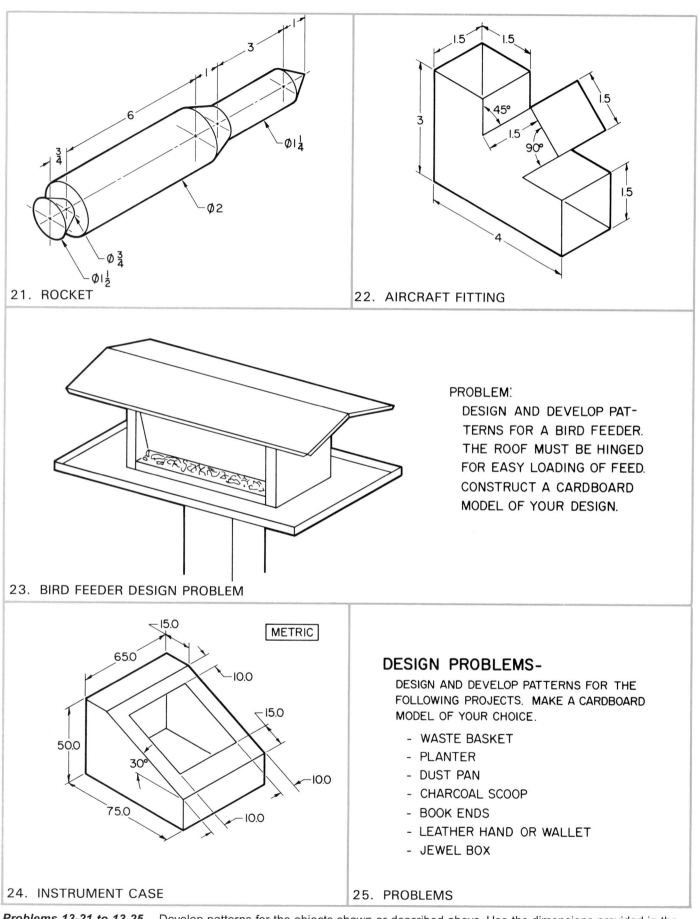

21. ROCKET

22. AIRCRAFT FITTING

PROBLEM:

DESIGN AND DEVELOP PAT-
TERNS FOR A BIRD FEEDER.
THE ROOF MUST BE HINGED
FOR EASY LOADING OF FEED.
CONSTRUCT A CARDBOARD
MODEL OF YOUR DESIGN.

23. BIRD FEEDER DESIGN PROBLEM

METRIC

DESIGN PROBLEMS-

DESIGN AND DEVELOP PATTERNS FOR THE
FOLLOWING PROJECTS. MAKE A CARDBOARD
MODEL OF YOUR CHOICE.

- WASTE BASKET
- PLANTER
- DUST PAN
- CHARCOAL SCOOP
- BOOK ENDS
- LEATHER HAND OR WALLET
- JEWEL BOX

24. INSTRUMENT CASE

25. PROBLEMS

Problems 13-21 to 13-25.   Develop patterns for the objects shown or described above. Use the dimensions provided in the problems or design appropriate size products for the end use.

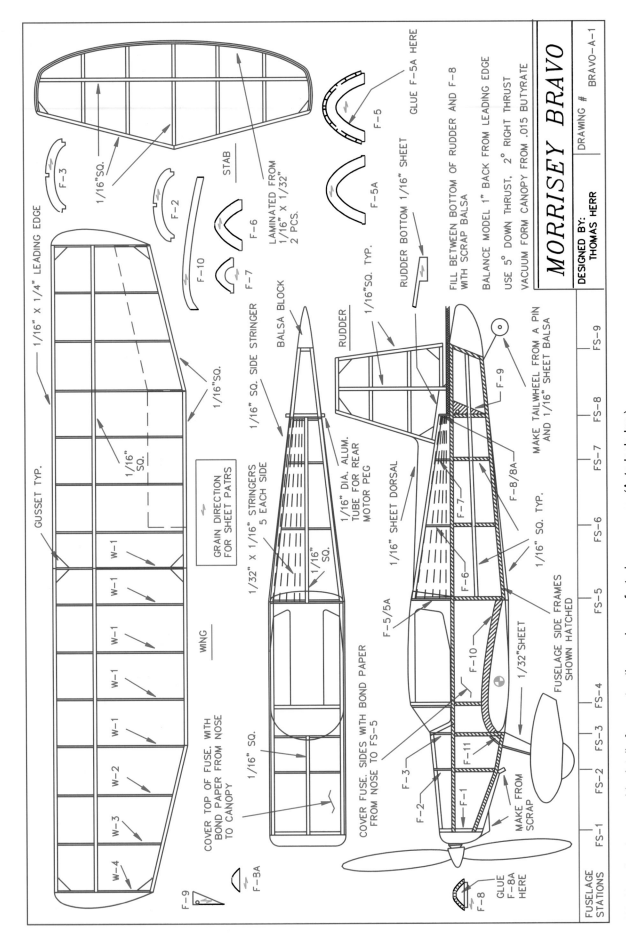

MORRISEY BRAVO

DESIGNED BY: THOMAS HERR

DRAWING # BRAVO–A–1

FILL BETWEEN BOTTOM OF RUDDER AND F–8 WITH SCRAP BALSA

BALANCE MODEL 1" BACK FROM LEADING EDGE

USE 5° DOWN THRUST, 2° RIGHT THRUST

VACUUM FORM CANOPY FROM .015 BUTYRATE

GLUE F–5A HERE

F–5

F–5A

RUDDER BOTTOM 1/16" SHEET

1/16"SQ. TYP.

RUDDER

1/16" SQ. SIDE STRINGER

BALSA BLOCK

LAMINATED FROM 1/16" X 1/32" 2 PCS.

F–6

F–7

F–10

F–2

F–3

STAB

1/16"SQ.

1/16" X 1/4" LEADING EDGE

GUSSET TYP.

1/16" SQ.

1/16" SQ.

W–1

W–1

W–1

W–1

W–1

W–1

W–2

W–3

W–4

WING

GRAIN DIRECTION FOR SHEET PATRS

COVER TOP OF FUSE. WITH BOND PAPER FROM NOSE TO CANOPY

1/16" SQ.

F–9

1/32" X 1/16" STRINGERS 5 EACH SIDE

1/16" SQ.

1/16" DIA. ALUM. TUBE FOR REAR MOTOR PEG

COVER FUSE. SIDES WITH BOND PAPER FROM NOSE TO FS–5

F–3A

1/16" SHEET DORSAL

F–5/5A

F–6

F–7

F–8/8A

F–9

MAKE TAILWHEEL FROM A PIN AND 1/16" SHEET BALSA

F–10

1/32"SHEET

FUSELAGE SIDE FRAMES SHOWN HATCHED

1/16" SQ. TYP.

F–3

F–2

F–1

F–11

MAKE FROM SCRAP

GLUE F–8A HERE

F–8

FUSELAGE STATIONS	FS–1	FS–2	FS–3	FS–4	FS–5	FS–6	FS–7	FS–8	FS–9

Working drawings provide details for construction and manufacturing purposes. (Autodesk, Inc.)

Working Drawings

After studying this unit and completing the assigned problems, you should be able to:

◆ Explain why working drawings are needed.

◆ Discuss the difference between detail drawings and assembly drawings.

◆ Identify the information provided on working drawings.

◆ Prepare simple detail and assembly drawings.

It would not be practical for industry to manufacture a product without using drawings which provide complete manufacturing details. Required drawings range from a single freehand sketch for a simple part, to many thousands of drawings needed to manufacture a complex product like the Space Shuttle. Working drawings tell the craftworker what to make, and establish the standards to which he or she must work, Figure 14-1.

Working drawings fall into two categories, ***detail drawings*** and ***assembly drawings***. A ***detail drawing*** includes a view (or views) of the product, with dimensions and other pertinent information required to make the part.

Refer to Figure 14-1 and note the following information on a detail drawing.

A. **Name of the object.** The name assigned to the part. This should appear on all drawings making reference to the piece.

B. **Quantity.** The number of units needed in each assembly.

C. **Drawing number.** Number assigned to the drawing for filing and reference purposes.

D. **Material used.** The exact material specified to make the part.

E. **Tolerances.** The permissible deviation, either oversize or undersize, from a basic dimension.

F. **Next assembly.** The name given to the major assembly on which the part is to be used.

G. **Scale.** Drawings made other than full size are called ***scale drawings.***

H. **Special information.** Information pertinent to the correct manufacture of the part that is not included on the various views of the object.

I. **Revisions.** Changes that have been made on the original drawing.

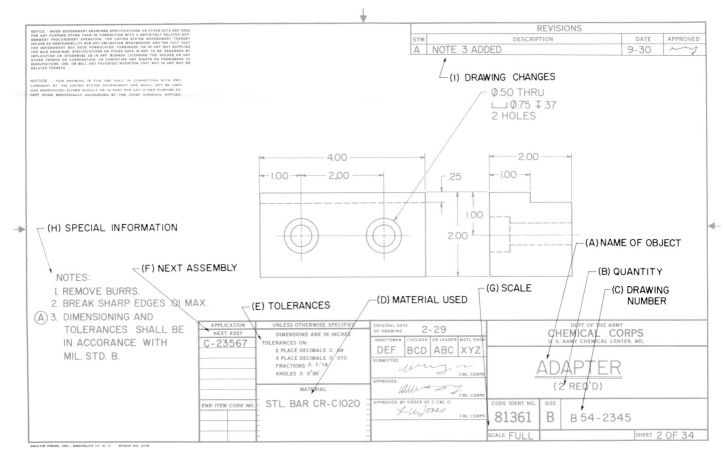

Figure 14-1. A typical working drawing with its components identified.

In most instances, the detail drawing provides information on a single part. See Figure 14-2. Can you identify all the information points in Figure 14-2? It is also permissible to draw all of the parts and the assembly of a small or simple mechanism on the same sheet, Figure 14-3.

An *assembly drawing* contains views that show where and how the various parts fit into the assembled product. Figure 14-4 is an example of a typical assembly drawing. They also show how the complete object will look.

When the object is large or complex it may not be possible to present all of the information on one sheet. A *subassembly drawing* illustrates the assembly of only a small portion of the complete object, Figure 14-5. A subassembly drawing solves the problem of not fitting all the drawing information on a single sheet.

Assembly and subassembly drawings may also be in pictorial form, Figure 14-6.

Under certain situations, a drawing may include a *parts list*, Figure 14-7, and/or a *bill of materials*, Figure 14-8. The information on the parts list is read from top down when the list is located at the upper-right corner of the drawing sheet, or, read from the bottom up when located above the title block. Either position is acceptable. The parts are usually listed according to size or in their order of importance.

An *identifying number* is placed on each sheet for convenience in filing or rapid location of the drawing, Figure 14-9. The drawing number is also usually the part number stamped or printed on the object after it is manufactured.

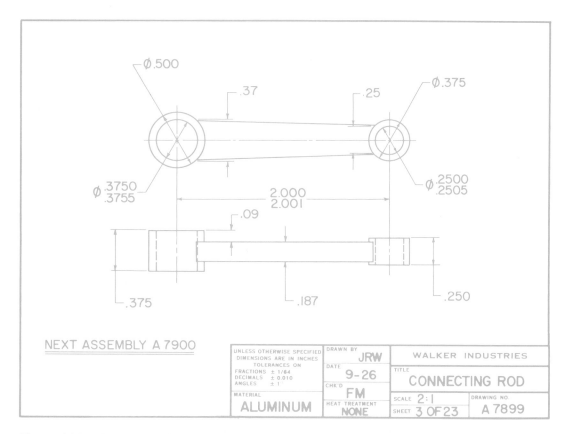

Figure 14-2. A detail drawing usually contains information for making one part.

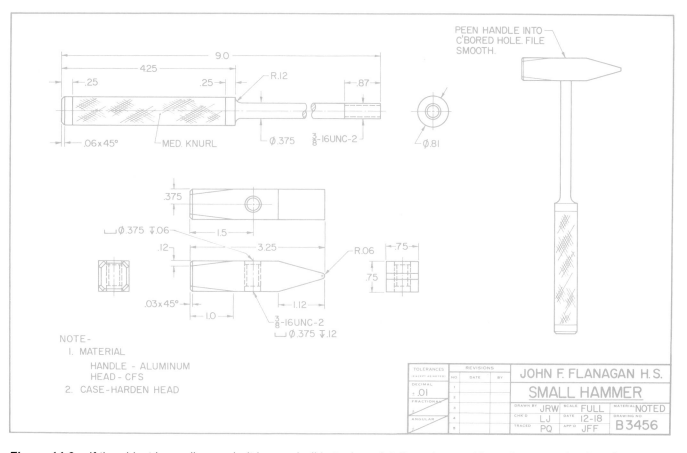

Figure 14-3. If the object is small enough, it is permissible to draw details and assembly on the same drawing sheet.

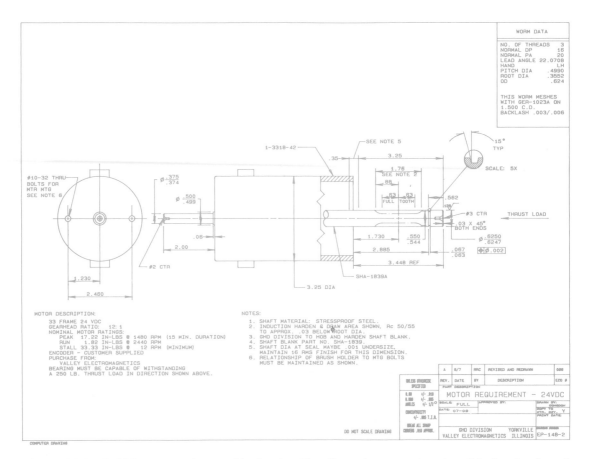

Figure 14-4. A CAD-generated assembly drawing. The dimensions are placed on this drawing for reference purposes for quality control. (T & W Systems, Inc.)

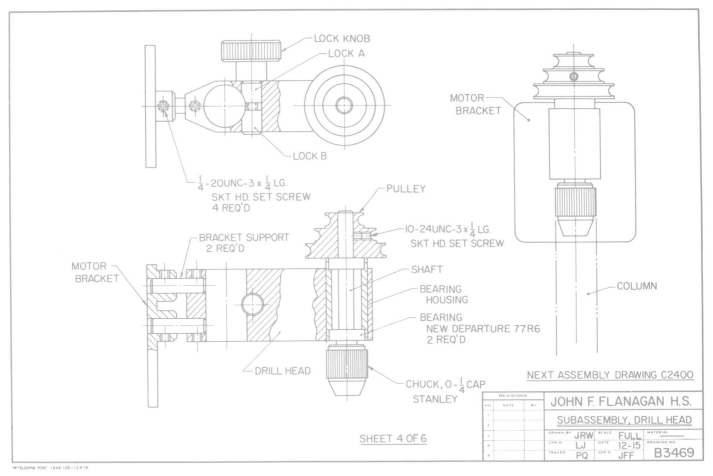

Figure 14-5. A conventional subassembly drawing. It shows the assembly of a small drill press head.

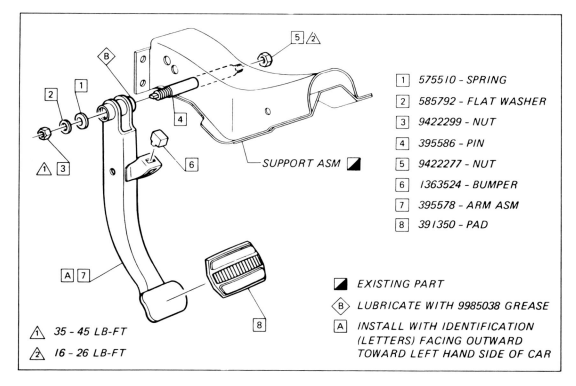

Figure 14-6. A pictorial subassembly drawing. This type of drawing is often used with semiskilled workers who have received a minimum of training in print reading. (General Motors Corp.)

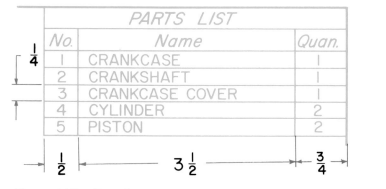

Figure 14-7. Parts list.

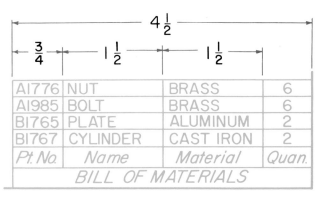

Figure 14-8. Bill of materials.

Figure 14-9. Each plate is given an identifying number for convenience in filing and locating drawings.

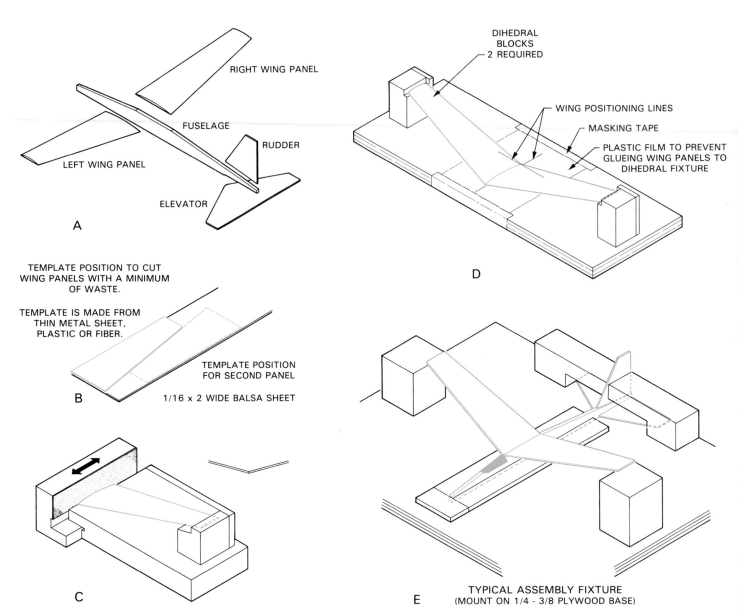

Figure 14-10. Jigs, fixtures, and other information helpful in mass producing the glider shown in Problem 14-2. A—The glider is made of five different parts. B—Plan carefully so parts can be cut with a minimum of waste. C—Jig for sanding dihedral angle where the wing panels meet. D—Fixture for holding wing panels while glue dries. The required dihedral (necessary for flight stability) is set automatically. E—Fixture for assembling glider. If properly made, all parts will be perfectly aligned.

Drafting Vocabulary

Assembly drawing	Parts list	Subassembly drawing
Bill of materials	Revision	Tolerances
Detail drawing	Scale drawing	Working drawing
Identifying number	Standards	
Manufacturing details	Subassembly	

Test Your Knowledge—Unit 14

Please do not write in the book. Place your answers on another sheet of paper.

1. Why are working drawings used?
2. The _____ drawing has a picture of the part with dimensions and other information needed to make the object.
3. The _____ drawing shows where and how the object described in the above drawing fits into the complete assembly of the object.
4. On large or complex objects, _____-_____ drawings are frequently used. These drawings show the assembly of only a small portion of the complete object.

The drawing described in question #2 includes information necessary to manufacture a part. Some of this information is described below. Match the lettered descriptions of each of these items with the numbered term.

5. Quantity
6. Tolerances
7. Material used
8. Scale
9. Revisions

A. The exact material specified to make the part.
B. Drawings made other than to full size.
C. The number of units needed in each assembly.
D. The permissible deviation, either oversize or undersize, from a basic dimension.
E. Changes that have been made on the original drawing.

10. Why are drawings given identifying numbers?

Outside Activities

1. From a local industry, try to obtain samples of detail drawings. Include samples of assembly drawings and samples of subassembly drawings. Prepare a bulletin board display using the theme *working drawings*.
2. Design a suitable title block for the drawing sheets of a company that you are planning to start. Make sure there are blanks to include all pertinent information required.
3. Secure the piston and rod assembly from a small single-cylinder gasoline engine. Prepare detail drawings for these parts. Indicate all dimensions as decimals. It may be necessary for you to learn how to read a micrometer to make accurate measurements.
4. Design a product with sales appeal. Prepare the drawings necessary to mass-produce it in the school laboratory. Make your drawings on tracing vellum so that a number of prints can be made.
5. Research the topic of Geometric Dimensioning and Tolerancing. Report to the class how industry uses this topic in preparing their assembly drawings. Show some examples.

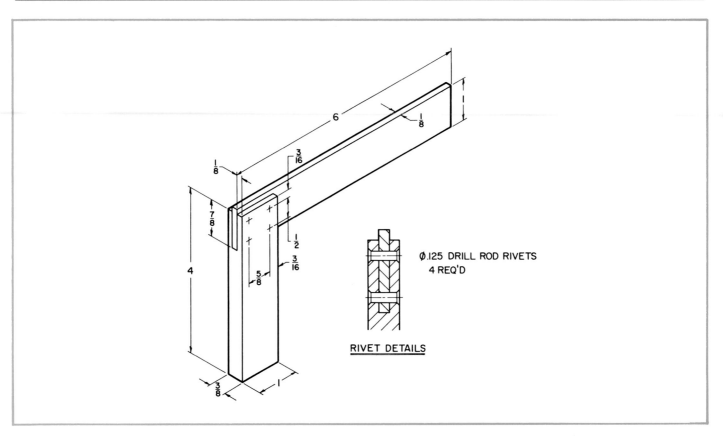

Problem 14-1. MACHINIST'S SQUARE.   Make detail and assembly drawings.

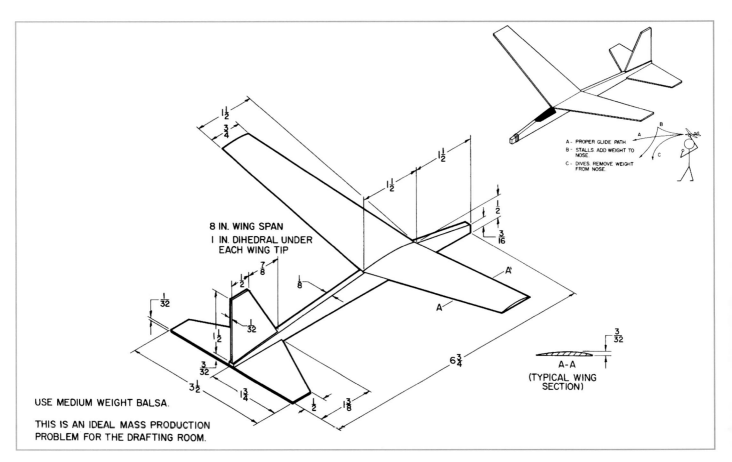

Problem 14-2. MODEL GLIDER.   Prepare detail and assembly drawings. Design the necessary patterns, fixtures, etc., needed to mass-produce the glider. Figure 14-10 shows jigs, fixtures, and helpful information for mass-producing the glider.

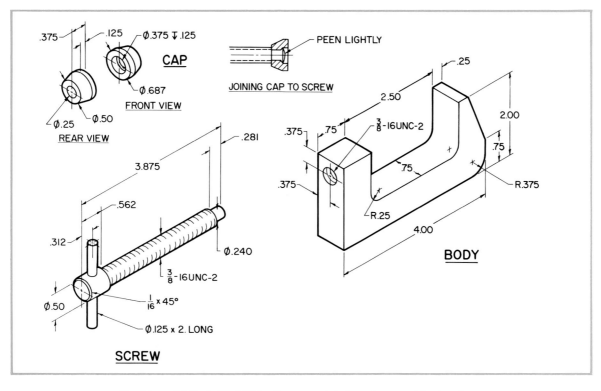

Problem 14-3. C-CLAMP.   Prepare detail and assembly drawings.

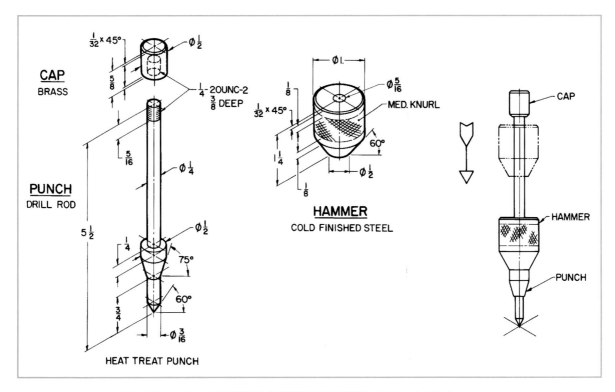

Problem 14-4. GRAVITY CENTER PUNCH.   Make detail drawings.

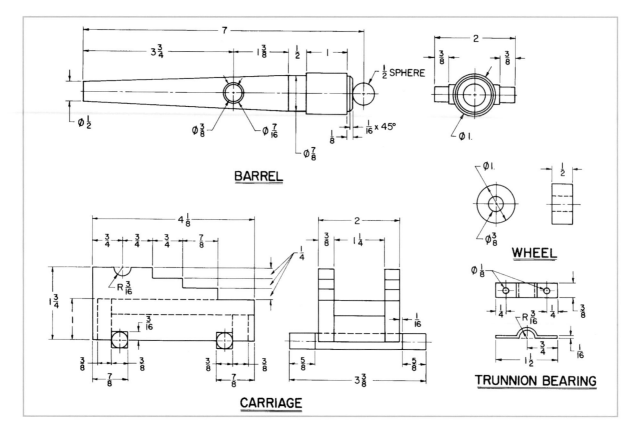

Problem 14-5. DECK GUN.   Prepare an assembly drawing.

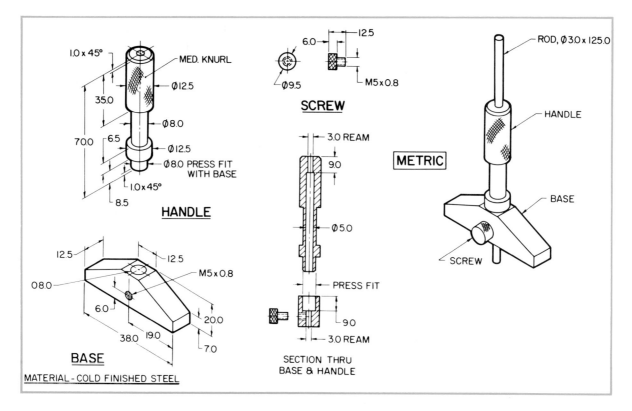

Problem 14-6. DEPTH GAUGE.   Make detail and assembly drawings.

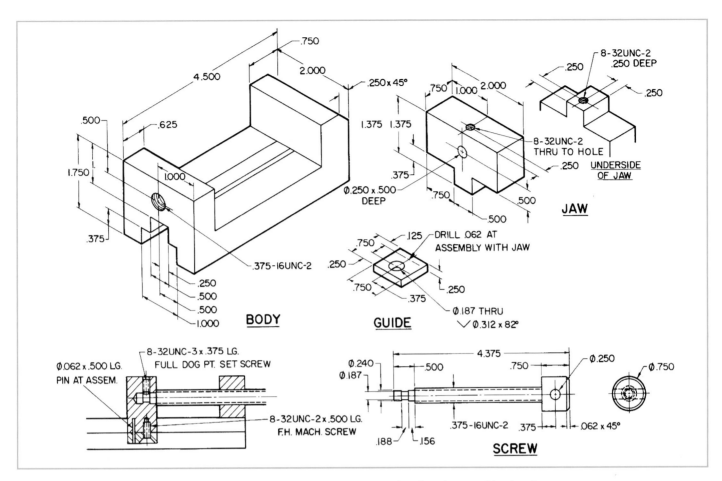

Problem 14-7. SMALL VISE.   Prepare detail and assembly drawings.

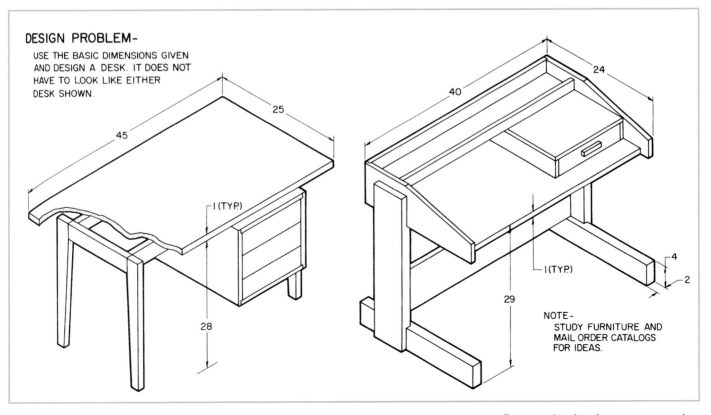

Problem 14-8. CONTEMPORARY DESK.   Design a desk using the dimensions given. Prepare the drawings necessary to make your design.

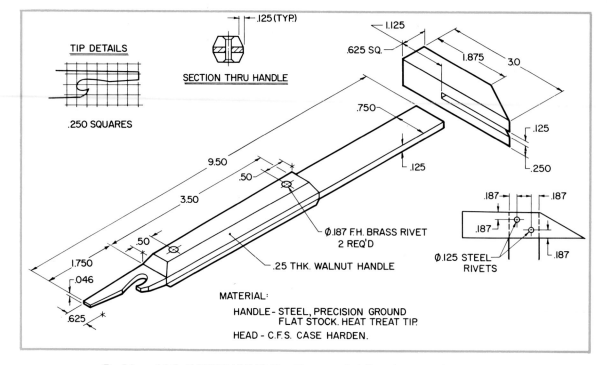

TIP DETAILS

.250 SQUARES

.125 (TYP.)

SECTION THRU HANDLE

.625 SQ.

1.125

1.875

3.0

.750

9.50

.50

3.50

.125

.125

.250

.50

Ø.187 F.H. BRASS RIVET
2 REQ'D

.25 THK. WALNUT HANDLE

1.750

.046

.625

.187

.187

.187

.187

Ø.125 STEEL
RIVETS

MATERIAL:

HANDLE- STEEL, PRECISION GROUND
        FLAT STOCK. HEAT TREAT TIP.
HEAD - C.F.S. CASE HARDEN.

Problem 14-9. HANDY HELPER. Prepare detail and assembly drawings.

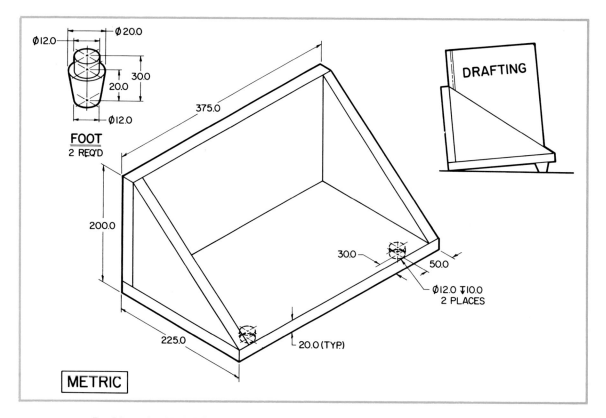

Ø12.0

Ø20.0

30.0

20.0

Ø12.0

FOOT
2 REQ'D

375.0

200.0

DRAFTING

30.0

50.0

Ø12.0 ▽10.0
2 PLACES

225.0

20.0 (TYP.)

METRIC

Problem 14-10. BOOK RACK. Make a fully dimensioned assembly drawing.

Print Making

After studying this unit, you should be able to:

◆ Identify reasons why prints are used in place of original drawings.

◆ Explain several print-making techniques.

◆ Demonstrate how to make a pencil or an ink tracing.

Original drawings are seldom found on a job site. They would soon become worn or soiled, and difficult to read. Also, workers at different locations producing the various parts of a product or structure need different sets of prints. This means that several sets of identical drawings are needed at the same time. Drawing up a set of plans for each person who needs them is impractical because of the costs. Creating new originals to replace plans that are damaged or ruined is also costly. Reproductions of the original drawings are used in these situations. These reproductions are called **prints.**

When several sets of prints are needed, the original drawings are reproduced or duplicated. The method used must produce accurate copies of the originals, must not destroy the original drawings, must be quick, and must be cost-effective. Several of the more common duplicating processes will be described in this unit.

Making Original Drawings

Before a print can be made, a high-quality original must first be made. It can be drawn using computer-aided drafting or manual techniques. Original drawings must have lines that are dense and uniform for good reproductions.

Computer-Aided Drafting

The growth in the computer field has had a direct and significant impact on the drafting field. Today, many of the drawings produced by drafting departments are created on computer-aided drafting (CAD) systems.

As computers become less expensive, so do the output devices attached to them. Personal computer packages are commonly purchased with a printer. In addition, the price of printers and plotters continues to drop as computers advance. Printers and plotters produce drawings, which have dense lines and other features.

Sometimes, printing or plotting multiple copies is the easiest and least expensive way to reproduce drawings. However, in situations where dozens of prints must be made, using one of the print-making processes is best.

Ink jet printers are inexpensive devices that produces good-quality output. Most are small enough to fit on a desktop, Figure 15-1. However, ink jet printers are available that print large-format drawings, Figure 15-2. Many ink jet printers can print colors.

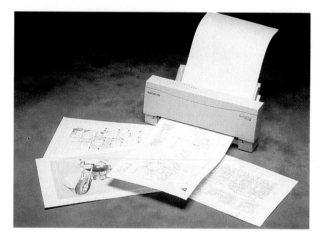

Figure 15-1. Ink jet printers are often small enough to fit on a desk. These printers can produce good-quality prints from a CAD system. (Calcomp)

Figure 15-2. Many ink jet printers are now available that can produce large-format prints from a CAD system. (Calcomp)

Laser printers and ***LED printers*** work on the electrostatic process. These printers typically produce a very high-quality image. Many laser printers are capable of printing large-size drawings, while maintaining a high-quality image, Figure 15-3. Color laser printers are also becoming common.

Pen plotters are the traditional way of making paper drawings from CAD systems, Figure 15-4. These devices duplicate the motion that a human hand might make while drawing. In other words, these devices draw lines instead of printing "dots" on the page. Pen plotters produce very high-quality drawings. However, these devices are not very good at drawing large quantities of text. Pen plotters can also create plots with color very easily.

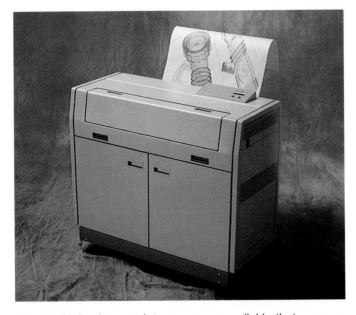

Figure 15-3. Laser printers are now available that can produce high-quality prints from a CAD system. Color laser printers are becoming more common as well. (Calcomp)

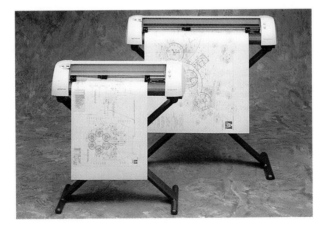

Figure 15-4. Pen plotters are the traditional way of producing prints from a CAD system. There are many different sizes of these plotters. These devices are also a very easy and inexpensive way of producing prints with color. (Calcomp)

Making a Tracing for Reproduction

Tracings are commonly made for manual drafting when prints are to be made. The first step in making a tracing is to draw it on a translucent (somewhat see-through) material. There are many types of this material available in sheet or roll form.

Tracing paper is frequently used because it is less expensive than film or cloth. Tracing paper is used for making preliminary plans or sketches.

Tracing film made from acetate, mylar, or polyester is ideally suited for work requiring print stability. Film will not expand or contract in heat, cold, or high humidity. Tracing film is tough and tear-resistant. It does not become discolored or brittle with age.

Pencil Tracings

Sharp, clear reproductions can be made from pencil tracings with modern reproduction equipment and materials. Many drafters make original drawings on the tracing material that will be used to produce the copies. For more complex jobs, the layout is usually "traced" from the original drawing. This is where the term *tracing* comes from.

Tracings for reproduction are made the same way as the conventional drawings you have been making. Complete the drawing, add notes, and dimension the drawing. Keep the lines uniformly sharp and dense to reproduce well.

Every effort must be made to keep the tracing clean. Work with clean hands. Also place a piece of clean paper under your hand when you letter, dimension, or add notes.

Inked Tracings

Inked tracings are more difficult to prepare than pencil tracings. Much patience, a knowledge of inking techniques, and lots of practice are required to produce acceptable inked tracings.

A dense, black, waterproof ink called *India ink* is used for inking. Lines are drawn with a *technical pen,* Figure 15-5. Technical pens are available in different widths. A technical pen must be held vertically to obtain the best line.

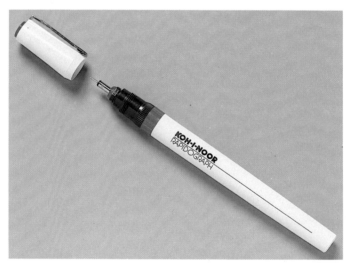

Figure 15-5. Technical pens are used to make ink tracings. Technical pens come in many different sizes. (Koh-I-Noor Rapidograph, Inc.)

Most of the difficulties you will encounter when starting to ink are shown in Figure 15-6. Many of these difficulties can be avoided if you keep the pen clean. Use great care so you do not accidentally slide your instruments or hand into freshly inked lines.

Reproduction Techniques

After an original drawing is completed, it may then be reproduced. New reproduction processes use computer and/or electronic techniques to create the prints. Many of the traditional processes employ a chemical reaction to produce prints.

CORRECTLY INKED LINE

INSTRUMENT ACCIDENTALLY PUSHED
INTO FRESHLY INKED LINE

PEN NOT HELD VERTICALLY

INK RAN UNDER EDGE OF
INSTRUMENT. PEN NOT HELD VERTICAL

Figure 15-6. A correctly inked line, and incorrectly inked lines with the probable causes.

Xerography (Electrostatic) Process

Xerography (pronounced ze-rog′-ra-fee) is a print-making process that uses an electrostatic charge to duplicate an original, Figure 15-7. Xerography process is commonly called *electrostatic process.* This process is based on the scientific principle that like electrical charges repel and unlike charges attract, Figure 15-8.

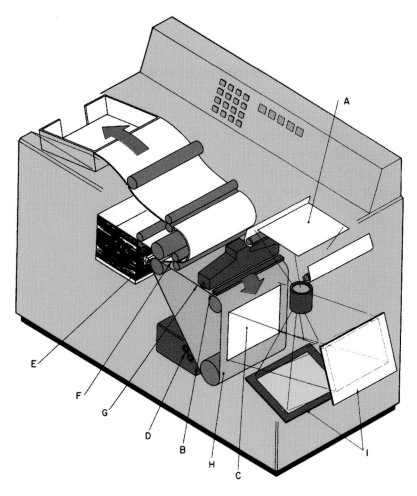

Figure 15-7. The various parts of an electrostatic (xerographic) printer. A—Original drawing. B—A positive charge is placed on the photoconductor. C—The positive charge is removed from the nonimage areas on the photoconductor. D—Negatively-charged toner ("ink") is placed on the photoconductor. E—The photoconductor presses against the copy paper. F—Rollers fuse the toner to the paper. G—Brushes and a vacuum remove remaining toner. H—The photoconductor belt. I—Mirrors and lens.

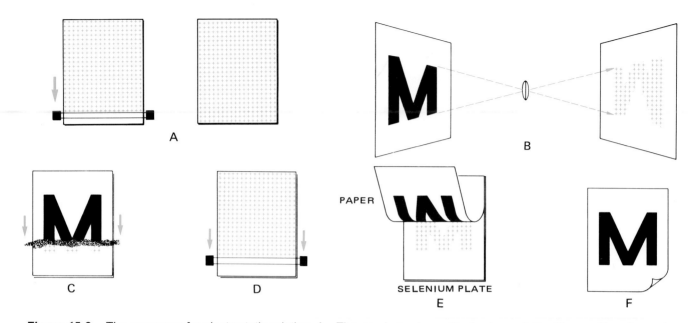

Figure 15-8. The sequence for electrostatic printing. A—The conductor is positively charged. B—The positive charge is removed from the conductor except where the image will be. C—Negatively-charged toner is passed over the conductor. D—The paper is positively charged. E—The paper is pressed against the conductor and the image is transferred. F—The toner is fused to the paper for the final product.

The electrostatic process exactly duplicates the original. The copy can be enlarged or reduced in size from the original if needed. Unlike traditional processes, the original drawing does not have to be on translucent material.

Large-format ***engineering copiers*** can quickly reproduce from original drawings. ***Document systems*** can be connected to computer networks, Figure 15-9. These systems receive drawing information from computer-aided drafting stations. The document systems then control the size of the reproduction, which drawings are reproduced first, and the use of color.

Figure 15-9. Document systems can be connected to computer networks. These electrostatic copiers can produce prints directly from CAD drawings. (Xerox)

Full-color copies can be made on color engineering copiers. Color is used very effectively in many different industries. On final prints or drawings, color can be used to identify different items, such as wires, piping, or other materials. Color is being used more often to represent specific items as more drawings are completed using computer-aided drafting.

The Microfilm Process

The *microfilm process* was originally designed to reduce storage facilities and to protect prints from loss. With this technique, the original drawing is reduced by photographic means to negative form. Finished negatives can be stored in roll form or on cards, Figure 15-10. Cards or film are easier to store than full-size drawings. To produce a working print, the microfilm image is retrieved from the files and printed on photographic paper. These prints are called *blow-backs.* The print is often discarded or destroyed when it is no longer needed. Microfilm can also be viewed on a *microfilm reader* to check details without making a blow-back.

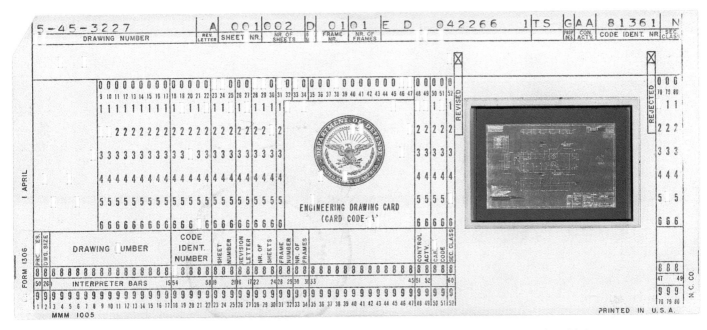

Figure 15-10.   Microfilmed drawings can be stored on cards. The drawings can be enlarged and printed later.

Traditional Print-Making Processes

There are two traditional print-making processes. These are called the diazo process and the blueprint process. As computers become more sophisticated and less expensive, more drafting is done on computers. This has caused a decline in the use of these traditional processes. However, you should be familiar with these processes. They are covered in the next sections.

Diazo Process

The *diazo process* is a copying technique for making direct positive prints, Figure 15-11. Positive prints are dark lines on a white background. The copies are produced quickly and inexpensively. In order to use this process, a tracing of the drawing must first be made on translucent material. This material is usually vellum or film.

A diazo print is made by placing the tracing in contact with light-sensitive paper or film, and exposing them to light. Light cannot penetrate the opaque lines drawn on the tracing. Exposure takes place when light strikes the light-sensitive coating of the print paper. After exposure, the print paper is developed by passing it through ammonia vapors.

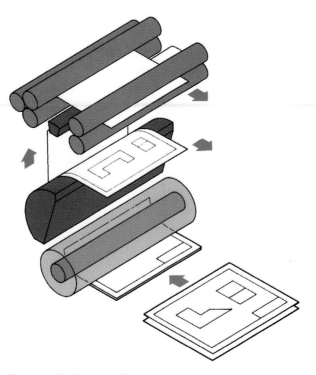

Figure 15-11. For diazo process prints, the original drawing is placed with light-sensitive paper into the machine. The two sheets are exposed to light, and ammonia vapors develop the exposed image.

Blueprints

The **blueprint process** is the oldest of the techniques used to duplicate drawings. A blueprint has white lines on a blue background. Many times, the term "blueprint" is used to refer to any print.

This process is similar to the diazo process. First, a tracing must be made. However, the print must be developed, "fixed" in a chemical solution, washed, and then dried. The blueprint process is not used much anymore.

Drafting Vocabulary

Blow-back	India ink	Pen plotter
Blueprint process	Ink jet printers	Print
Diazo process	Laser printers	Technical pen
Document systems	LED printers	Tracing
Electrostatic process	Microfilm process	Tracing film
Engineering copier	Microfilm reader	Tracing paper

Test Your Knowledge—Unit 15

Please do not write in the book. Place your answers on another sheet of paper.

1. Why are prints used instead of the original drawings?
2. For traditional print-making processes, a(n) _____ is needed.
3. Name two methods of creating tracings.

4. Name three devices used to make prints from drawings created on a CAD system.

5. The ink used to make an ink tracing is called _____ ink.

6. How is color used in CAD drawings?

7. How can color be reproduced in prints?

8. Enlarged prints made from _____ are called blow-backs.

9. The _____ is based on the scientific principle that like electrical charges repel each other, and unlike charges attract each other.

Outside Activities

1. Make a pencil tracing of one of your drawings and make a print. Make an ink tracing of the same drawing and make a print using the same process. Compare the results. How do they differ?

2. Make a CAD drawing of the drawing from number 1. Make a print using a pen plotter and a laser printer (or ink jet printer). Compare the results. How do these prints compare to the prints you made in number 1?

3. Get a print made using each of the reproduction processes described in this unit. Get photos of the devices used for each of the processes. Prepare a bulletin board display. Label each method.

4. Obtain an example of a microfilm aperture card and a print made from it. How does the quality compare to a full-size reproduction?

5. Survey industries in your area. What type of reproductions are used by architects, contractors, and manufacturers?

A high degree of technical "know-how" is needed to successfully design most products today. (American Foundrymen's Society, Inc.)

After studying this unit and completing the assigned problems, you should be able to:

◆ Describe the role of the industrial designer in industry.

◆ Explain why product design requires so much research and development (R&D).

◆ Recognize that a design may have several solutions.

◆ Know how to apply the basic design guidelines when developing your projects.

◆ Cite the qualities that help identify good design.

Design is a plan for the simple and direct solution to a technical problem. Good design is the orderly and interesting arrangement of an idea to provide certain results and/or effects, Figure 16-1.

A well-designed product is functional (does what it is supposed to do and does it well), efficient, and dependable. Such a product is less expensive than a similar poorly designed product that does not function properly and must constantly be repaired.

The men and women doing this planning for industry are called *industrial designers*, Figure 16-2.

Figure 16-1. A design is a plan to solve a problem. The concern for environmental protection and energy conservation has required the redesign of many products. The experimental vehicle shown uses a gas turbine to drive a generator to charge the batteries. A gas turbine has lower polluting emissions than a piston engine. The car has three modes of operation. The first mode is pure electric for pollution-free city driving. In the second (hybrid) mode, the turbine automatically activates to drive the generator when battery power declines to a certain level. The third mode uses the turbine alone when maximum power is required in emergency situations. (Volvo Cars of North America, Inc.)

Figure 16-2. Industrial designers, in addition to having a creative imagination, must have a knowledge of engineering, production techniques, tools, machines, and materials. (Ford Motor Company)

245

In addition to a creative imagination, skilled designers must have a knowledge of engineering, production techniques, tools, machines, and materials to design a new product for manufacture, or to improve an existing product, Figure 16-3.

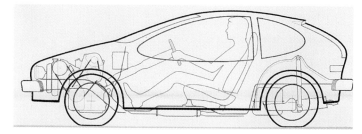

Figure 16-4. After the design problem is identified, many techniques are used to put the design into visual form for further study. In the automotive industry, full-size tape drawings are made to see if the idea is feasible after design sketches have been made. (Design sketches were discussed in Unit 3.) Shown is a tape outline of a small town car that is under study. (General Motors Corp.)

Figure 16-3. Computer-controlled flight simulator used to introduce flyers to new aircraft. Its movements, sounds, and visual simulations are so realistic that the aviators have no problem when they fly real aircraft. Much creative imagination went into its design and construction. (Sperry Div., Sperry Corp.)

Figure 16-5. While this trademark or logo appears to be simple, it was a challenging problem for its designer. The "mark" has great commercial value. People may not remember the company name but they remember its trademark. Can you name familiar trademarks?

Design Guidelines

Product design requires much research and development. It starts out as an idea, Figure 16-4. Many concepts of that idea must be studied, tried, refined, and then either used or discarded. The same technique is followed whether designing a trademark (very important to a company), Figure 16-5, or a complex product like a ship, Figure 16-6, or a student learning experience for this class.

Figure 16-6. Product design requires much research and development. Many people are responsible for the design of components, no matter how complex. Proper product design will help increase productivity and reduce cost, while improving quality. (General Electric)

The design process is not an exact science. There is no special formula that, if followed, 1, 2, 3 . . . will guarantee a well-designed product. However, there are certain qualities that can help you identify good design.

Function. A well-designed product "works." It does what it is designed to do and does it efficiently. Function often dictates product shape. Some design problems have several solutions, Figure 16-7.

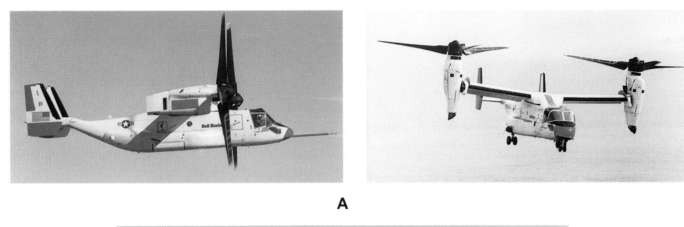

A

B

C

Figure 16-7. A design problem may have several solutions. Here are how three companies designed a vertical take-off and landing (VTOL) aircraft. Each solved the problem. Can you explain how each aircraft operates? (A—Bell Helicopter Textron/Boeing Helicopters; B—McDonnell Douglas; C—Robert Walker)

Honesty. The qualities of the materials (strength, weight, texture, etc.) used in the design are emphasized. One material is not made to look like another material, Figure 16-8.

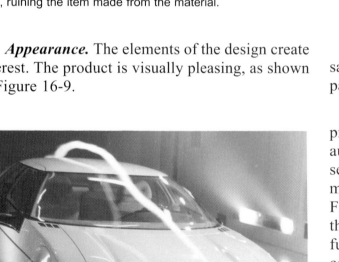

Figure 16-8. A well-designed product does not have one material masquerading as another material. Shown is a table made of cherry wood. Included on the table top is a material made of compressed sawdust and covered with a thin plastic sheet on which a wood grain pattern has been printed. It has been scratched to show how thin the imitation wood covering is. The scratch cannot be sanded out and the finish cannot be repaired. The plastic sheet also has a tendency to blister and peel, ruining the item made from the material.

Appearance. The elements of the design create interest. The product is visually pleasing, as shown in Figure 16-9.

Figure 16-9. The elements of good design create interest and the product is visually pleasing. The Probe IV is considered a very attractive automobile by many people. (Ford Motor Company)

Reliability. A well-designed product is not always malfunctioning (breaking down) or failing. It should be easy to service and economical to maintain. See Figure 16-10.

Figure 16-10. A well-designed product is dependable; it is not always in need of repair. Here an auto body from the production line is being checked to be sure it meets design specifications. Production units are selected at random and actually "torn apart" to check whether they have been welded properly. (Ford Motor Company)

Safety. When used properly, the product is safe. Protection from possible injury to the user is part of the basic design, Figure 16-11.

Color. Color plays an important role with some products and is included in the design process. The automotive industry does much research before selecting the interior and external colors and the materials for interior details for new models, Figure 16-12. A well-designed product can be further enhanced in appearance when finished in carefully selected colors. However, the addition of color, no matter how carefully chosen, will not improve a poorly designed product.

Quality. Quality is a basic part of good design. It cannot be added after the product is made. Good design should be a guarantee of quality, Figure 16-13. Quality is designed and built into items.

Simply stated, good design is characterized by certain qualities. All of them are necessary. Overlooking or eliminating one can potentially destroy the entire design.

Figure 16-11. A well-designed product is safe. This "thrill ride" was designed to provide certain results and effects. What things do the designers have to consider when planning such rides to make them safe?

Figure 16-13. Quality is a basic part of good design. This hardwood (mahogany) stool is well-designed and constructed, and it looks that way.

Figure 16-12. Color plays an important role in the design of some products. Colors must be carefully chosen to attract buyers. (Ford Motor Company)

Designing a Project

How should *you* go about designing a project? One way is to employ a method followed by professional designers. Think of your project as a *design problem*:

1. **The idea.** What kind of a project do you want to make? Keep your first few design problems simple. How will it be used? Will its use affect the shape of the project? Are there cost limitations?

2. **Develop your ideas.** What must your project do? How can it best be done? Research (study) similar products. How did other designers solve the problem? Make sketches of your ideas. Select the best of the ideas. When possible, submit your design for class critique (review).

3. **Make models.** Construct models of your best ideas, Figure 16-14. Evaluate them. Select the best. You may want to make refinements (changes) in your design when you see it in three dimensions.

4. **Make working drawings.** When satisfied that you have done the best you can to solve the

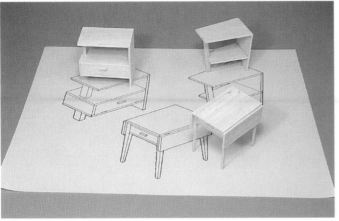

Figure 16-14. Develop your ideas by making many sketches. Construct models of your best ideas to determine how they will look in three dimensions. The sketches shown were made with instruments and ink so they would photograph well. Your drawings can be drawn in pencil.

Figure 16-15. The proof of a design is how it looks when it is constructed and put into use. This student designed and made table is nicely designed and has been constructed using approved cabinetmaking techniques. The only metal fasteners were the screws used to hold the top to the body.

problem, prepare working drawings for the project. If practical, make the drawings full size.

5. **Construct the project.** Make the project to the best of your ability. Do a job you will be proud to show to others, Figure 16-15. Do not be afraid to ask for help from your instructor if you have a problem.

Developing a well-designed project takes time. Do not be discouraged if your first attempts fall short of your goals. It takes practice and effort to acquire the skill and ability to solve design problems. The only way to succeed is by doing.

Keep a notebook of your design efforts. Include your sketches, drawings, and photos of projects you have designed. It will make it easier to evaluate your progress.

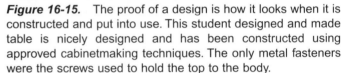

Drafting Vocabulary

Appearance	Function	Refinements
Color	Honesty	Reliability
Creative imagination	Industrial designer	Research
Dependable design	Limitations	Safety
Design problem	Quality	Solution

Test Your Knowledge—Unit 16

Please do not write in the book. Place your answers on another sheet of paper.

1. Good design is a(n) _____.
   A. orderly and interesting arrangement of an idea to produce certain results and/or effects
   B. carefully thought out plan to produce a product that is functional and pleasing to the eye and/or touch
   C. creative process
   D. All of the above.
   E. None of the above.

2. The men and women who do design work in industry are called _____ _____.

3. They must have working knowledge of _____ _____ _____ _____.

4. List the characteristics that make up good design.

5. List the steps you would follow to solve a design problem. Briefly describe each.

Outside Activities

The design problems listed below are present to give you practice in design activities. You may also originate your own products to be designed.

A. Hand-launched glider.

B. Jet-propelled (CO_2 cartridge engine) model automobile.

C. Book rack.

D. Box to store compact disks or tapes.

E. Tool box.

F. Turned bowl.

G. Clock.

H. Night stand table.

I. Stereo cabinet.

J. Bicycle rack.

K. Storage container for video cassettes.

L. Display cabinet for awards and trophies.

M. Fiberglass model boat.

N. Cutting board.

O. Metal base for a table lamp.

P. Coffee table.

Q. Tray.

R. Turned wood lamp.

S. Plastic soft food spreader.

T. Salad server (laminated wood, wood, metal, plastic).

U. Wall shelf.

V. Shoeshine box.

W. Disposable waste basket (made from corrugated cardboard).

X. Desk top pencil holder.

Y. Folder to store drafting plates.

Z. Carrying case for drafting tools.

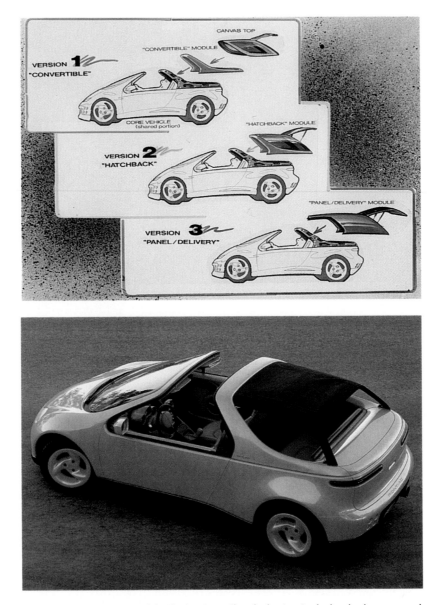

Concept cars are used in the automotive industry to help designers and engineers envision a design. Shown here are three design sketches of a concept car, along with a prototype. (Pontiac)

Unit 17
Models, Mockups, and Prototypes

After studying this unit and completing the assigned problems, you should be able to:

◆ Explain why industry uses models, mockups, and prototypes.

◆ Define the terms—model, mockup, and prototype.

◆ Describe how some industries use models, mockups, and prototypes.

◆ Construct simple models.

◆ Use model-making equipment and supplies safely.

To many people, model making is an interesting and challenging hobby. You have probably made models of famous planes, boats, or cars from wood or plastic.

Industry makes extensive use of three modeling activities—***models***, ***mockups***, and ***prototypes***. They are used for engineering, educational, and planning purposes, as well as for merchandising. Models have proven to be very helpful in solving design problems and to check the workability of a design or idea before it is put into production.

Model

Industry's definition of a ***scale model*** is a replica of a proposed, planned, or existing product. Even though CAD is a great help in developing a new product design, a model may be constructed to see how the product will look in three-dimensions, to check out scientific theory, to demonstrate ideas, to aid in training, or to use for advertising purposes. See Figure 17-1.

Computer-generated models. Computers are commonly used to create models in industry. Three types of models are created—***wireframe models, surface models,*** and ***solid models.***

Wireframe models, Figure 17-2, looks like they are made of wire. You can easily see through them. These models are sometimes hard to visualize since you cannot tell which is the front or the back.

Surface models, Figure 17-3, are easier to visualize. These models are basically a wireframe with a thin "skin." Many times they look like solid models, but surface models are *not* solid.

As their name implies, ***solid models,*** are "solid" all the way through. When a section is made through a solid model, the interior features can be seen, Figure 7-4.

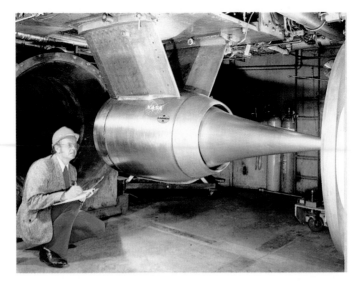

Figure 17-1. A working scale model of an experimental scramjet engine capable of sustaining flight at speeds in excess of Mach 6. It is fueled by liquid hydrogen, which also cools the engine's combustion chamber components to keep them from melting. (NASA)

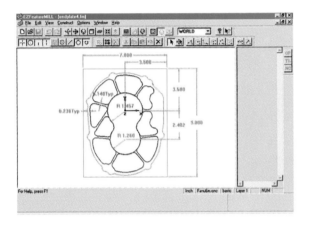

Figure 17-2. With some CAD/CAM software, the geometry of the part to be machined can be created at the machine tool or imported from files.

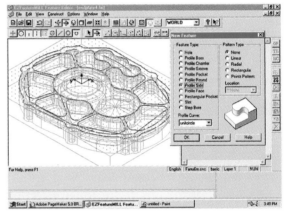

Figure 17-3. The CAD/CAM software defines the features of the part.

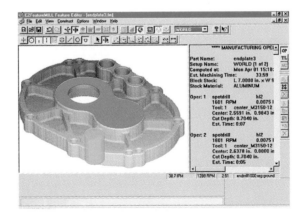

Figure 17-4. The CAD/CAM software automatically produces ready-to-use operations sheets and tool lists for the machinist.

Mockup

A *mockup* is a full-size, three-dimensional copy of a product, Figure 17-5. The mockup is usually made of plywood, plaster, clay, or a combination of materials.

Figure 17-5. Full-size mockup of the front seat and controls of a concept car. Such design features often find their way into future production vehicles. (Pontiac Div., General Motors Corp.)

Prototype

A *prototype* is a full-size operating model of the production item, Figure 17-6. It is often handcrafted to check out and eliminate possible design and production "bugs."

Figure 17-6. When designing this 200 mph (320 km/h) train, it was important to know how it would handle actual operating conditions. Before it was put into production, a prototype was constructed and run more than a million miles. Only a few changes had to be made. (French National Railroad)

How Models, Mockups, and Prototypes are Used

Many industries employ models, mockups, and prototypes for design tools. A few of the more important applications are automotive, aerospace, architecture, ship building, city planning, and construction engineering. See Figure 17-7.

Figure 17-7. Cut-away of a turbofan jet engine. It is used for sales and training purposes. (Pratt & Whitney Canada)

Automotive Industry

The automotive industry places great importance on the use of models, mockups, and prototypes. Mistakes in marketing and production can be very costly. An automobile starts "life" as a series of sketches, Figure 17-8. These are usually developed around specifications supplied by management. Promising sketches are usually drawn full size for additional evaluation. Clay models are employed for three-dimensional studies. In some cases, three-dimensional computer-generated models are created to evaluate the design, Figure 17-9.

Figure 17-8. The automotive industry uses sketches to develop ideas. Scale models are made of promising designs for future study and development. (Pontiac Div., General Motors Corp.)

Figure 17-9. A computer-generated model of the suspension of a road-rally car. (Autodesk, Inc.)

Upon completion of further design development, a full-size mockup is constructed, Figure 17-10. If it is decided that the design has potential, a fiberglass prototype is generally fabricated, Figure 17-11. This is the concept car you see at auto shows where they are displayed for evaluation by the public. If the comments are favorable, the concept car may be put into production. After many more months of development and production planning, the vehicle is manufactured and made available to the public, Figure 17-12. ***Production fixtures*** (devices to hold body panels while they are welded together), dies to shape other parts, and other tools are developed from the mockup.

Figure 17-10. Final smoothing of the vinyl sheets coating the full-size clay model of this automobile takes place before review by company officials. Modeled exactly to scale from the full-size line drawing on the wall, this clay body eventually will be destroyed after a two-part epoxy/fiberglass mold is made. The resulting mold will be used to shape the body for the prototype vehicle. (Ford Motor Company)

Figure 17-11. Fiberglass prototype of the Pontiac Trans Sport. While only a few "idea" vehicles go into production, they are used for design studies. Many proposed innovations are utilized on production vehicles. (Pontiac Div., General Motors Corp.)

Figure 17-12. Some "idea" vehicles actually go into production with minor changes. Shown is the production version of the Trans Sport. With few exceptions (including the color), it bears a good resemblance to the concept vehicle shown in Figure 17-8. (Pontiac Div., General Motors Corp.)

Aerospace Applications

The tremendous cost of aerospace vehicles makes it mandatory that they first be developed in model and mockup form, Figure 17-13. Flight characteristics can be determined with considerable accuracy, without endangering human life, by testing complicated models in a wind tunnel, Figure 17-14. Prototype aircraft are the first two or three production planes (usually handcrafted) that are flown to check the data obtained from wind tunnel and computer research, Figure 17-15.

Figure 17-13.   Model of advanced United States Air Force aircraft used for design studies. It was constructed employing information developed by CAD. (Grumman Aerospace Corp.)

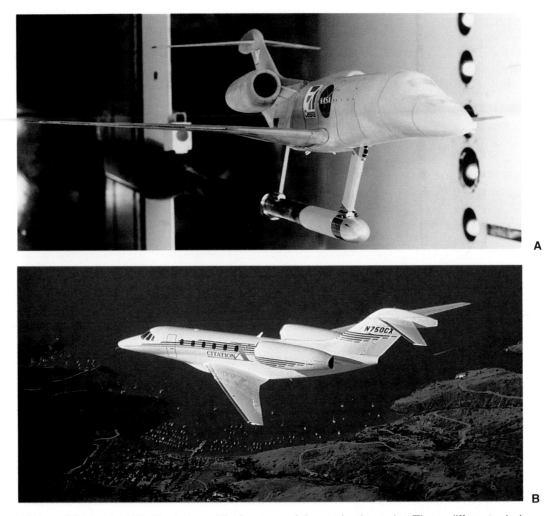

A

B

Figure 17-14. A—Wind tunnel model of a state-of-the-art business jet. Three different wind tunnel models were used; a high-speed model for the clean configuration (gear and flaps up) for testing at cruise Mach numbers and beyond; a low-speed model for landing and take-off; and a flutter model to determine proper weight distribution and stiffness at high Mach numbers. The flutter model is shown. B—Prototype of the Citation "X" shown as wind model in Figure 17-14A. Flight tests were almost exactly as predicted by wind tunnel models. (Cessna Aircraft Company)

Figure 17-15. Prototype of the advanced Boeing 777. It is used for flight testing and crew and maintenance technician training. Commercial aircraft must undergo hundreds of hours of rigorous testing before it can be put into passenger service. (The Boeing Company)

Architecture

You have seen photos of proposed buildings in the real estate section of your newspaper. Some of these illustrations are models that are very accurate replicas of the proposed buildings, Figure 17-16. Many people use models when planning new homes, Figures 17-17 and 17-18. This helps them visualize how the house will look when completed. The model enables the owner to see the completed design in three dimensions and how paint colors and shrubbery plantings will look. Models of buildings are also used for advertising purposes, Figure 17-19. Photos of the models are used in magazines, newspapers, and brochures.

Figure 17-16. Computer-generated architectural model of a proposed house. (Autodesk, Inc.)

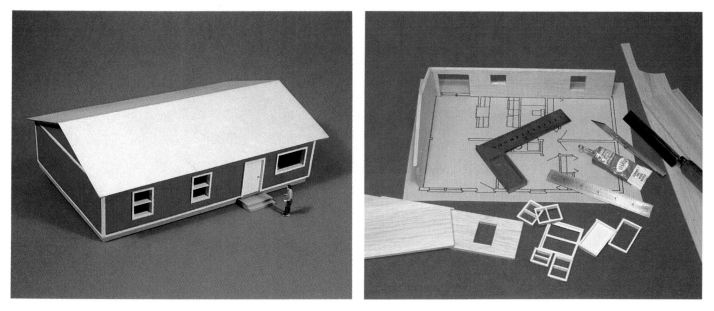

Figure 17-17.   Many potential home builders make scale models so they can see how their home will look before starting to build.

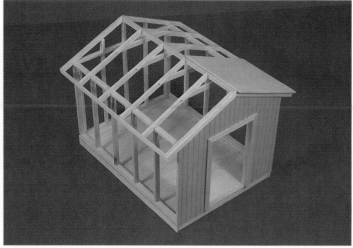

Figure 17-18. Computer-generated model of the interior of a room in a house. Note how the owners have positioned the furniture to take advantage of the scenery outside. (Autodesk, Inc.)

Figure 17-19. Student-designed and constructed storage shed. Models like this are often used for advertising purposes.

Shipbuilding

Ship hulls are tested in model form before designs are finalized and construction begins, Figure 17-20. Specially designed equipment tows the model hull through a water test basin. The model hull behaves like the full-size ship so design faults can be located and corrected. The models test a proposed ship's seaworthiness, power requirements, resistance as it travels through water, and stability characteristics.

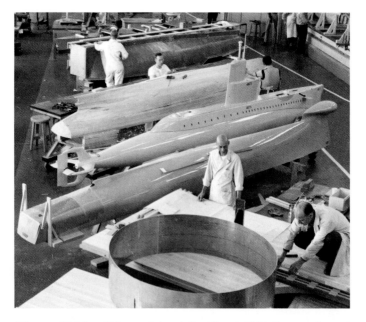

Figure 17-20. Ship models at the Naval Ship Research and Development Center used to determine a proposed ship's characteristics. The model is tested by towing the model hull with instrumentation in the 140 foot test basin at the Center.

City Planning

Most large cities use scale models to show city officials and planners how proposed changes and future developments will look, Figure 17-21. Models, while costly, permit intelligent decisions to be made before large sums of money are spent acquiring land and before existing buildings are torn down. The models can be used to plan for the movement of vehicles and people to avoid congestion after the plan is built. The models can be used to "help sell" the future development so financing can be obtained.

Figure 17-21. Models showing the details of the tip of Manhattan and a portion of Brooklyn. The model is used for planning purposes. Many drafters will be needed to plan the future expansion of the city. (Lester Associates, Inc.)

Construction Engineering

Many construction projects are designed from carefully constructed models, Figure 17-22. By working from models, engineers can see how space can best be utilized. In some instances, they can determine how the proposed project will affect surrounding communities and the environment. This helps minimize field problems and changes during construction.

Constructing Models

Model-making materials are readily available commercially, Figure 17-23. Many products made for the model railroader are ideally suited for making architectural models. Kits are available for the small home builder who wants to design his or her own home. A professional touch can be added to models by using accurately scaled furniture, automobiles, and figures that can be purchased at toy and hobby shops. Preprinted sheets of brick and stone can be glued to suitable thicknesses of balsa wood for walls and partitions.

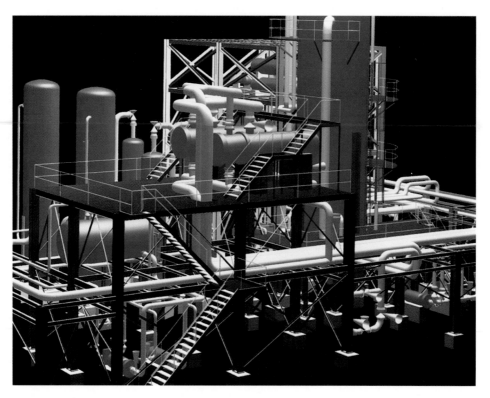

Figure 17-22. A computer-generated model such as this can be used to determine the best way to utilize space.

Figure 17-23. A few of the model-making materials available at hobby shops.

Various types of abrasive paper are suitable for roofing, driveways, and walkways. Simulated window glass can be made from transparent plastic sheet. Several different scale sizes of window and door frames are available molded in plastic. Most model shops can supply brushes, paints, and trees in various types and sizes.

Other types of models—autos, planes, and boats—are made from basswood, mahogany, balsa wood, metal, plaster, and various kinds of plastics, Figure 17-24. Regular model-making paint is produced in hundreds of colors and is ideal for painting all types of models, Figure 17-25. However, care must be exer-

Figure 17-24. Plastic model of helicopter to be used as a basis for constructing a hangar.

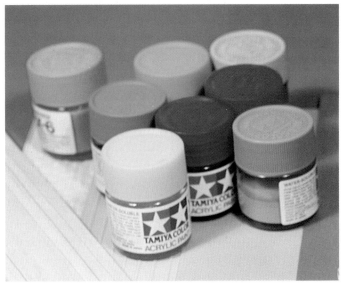

Figure 17-25. Regular model-making paints are available in hundreds of colors and are ideal for painting all types of models. Painting must be carefully done.

cised when painting models that have plastic in their construction. Be sure the paints are designed for plastics. If you are not sure, paint a scrap portion of similar plastic or a small portion of the plastic that is hidden from view. This will help determine whether the paint is compatible with the material.

Safety Note: Carefully read the instructions on the paint and adhesive containers. Use them only as specified. Wear a dust mask if you must do much sanding of wood or plastic. Wash your hands after cleaning brushes and work area. Models may be assembled with model airplane cement, Elmer's Glue, Titebond™, cyanoacrylates, or epoxies. Use care when using them and follow all safety directions according to the manufacturer's directions.

Drafting Vocabulary

Fabricated	Mockup	Scale Model
Fiberglass	Model	Wind Tunnel
Fixture	Preproduction	
Handcrafted	Prototype	

Test Your Knowledge—Unit 17

Please do not write in the book. Place your answers on another sheet of paper.

1. Industry uses models, mockups, and prototypes for what purposes?
2. A model is _____.
3. A mockup is _____.
4. A prototype is _____.
5. List four industries that employ models, mockups, and prototypes as design tools. Briefly describe how each industry listed used them.

6. Walls and partitions in a model home are usually constructed of _____ wood.

7. Roofs, driveways, and walkways can be made from different kinds and grades of _____.

8. Models of cars, planes, and boats may be made from what kinds of materials?

Outside Activities

1. Visit a model or hobby shop, then prepare a list of available materials which may be used for building architectural models.

2. Review technical magazines and clip illustrations that show models, mockups, and prototypes being used for engineering, educational, planning or other purposes. (Do not cut up library copies.)

3. Make a collection of materials suitable for building model homes.

4. Visit a professional model maker in your community. With permission, make a series of slides showing examples of various models and how these models are made. Then, give a talk to your class on professional model making.

5. Make a scale model of your drafting room. Discuss alternate layouts.

6. Construct a model helicopter similar to the one shown in Figure 17-24. Design and construct from balsa wood a minimum building that would protect the helicopter from the elements.

After studying this unit and completing the assigned problems, you should be able to:

◆ Identify the components of a map.

◆ Explain why maps are used.

◆ Describe various types of maps.

◆ Read a map.

A ***map*** is a graphic representation, usually on a flat plan, of a portion of the earth's surface, Figure 18-1. Maps are used to communicate ideas and concepts, as well as provide direction. A typical map of a community is shown in Figure 18-2.

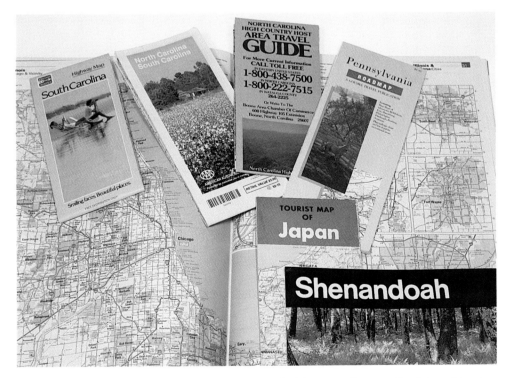

Figure 18-1. Maps are important to many people. They use them on the job and to plan business trips and vacations. Many types of maps are available.

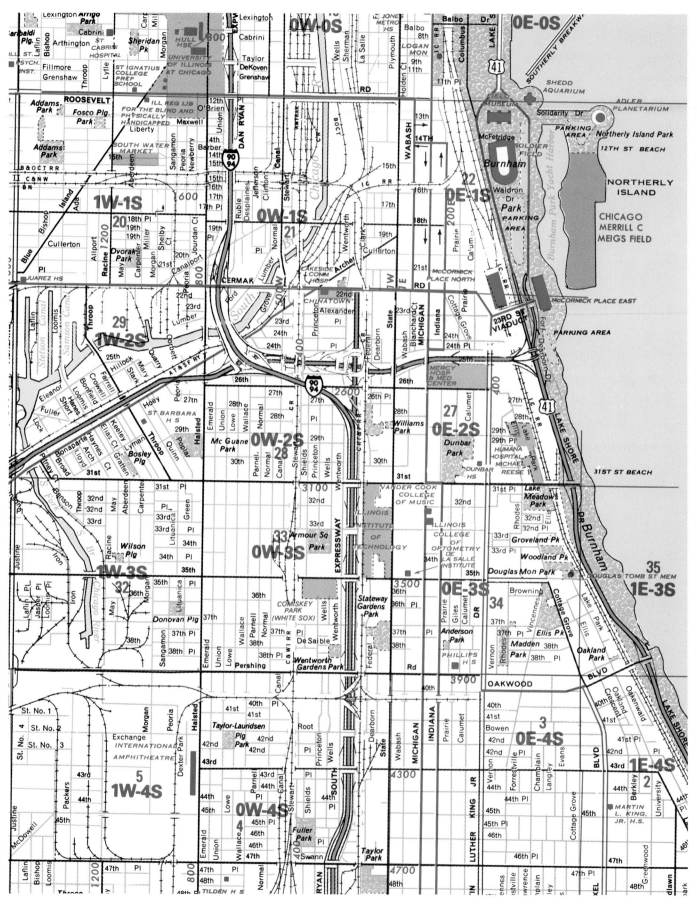

Figure 18-2. Well-designed map showing the downtown area of Chicago, Illinois. (Map ©1995 by Rand McNally, R.L. 95-S-125)

General reference maps show elevation and political features (cities and towns) of an area. United States Geological Survey maps show the exact location of features of the earth. Surveyors rely on these precision maps whenever they need data or facts.

Aeronautical and marine maps (often called *charts*) provide navigational information for those using the air and sea lanes. Weather maps indicate meteorological conditions for a given period of time, Figure 18-3.

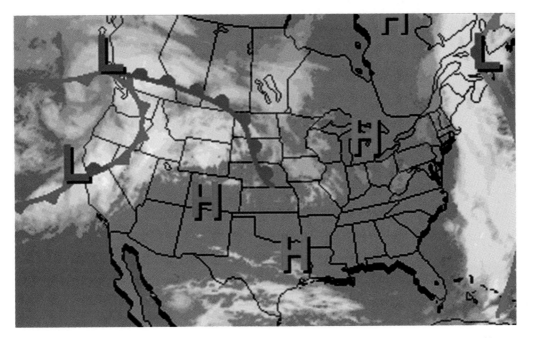

Figure 18-3. Weather map. Most weather maps are now generated by computer. (Accu-Weather Inc.)

Road maps show the roads and highways in a region and aid in planning business and vacation trips. Other maps show or emphasize some particular features or surface characteristics of the earth (or planets).

A professional map maker is called a *cartographer*. He or she has the specialized skills to combine data from many sources to design and prepare accurate maps.

There are times when drafters are called upon to draw maps. This unit will describe the maps they would be expected to prepare.

Preparing Maps

Aerial photography, space satellites, and the computer have reduced the time needed to map large areas of land and sea. Accuracy has also been greatly improved. However, a great deal of "leg work" is still necessary to map small areas. A *surveyor*, with an instrument called a *transit*, establishes the boundaries of the area being mapped, Figure 18-4. Distances are measured with a *tape*. The *bearings* (angles the lines make with due north and south) are recorded in the form of *field notes*. With this information, a map or *plat plan,* such as the one shown in Figure 18-5 is developed and drawn.

In metricated surveying, the areas of the surveyed land is given in *hectares (ha)* instead of acres, Figure 18-6. Distances are given in meters and kilometers.

Figure 18-4. Surveyor using a transit to establish boundaries.

Maps are always drawn to scale because it is not possible to draw them full size. The size of the land area being mapped determines the scale.

In most towns, cities, and counties/parishes, a drawing of a building lot (called a ***plat***) or farm may be drawn 1 in. = 30 ft. In places where metric units have been adopted, they will be drawn to a scale of 1:300. That is, 1.0 mm on the drawing equals 300.0 mm of the lot.

The ***United States Geological Survey*** has started mapping the United States in metric. Each map will cover 4157 square kilometers. The maps are printed to a scale of 1:25,000 (4.0 cm = 1.0 km). Such maps not yet metricated are drawn to a scale of 1:62,500 (approximately 1 in. = 1 mile).

Many maps use a graphic scale to indicate map scale, Figure 18-7.

Maps Drawn by Drafters

The following maps are examples of a few of the many types a drafter is often requested to prepare. An important point to remember is to include the proper scale.

Plot Plan

The map of a lot on which a house or building is to be constructed is called a ***plot*** or ***site plan***. It shows site dimensions, the exact location of the lot, as well as where the structure is to be situated on the lot. It may also show topographical features, Fig. 18-8.

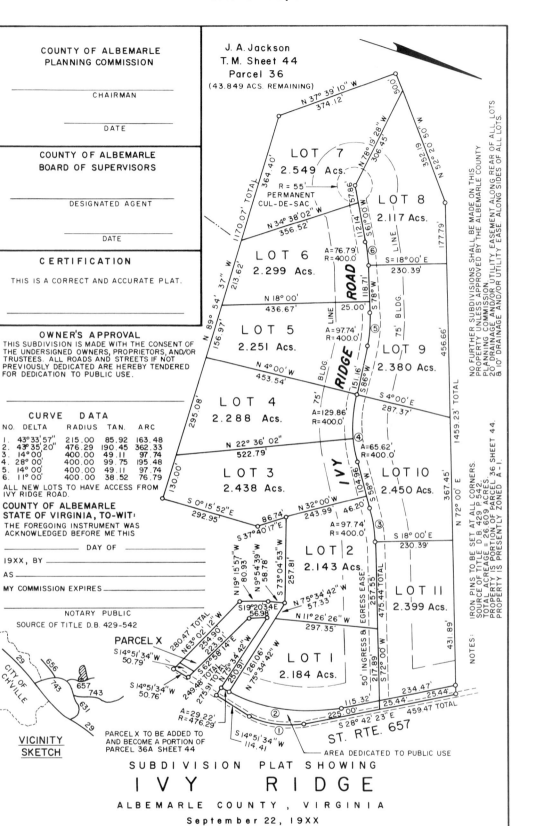

Figure 18-5. A plat plan. It is a map typical of those prepared from information gathered by a surveyor. This is a plat showing a housing subdivision.

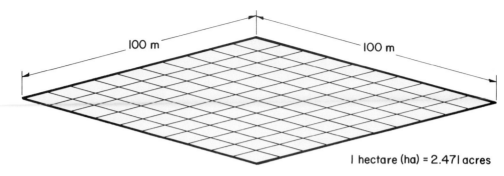

Figure 18-6. A hectare (ha) is an area that is 100 meters long and 100 meters wide. (It equals about 2.5 acres.)

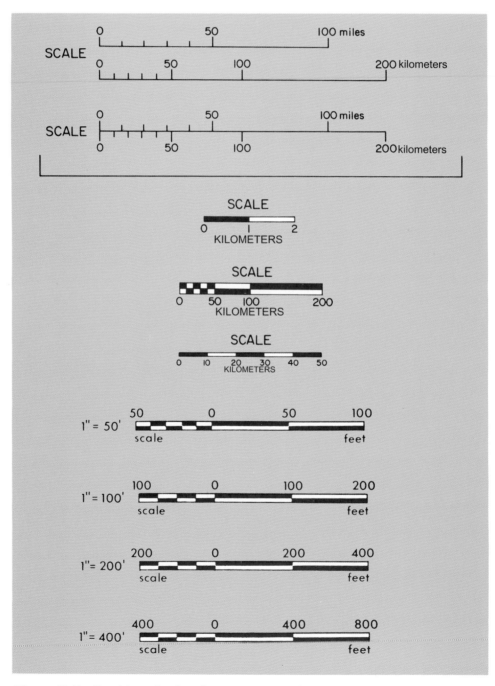

Figure 18-7. Graphic scales found on many types of maps.

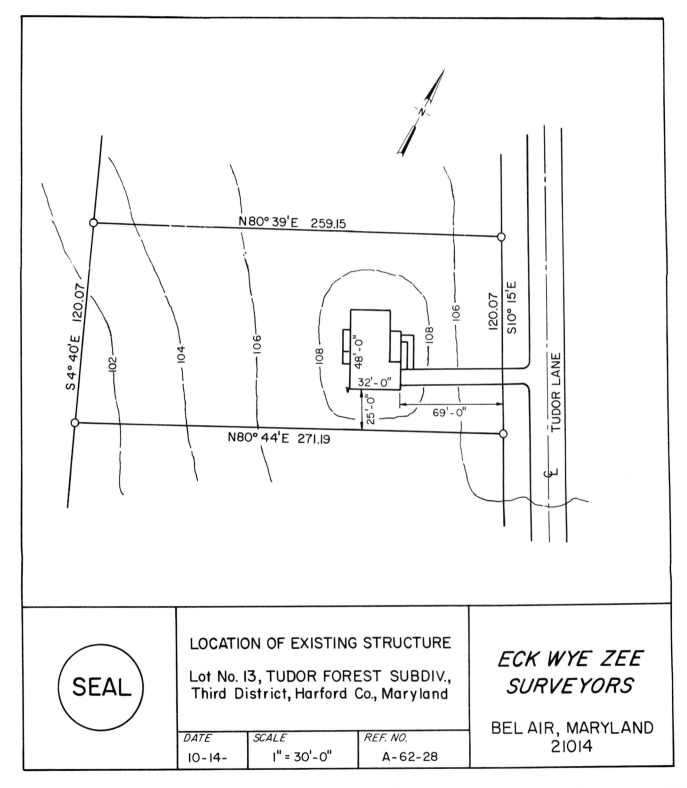

Figure 18-8. Plot plan. The elevation or height of the land above sea level is shown by the series of irregular lines (called **contour lines**). They are drawn on the plan at predetermined differences in elevation.

City Map

Street and lot layouts are shown on a *city map*, Figure 18-9.

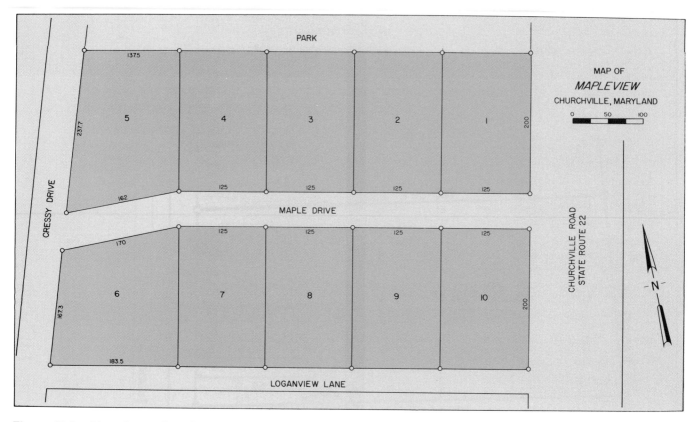

Figure 18-9. Map of a section of a city showing the layout of streets and lots.

Special Maps

Special maps such as *location* or *vicinity maps* help you find your way around a specific area, Figures 18-10 and 18-11. Other special maps aid in determining travel time and calculating distances. These maps are used to plan trips when traveling on vacations or for business, Figure 18-12.

Topographic Maps

Information about physical features is included on a *topographic map.* Crops, streams, airports, roads, bridges, buildings, and other land characteristics are shown. Contour lines are often included on topographic maps.

Other Types of Maps

Other types of maps include *real estate maps,* Figure 18-13, which show available home sites to potential buyers.

Special maps are often located at historical sites to guide visitors to the various points of interest, Figure 18-14.

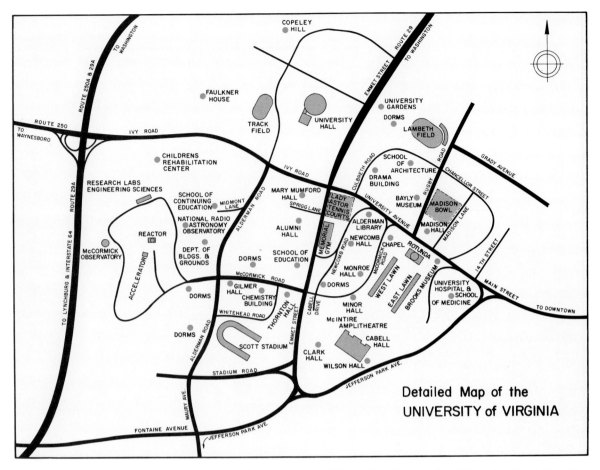

Figure 18-10. Map showing the location of the various facilities of the University of Virginia, Charlottesville, Virginia.

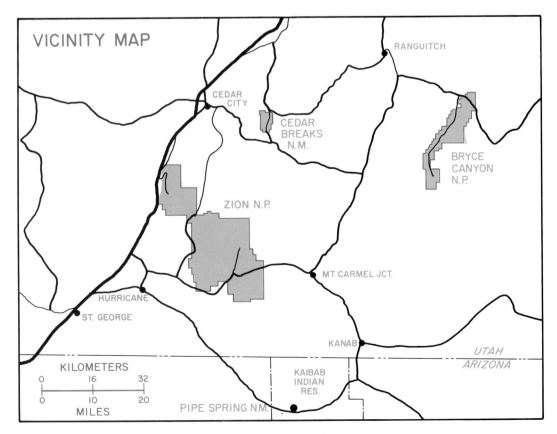

Figure 18-11. Vicinity map of Zion National Park in Utah.

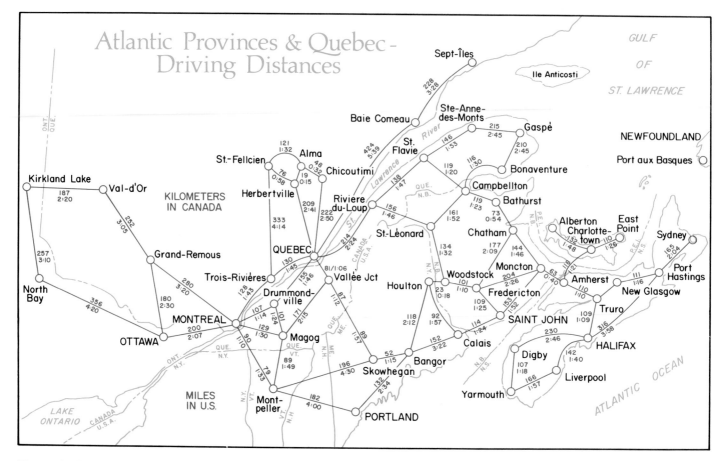

Figure 18-12. Map to aid in determining travel time and calculating distances when traveling.

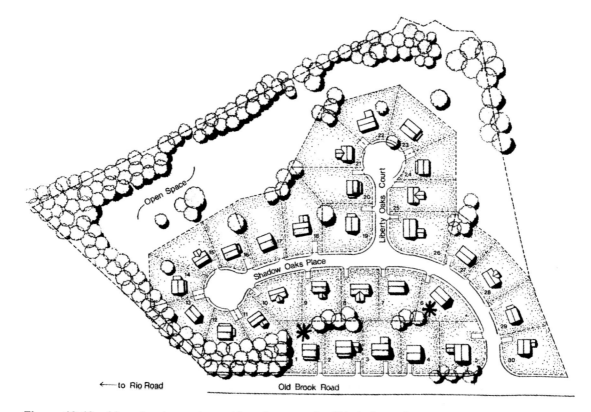

Figure 18-13. Map showing a planned housing area. It will help home buyers select sites for their new homes.

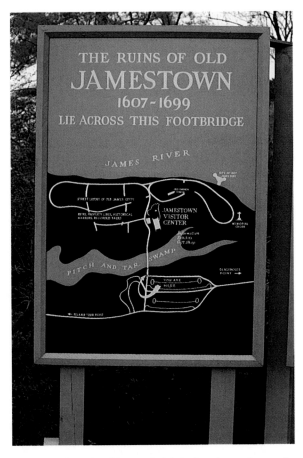

Figure 18-14. Map of historic Jamestown in Virginia is posted at the entrance to the National Park. It shows points of interest.

Map Symbols

By using *symbols* such as those shown in Figure 18-15, it is possible to include a large amount of information on a map. The symbol is often a stylized drawing (has a resemblance) of the feature it represents.

Figure 18-15. A few of the hundreds of symbols used on maps. How many can you identify?

Drafting Vocabulary

Acres
Bearings
Boundaries
Cartographer
Charts
City map
Contour lines
Elevation
Field notes

Hectares
Location map
Map
Meteorological
Metricate
Navigational
Plat plan
Plot
Real estate map

Surveyor
Symbols
Tape
Topographical map
Transit
United States Geological
  Survey
Vicinity map

Test Your Knowledge—Unit 18

Please do not write in the book. Place your answers on another sheet of paper.

1. What is a map?
2. List some uses you have made of maps.
3. A professional map maker is called a(n) _____.
4. A surveyor uses an instrument called a(n) _____ to gather data to prepare a map.
5. Measurements made by a surveyor are recorded in the form of _____ _____.
6. As surveying is metricated, the area of the surveyed land is given in _____ instead of acres.
7. Why must maps be drawn to scale?
8. A plot or plat plan is a map of _____.
9. The elevation or height of land above sea level is often shown on a map with _____ lines.

Outside Activities

1. Secure samples of special-purpose maps and make a bulletin board display.
2. Prepare a map showing the school grounds. Label all athletic fields.
3. Make a plat plan of the property on which your home is located. Show the street and make sure North is up. Label all businesses, churches, and schools.
4. Prepare a map of your neighborhood.
5. Draw a map showing the route you take in coming to school.
6. **Class Project:** Draw a map of the surrounding community and have each student draw in the route traveled in coming to school.
7. Secure a road map of your state. Plot the shortest route between your home town and the capitol of your state. If you live in the capitol city, plot the shortest route between it and the next largest city in the state.
8. Prepare a special map that will show the locations of the various schools in your school district.
9. **Multigroup Project:** If survey equipment is available, have each group survey the school grounds and compare their results. Compare the results with an actual survey of the grounds.

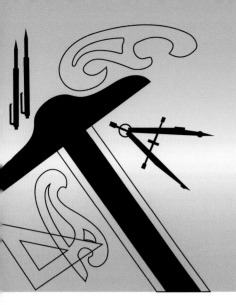

Unit 19
Charts and Graphs

After studying this unit and completing the assigned problems, you should be able to:

◆ Explain why charts and graphs are used.

◆ Create simple charts and graphs to show trends, make comparisons, and measure progress.

◆ Describe the most frequently used charts and graphs.

◆ Explain the difference between a chart and a graph.

◆ Cite the advantage of using a computer to generate graphs.

Industry and education have many uses for charts and graphs. A few of the more widely employed types are described and shown in this unit.

Charts and graphs make it possible to show trends, make comparisons, and measure progress quickly without studying a mass of statistics (numbers and information).

A drafter is often asked to prepare charts and graphs. Drafters take the statistical information and decide upon the most effective way to present this material in an interesting and easily understood manner.

Graphs

Graphs most frequently used are:

1. Line graph.
2. Bar graph.
3. Circle or pie graph (sometimes called an area graph).

Line Graph

The *line graph* may be used to make comparisons, Figure 19-1. Lines that present the information are called *curves*. When only one curve appears on the graph, it should be drawn as a solid line.

When more than one curve is employed, each line should be clearly labeled and a *key* included with the graph to show what each curve represents.

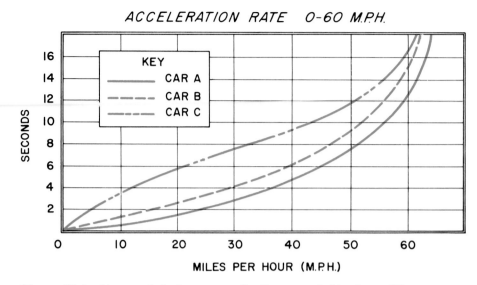

Figure 19-1. Line graph that compares the times needed by three different cars to reach 55 mph.

A line graph may be used to show trends, that is, what has happened or what may happen. A line graph is illustrated in Figure 19-2. In designing line graphs, be sure to use appropriate scales for each of the axes.

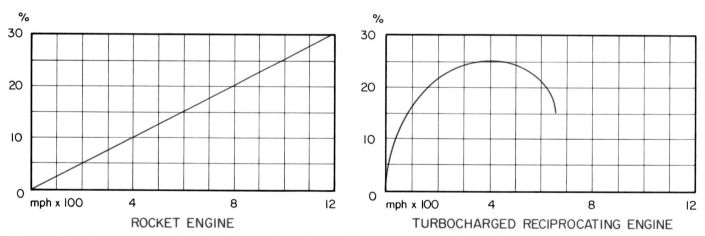

Figure 19-2. Line graph showing that rocket engine efficiency increases the faster it travels while a turbocharged internal combustion engine's efficiency drops off at high speed.

Bar Graph

Comparisons between quantities or conditions can also be made with **bar graphs**. Several different forms of the bar graph are available to the graph maker.

The **horizontal bar graph** presents information on a horizontal plane, Figure 19-3.

With the **vertical bar graph** information is given in a vertical or upright position for easy comparisons, Figure 19-4.

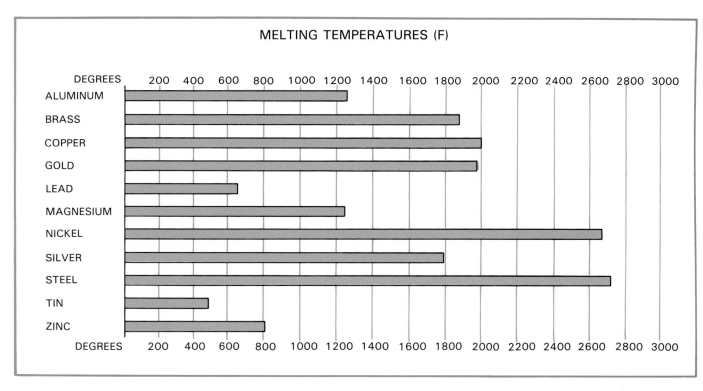

Figure 19-3.   Horizontal bar graph used to indicate melting temperatures of metal.

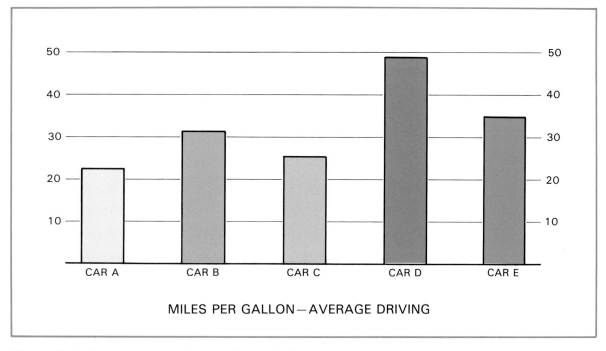

Figure 19-4.   Car economy is shown in this vertical bar graph.

The **composite bar graph** can be drawn in either a vertical or horizontal position, Figure 19-5. This type of graph compares several items of information in the same graph.

A **100 percent bar graph** consists of a singular rectangular bar, Figure 19-6. Information is presented on a percentage basis.

A **pictorial bar graph** is a variation of the bar graph. It employs pictures to represent the information instead of bars, Figure 19-7. Pictures tend to make a graph more interesting.

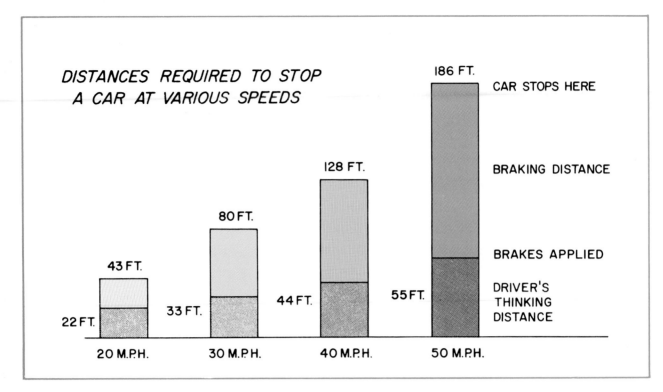

Figure 19-5. Composite bar graph shows stopping distances at various speeds.

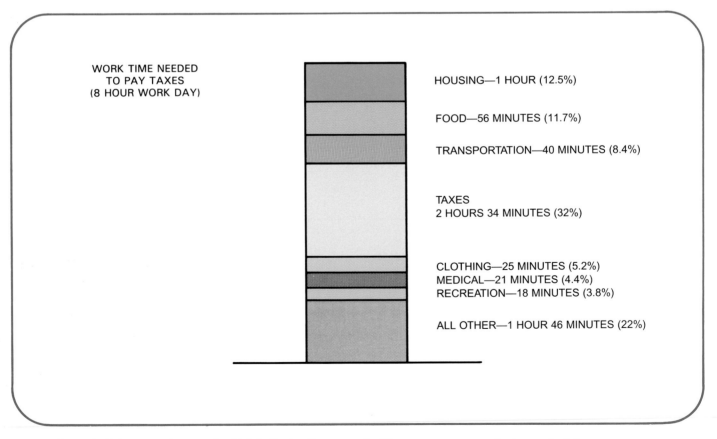

Figure 19-6. A 100 percent bar graph which indicates the amount of time one person works to pay their taxes.

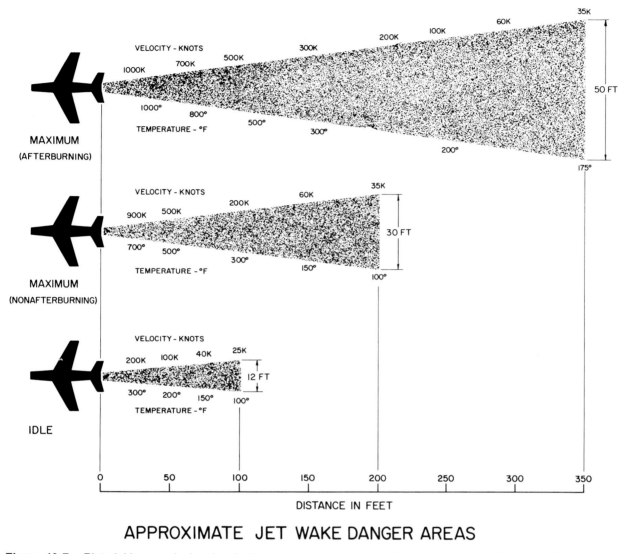

APPROXIMATE JET WAKE DANGER AREAS

Figure 19-7.   Pictorial bar graph showing the jet wake danger areas and the temperature and speed of the exhaust gases.

Circle or Pie Graph

The *circle graph* or *pie graph* is shown in Figure 19-8. It is composed of a segmented circle and shows the entire unit divided into comparable parts. It is also known as an *area graph*.

Charts

Charts are another means employed to convey information rapidly. The most familiar of these is the *organization chart*, Figure 19-9. It shows the order of responsibility and the relation of persons and/or positions in an organization.

The *pictorial chart* is an ideal way of presenting comparisons in an interesting and easily understood manner, Figure 19-10.

Flow charts may be used to show the sequence, or order of operations on how a product is manufactured and/or distributed, Figure 19-11. A flow chart is often employed to show a cycle of events in the order they occur.

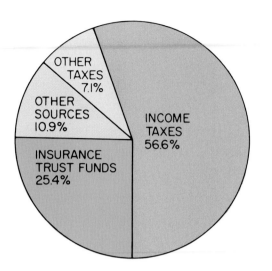

Figure 19-8. The circle or pie graph is sometimes called an area graph. The graph may be as simple as the above which shows where the federal government gets the money to operate, or it can be made more complex.

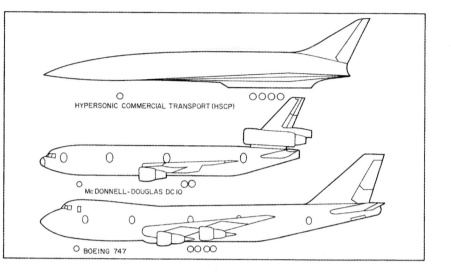

Figure 19-10. Pictorial chart showing how the size of a hypersonic commercial transport compares with present-day commercial aircraft. (Pratt & Whitney Aircraft)

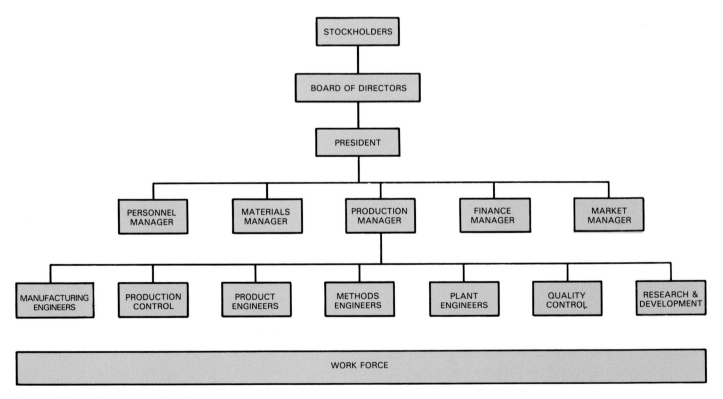

Figure 19-9. An organization chart.

XYZ CORPORATION
FLOW CHART

PRODUCT: #610 Birdhouse   **PART:** Side

TASK	DESCRIPTION	DISTANCE	RATE
○⇨□▼◖	Move to cut-off station.	10 feet	
○⇨□▼◖	Cut part to length.		50 min/100
○⇨□▼◖	Move to ripping station.	12 feet	
○⇨□▼◖	Rip to width.		55 min/100
○⇨□▼◖	Move to sanding.		
○⇨□▼◖	Sand sides and edges.		95 min/100
○⇨□▼◖	Move to inspection.		
○⇨□▼◖	Inspect size and sanding.		20 min/100
○⇨□▼◖	Move to storage.	20 feet	
○⇨□▼◖	Store for assembly.		

SYMBOLS

○ **OPERATION**   Object is changed in its chemical or physical makeup. It is assembled or disassembled.

⇨ **TRANS-PORTATION**   Object is moved from one place to another.

◖ **DELAY**   Object is held awaiting the next operation.

□ **INSPECTION**   Quality of the object is checked.

▼ **STORAGE**   Object is placed in a protected location.

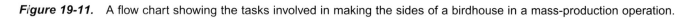

Figure 19-11.   A flow chart showing the tasks involved in making the sides of a birdhouse in a mass-production operation.

Preparing Charts and Graphs

There are many variations of the graphs presented in this unit. It is up to the ingenuity and creativity of the person preparing the material to decide which type will be best for the job.

Do *not* start a graph or chart until *all* of the information has been gathered. A rough layout should be prepared first to best determine what size and proportion presentation will be best.

Use color whenever possible to brighten your presentation and to make the information stand out. Hundreds of styles of transfer lettering, symbols, and figures are available, Figure 19-12. They are useful in making the chart or graph more interesting.

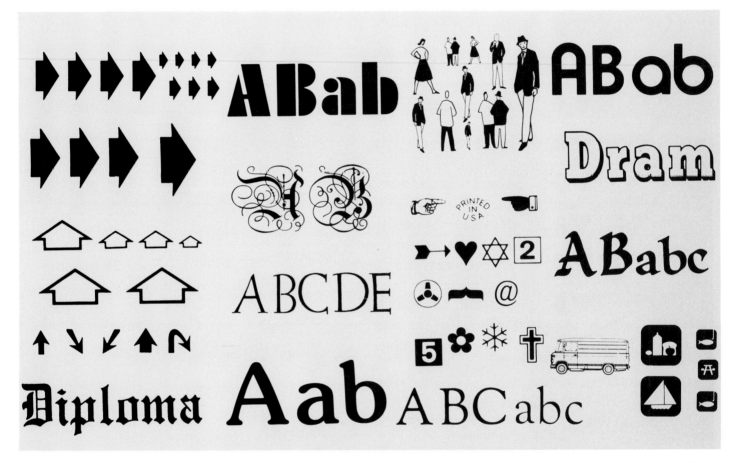

Figure 19-12. A few of the commercially available transfer lettering and symbols that can be employed to make professional looking charts and graphs.

Computer-Generated Graphs and Charts

Computer-generated graphics has greatly decreased the amount of time spent preparing graphs and charts, Figure 19-13. With the proper program, any type graph or chart can be developed in two- and three-dimensional forms. Multicolor presentations are also possible. See Figure 19-14.

Figure 19-13. With the computer, it is possible to graph engine emissions while testing is underway. (Ford Motor Company)

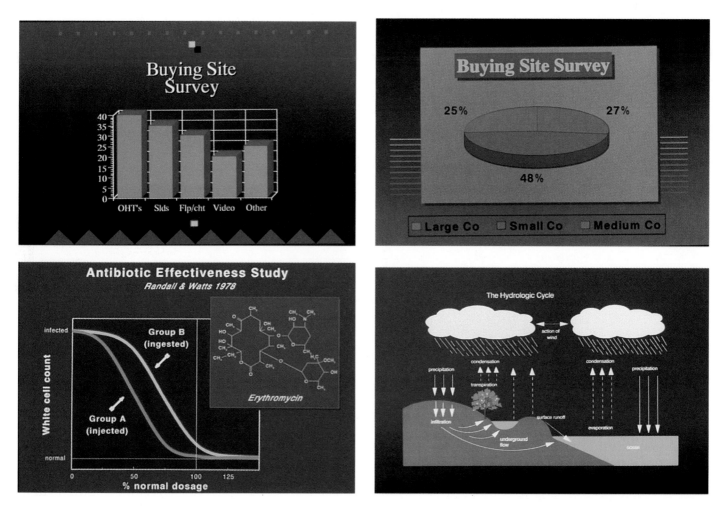

Figure 19-14. Computer-generated graphics are being more widely used by industry in sales and in production planning and evaluation. With the computer, information can always be up-to-date. How many of the charts and graphs can you identify? (Polaroid)

Drafting Vocabulary

Area graph	Graph	Pie graph
Bar graph	Horizontal bar graph	Statistics
Chart	Key	Transfer lettering
Circle graph	Line graph	Trends
Composite bar graph	Organization chart	Variations
Curve	Pictorial bar graph	Vertical bar graph
Flow chart	Pictorial chart	100 percent bar graph

Test Your Knowledge—Unit 19

Please do not write in the book. Place your answers on another sheet of paper.

1. Why are charts and graphs used?
2. List the three most commonly used kinds of graphs.
3. The line that presents the information on a(n) _____ graph is called a(n) _____.
4. Of what use is the key when more than one line is used on a graph?
5. Briefly describe each of the following graphs:
    A. Horizontal Bar Graph.
    B. Vertical Bar Graph.
    C. Composite Bar Graph.
    D. 100 Percent Bar Graph.
    E. Pictorial Bar Graph.
6. The _____ or _____ graph is also known as an area graph.
7. Briefly describe each of the following charts:
    A. Flow Chart.
    B. Organization Chart.
    C. Pictorial Chart.
8. Why should color be used on a chart or graph?

Outside Activities

1. Develop a pie graph of how you spend your allowance.
2. Prepare a bar graph to show the approximate increase in horsepower in a particular make of automobile from 1940 to the present time. Use five year steps.
3. Prepare a line graph showing how the price of the automobile used in Activity 2 has risen in the same periods.
4. Make a 100 percent bar graph showing a breakdown of the cost of a gallon of gasoline including cost of the gasoline, state, federal, and local taxes.
5. Draw a pictorial bar graph showing how far an automobile will travel after the brakes are applied at 25 mph, 45 mph, 55 mph, and 65 mph.

6. Make an organization chart of the pupil personnel system used in your drafting room or Industrial Technology Labs.

7. Develop a picture chart showing how the size of a particular make of auto has increased or decreased in the past 20 years.

8. Make a picture graph showing the enrollment in each grade of your school. Let each symbol represent 25 students.

9. Design a flow chart showing how a simple product could be manufactured in the Industrial Technology lab.

Welding was used to fabricate the fixtures that hold the Space Shuttle during preparation for its next flight. Welders had to correctly read and interpret the welding drawings since every weld was carefully checked. (NASA)

After studying this unit and completing the assigned problems, you should be able to:

◆ Describe the welding process.

◆ Explain why weld symbols are used on drawings.

◆ State what the various weld symbols indicate.

◆ Use weld symbols on drawings to provide exact weld specifications.

Welding is a widely used industrial technique for *fabricating* (joining) metal pieces. It is a method of joining by heating metals to a high temperature which causes them to melt and fuse together. The high temperatures are generated by electricity (arc welding, for example) or by burning gases (oxygen and acetylene). See Figure 20-1.

Some method had to be devised to tell the welder what type of weld the engineer wanted on the job. The American Welding Society (AWS) developed and standardized the basic weld symbols shown in Figure 20-2. These symbols provide a means of giving complete and specific welding information on the drawing of the part to be welded.

The welding symbol placed on the drawing is made up of basic and supplementary weld symbols that have been selected to describe the required weld, Figure 20-3.

Using Welding Symbols

Before welding symbols can be employed effectively, the drafter must be familiar with the various types of joints used in welding, Figure 20-4.

The location of the arrow with respect to the joint is important when using the welding symbol to specify the required weld. The side of the joint indicated by the arrow is considered the *arrow side*. The side opposite the arrow side of the joint is considered the *other side* of the joint.

When the weld is to be made on the *arrow side* of the joint, the weld symbol is placed on the reference line so it is *toward the reader*, Figure 20-5.

To specify a weld on the *other side* of the joint, the weld symbol is placed on the reference line *away from the reader*, Figure 20-6.

Welds on both sides of the joint are specified by placing the symbol on *both sides* of the reference line, Figure 20-7.

A weld that is made all of the way around the joint is specified by drawing a circle at the point where the arrow is bent, Figure 20-8. *Field welds* (welds not made in the shop) are specified as shown in Figure 20-9.

Weld size is placed to the left of the symbol. The figure indicating the length of the weld is placed to the right of the symbol, Figure 20-10.

Figure 20-1. There are many welding processes. All welding techniques join metals by heating them to a suitable temperature where they melt and fuse together. (U.S. Army and American Welding Society)

FILLET	PLUG OR SLOT	SPOT PROJEC- TION	SEAM	GROOVE							BACK OR BACKING	SURFACING	FLANGE	
				SQUARE	"V"	BEVEL	"U"	"J"	FLARE "V"	FLARE BEVEL			EDGE	CORNER

BASIC ARC AND GAS WELD SYMBOLS

WELD ALL AROUND	FIELD WELD	MELT-THRU	CONTOUR		
			FLUSH	CONVEX	CONCAVE

SUPPLEMENTARY WELD SYMBOLS

Figure 20-2. Basic weld symbols.

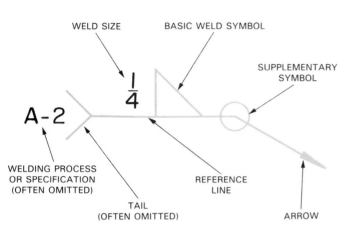

Figure 20-3. A welding symbol is made up of basic and supplementary weld symbols. It gives complete and specific welding information to the welder.

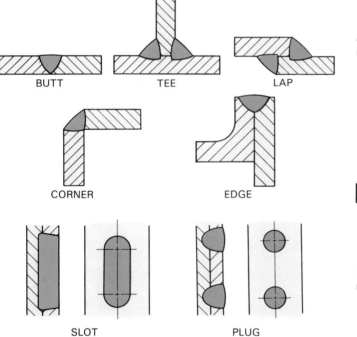

Figure 20-4. Basic joints used in welding.

Drawings for Assemblies to be Fabricated by Welding

Items that are to be assembled by welding are usually composed of several pieces, Figure 20-11. Since the individual pieces are seldom cut to size, welded, and machined in the same general shop

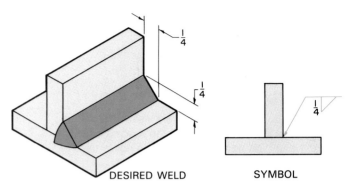

Figure 20-5. Welding symbol that indicates the weld is to be made on the *arrow side* of the joint. Note the weld symbol is *toward* the reader.

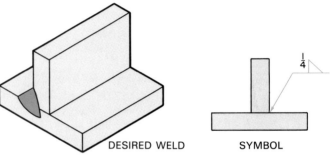

Figure 20-6. Welding symbol that indicates the weld is to be made on the *other side* of the joint.

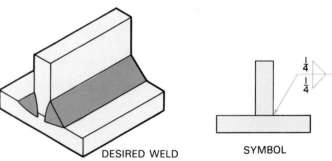

Figure 20-7. Welding symbol that indicates the weld is to be made on *both sides* of the joint.

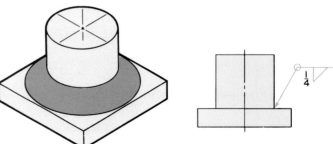

Figure 20-8. Welding symbol that indicates the weld is to be made *all around* the joint.

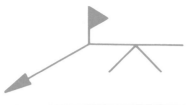

Figure 20-9. Field welding symbol. This indicates the welding is to be done on the job and not in the fabrication shop.

area, several drawings are often required. Each drawing will provide information on a specific operation. For simple jobs, all of the needed information can be included on a single drawing, Figure 20-12.

DESIRED WELD

WELD LENGTH

WELD SIZE

SYMBOL

Figure 20-10. How weld size and length are indicated.

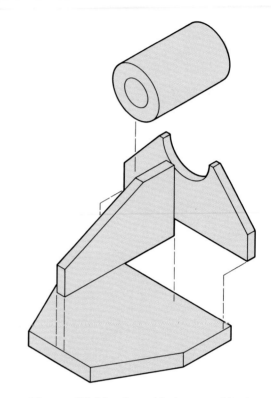

Figure 20-11. A welded assembly is composed of several different pieces.

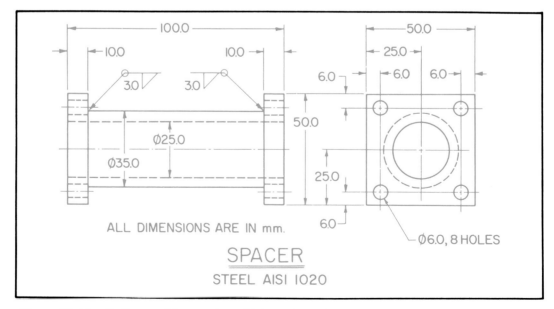

ALL DIMENSIONS ARE IN mm.

Ø25.0

Ø35.0

Ø6.0, 8 HOLES

SPACER

STEEL AISI 1020

Figure 20-12. Cutting, welding, and machining information is included on this drawing. Note the dimensions are in millimeters.

Drafting Vocabulary

American Welding Society
 (AWS)
Arrow side
Fabricating

Field weld
Fuse
Other side
Weld symbol

Welding
Welding symbol

Test Your Knowledge—Unit 20

Please do not write in this book. Place your answers on another sheet of paper.

1. How are the high temperatures needed for welding generated?
2. Welding symbols were developed to _____.
   A. lessen the possibilities of the wrong type of weld being made
   B. tell the welder exactly what to do
   C. eliminate "hit or miss" welds
   D. All of the above.
3. Sketch the symbol for a fillet weld on the arrow side of the joint.
4. Sketch the symbol for a fillet weld on the other side of the joint.
5. Sketch the symbol for a fillet weld on both sides of the joint.
6. Sketch the symbol for a fillet weld all of the way around the joint.
7. What is a field weld?
8. Why are several different drawings needed for a job that is made up of several pieces and assembled by welding?

Outside Activities

1. Secure samples of welded joints and mount them on a display board. Label the samples and add the proper drafting symbol to the display.
2. Visit a local industry that makes extensive use of welding and get samples of the work they produce.
3. Invite a professional welder to the school shop to demonstrate the safe and proper way to gas weld and to electric arc weld.
4. Report for the class on the welding techniques known as Gas Tungsten Arc Welding and Gas Metal Arc Welding.

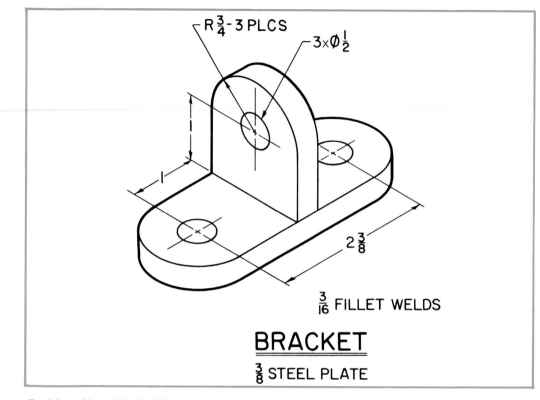

BRACKET

3/8 STEEL PLATE

3/16 FILLET WELDS

Problem 20-1. BRACKET. Prepare a drawing showing the necessary views to fabricate it. Use the appropriate welding symbol.

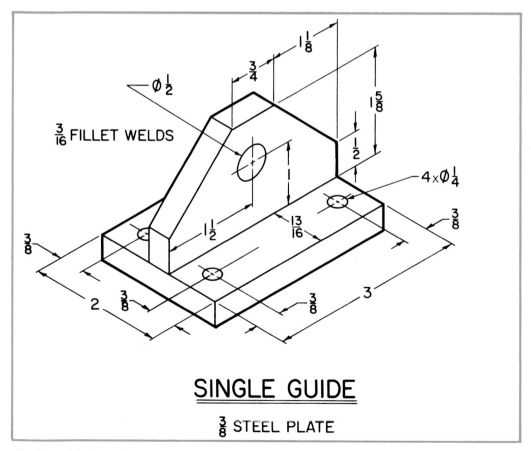

SINGLE GUIDE

3/8 STEEL PLATE

3/16 FILLET WELDS

Problem 20-2. SINGLE GUIDE. Prepare a drawing with the information necessary to fabricate it by welding. Welds are to be made on both sides of the joint.

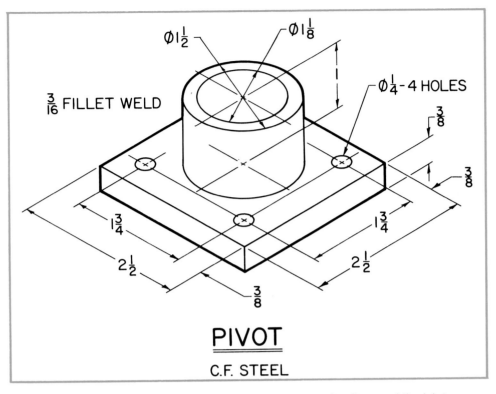

PIVOT
C.F. STEEL

Problem 20-3. PIVOT. The fillet weld is to be made all around the joint.

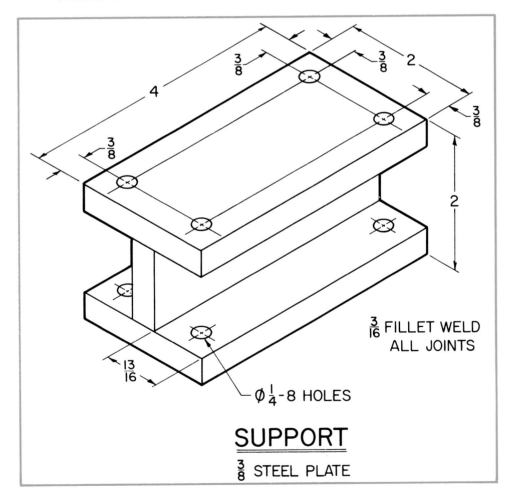

SUPPORT
$\frac{3}{8}$ STEEL PLATE

Problem 20-4. SUPPORT. Prepare a drawing showing the views necessary to fabricate the object. Welds are to be made on both sides of each joint.

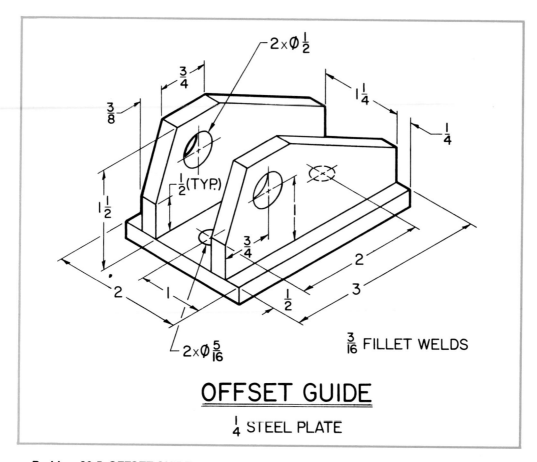

OFFSET GUIDE

$\frac{1}{4}$ STEEL PLATE

Problem 20-5. OFFSET GUIDE. The weld is to be made on both sides of each vertical piece.

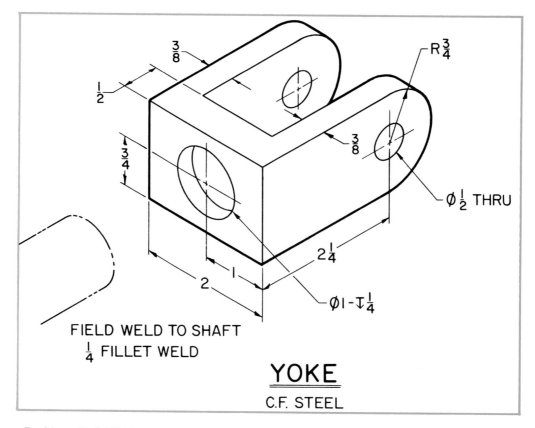

FIELD WELD TO SHAFT
$\frac{1}{4}$ FILLET WELD

YOKE

C.F. STEEL

Problem 20-6. YOKE. This part is to be welded to a 1 inch diameter shaft at the job site.

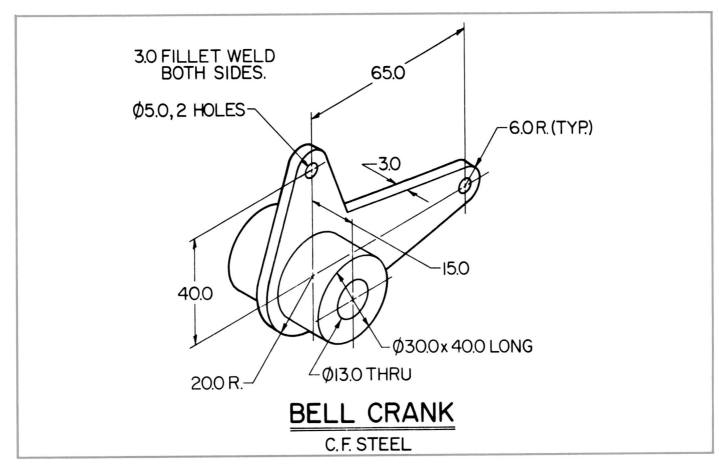

3.0 FILLET WELD
BOTH SIDES.

Ø5.0, 2 HOLES

65.0

6.0 R. (TYP.)

3.0

15.0

40.0

Ø30.0 x 40.0 LONG

Ø13.0 THRU

20.0 R.

BELL CRANK
C.F. STEEL

Problem 20-7. BELL CRANK.   A 3.0 mm fillet weld is specified on both sides of the joint.

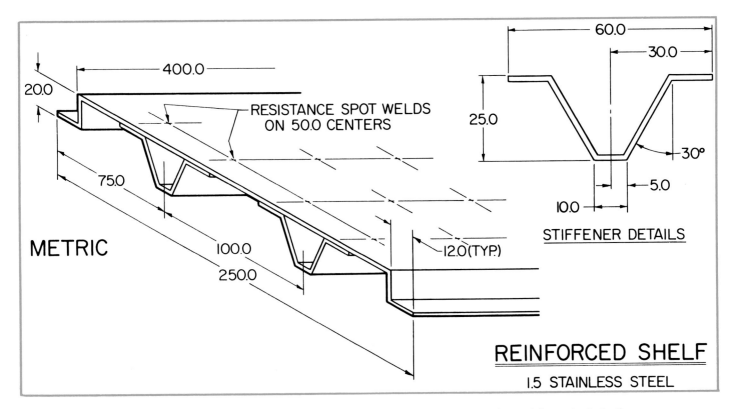

20.0

400.0

RESISTANCE SPOT WELDS
ON 50.0 CENTERS

75.0

METRIC

100.0

250.0

12.0 (TYP.)

60.0

30.0

25.0

30°

5.0

10.0

STIFFENER DETAILS

REINFORCED SHELF
1.5 STAINLESS STEEL

Problem 20-8. REINFORCED SHELF.   Stiffeners are spot welded to the stainless steel shelf.

Hundreds of different fasteners are used in the manufacture of this vehicle. (Ford Motor Company)

Unit 21
Fasteners

After studying this unit and completing the assigned problems, you should be able to:

- Describe the function of a typical fastener.
- Identify a variety of fasteners.
- Indicate several applications for screw threads.
- Specify inch-based and metric-based threaded fasteners on drawings.
- Explain why inch-based and metric-based threaded fasteners are not interchangeable.
- Demonstrate three approved methods for drawing threads for nuts and bolts.

Manufactured products are assembled by many different kinds of fasteners, Figure 21-1. A *fastener* is a device used to hold two or more parts together. They include screws, nuts and bolts, rivets, nails, etc. Since fasteners are so important to industry, drafters, engineers, and designers must be familiar with them, know what types to use for a particular application, and how to draw them.

Figure 21-1. A few of the thousands of different types and sizes of fasteners used by industry. How many can you identify?

Threaded Fasteners

Threads have many applications. They are employed to:

1. Make adjustments.
2. Transmit motion.
3. Assemble parts.
4. Apply pressure.
5. Make measurements.

How many examples of each can you name?

Threaded fasteners employ the wedging action of the screw thread to hold items together. They are found extensively in manufactured products. See Figure 21-2. Threaded fasteners vary in cost from several thousand dollars each for special bolts that attach the wings to the fuselage of large aircraft, to the small machine screw that costs a fraction of a penny. (One auto manufacturer uses more than 11,000 fasteners of various types and sizes.)

Until recently, industry usually worked only with the inch-based *unified thread series*. These fasteners are made in both coarse (identified by UNC) and fine (identified by UNF) thread series, Figure 21-3. Now, industry must be familiar with metric-based threaded fasteners.

Figure 21-3. It is easy to see the difference between a UNF (fine) and UNC (coarse) thread series. The bolts have the same diameter and are the same length.

Inch-based UNC and UNF threads and metric-based threads have the same basic profile (shape), Figure 21-4. While some inch-based and metric-

Figure 21-2. Many types of fasteners and adhesives are used in manufacturing boats. When selecting fasteners, what important factor had to be considered?

based threaded fasteners may appear to be the same size, *they are not interchangeable*. A metric-based bolt **cannot** be used with an inch-based nut that appears to have the same thread size as the bolt.

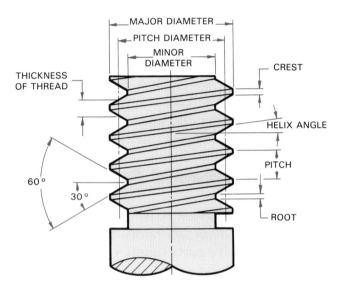

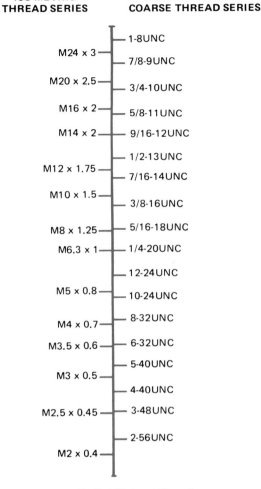

Figure 21-4. Metric- and inch-based threads have exactly the same shape. The parts of a screw thread are shown.

ISO and Unified National Thread Series ARE NOT INTERCHANGEABLE

Figure 21-5. A comparison of ISO metric coarse and Unified Coarse (UNC) inch-based thread sizes. Even though several of them seem to be the same size, they are not interchangeable (one cannot be substituted for the other).

More than two million different kinds, shapes and sizes of inch-based threaded fasteners are made. Add to this number, the metric-based threaded fasteners. Both inch-based and metric-based fasteners will have to be kept in inventory for many, many years. Both inch-based and metric-based wrenches will also be needed.

A comparison of metric coarse pitch and UNC inch-based thread sizes is shown in Figure 21-5.

A user may have trouble telling a metric bolt from a similar inch-based fastener. Some simple way will have to be devised to identify the metric fastener. At present in the United States, the imprinted hex head and a unique 12-element spline head are being considered. See Figure 21-6.

Drawing Threads

It is very time consuming to show threads on a drawing as they would actually appear. For this reason, either the *schematic* or *simplified* represen-

tation is used. See the approved methods of thread representation in Figure 21-7.

The *detailed representation*, Figure 21-7 Top, looks like the actual screw thread. It is sometimes employed where confusion might result if the simplified representation were used.

The *schematic representation* of a screw thread, Figure 21-7 Center, is easier to draw. It should not be used for hidden threads or sections of external threads.

The *simplified representation* of a screw thread, Figure 21-7 Bottom, is a fast and easy method used to draw threads. For this reason, it is

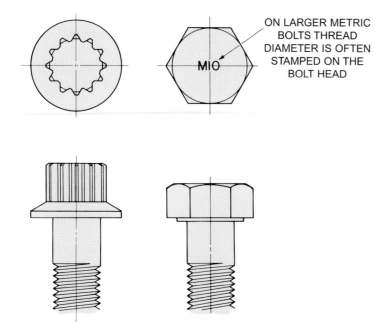

ON LARGER METRIC BOLTS THREAD DIAMETER IS OFTEN STAMPED ON THE BOLT HEAD

M10

Figure 21-6. Much study is being done to devise an easy way to identify metric-based fasteners from inch-based fasteners. The 12-element spline head and imprinted hex head (thread diameter is stamped on the head) are two methods being considered.

widely used in drafting. It should be avoided where there is a possibility of this representation being confused with other details on a drawing.

Avoid mixing the various methods on the same drawing.

Regardless of which thread representation the drafter decides to draw, the thread size must be shown on the drawing. See Figure 21-8 for the accepted way to present thread information.

Threads are understood to be right-hand threads. Left-hand threads are represented by the letters LH following the class of fit (inch-based thread) or thread tolerance (metric-based thread).

Types of Threaded Fasteners

The fasteners described in this unit can usually be found in the school laboratory, typical hardware store, or automotive parts supply store. The major-

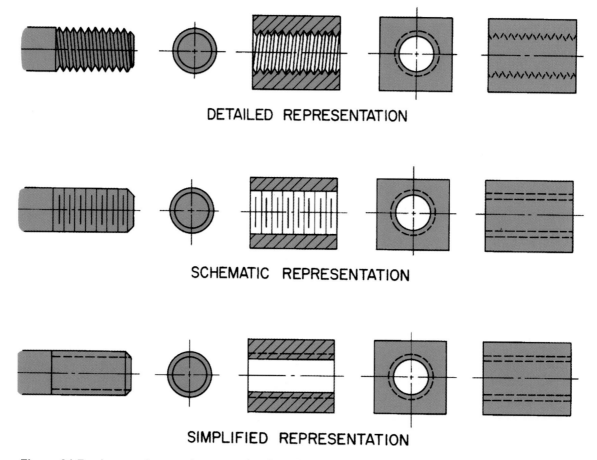

DETAILED REPRESENTATION

SCHEMATIC REPRESENTATION

SIMPLIFIED REPRESENTATION

Figure 21-7. Approved ways of representing threads on drawings.

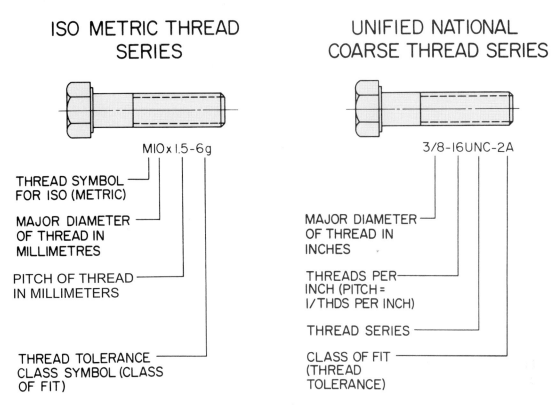

Figure 21-8. How thread size is noted and what each term means.

ity of them are made of steel, brass, or aluminum. Special applications may require them to be made of other materials.

Machine Screws

Machine screws are available with single slotted, cross slotted (Phillips), or hex-style heads, and round, flat, fillister, pan and oval head shapes, Figure 21-9. Nuts (square or hexagonal) are not furnished with machine screws and must be purchased separately. They have many applications in general assembly work where screws less than 1/4 in. (6.0 mm) in diameter are needed. Machine screws are available in many sizes starting as small as #0-80UNC (0.06 in. diameter) inch-based and 1.4 × 0.3 metric-based.

Machine Bolts

Machine bolts are manufactured with hexagonal and square heads, Figure 21-10. They are used to assemble machinery and other items that do not require close-tolerance fasteners which are more expensive. Machine bolts are secured by tightening the matching nut.

Cap Screws

Cap screws are used when the assembly requires a stronger, more precise, and better appearing fastener, Figure 21-11. They are primarily employed to bolt two pieces or sections together. The screw passes through a *clearance hole* in one part and screws into a *threaded hole* in the other part, Figure 21-12. Machine tools are usually assembled with cap screws.

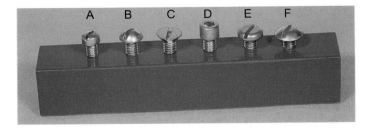

Figure 21-9. A few of the many styles of machine screw head types available. A—Fillister. B—Round. C—Flat. D—Socket. E—Pan. F—Truss.

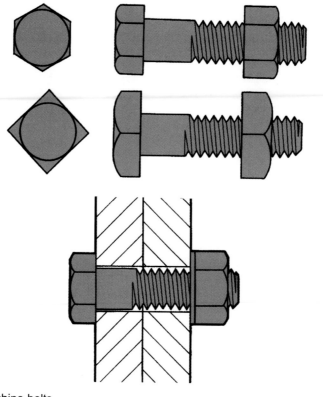

Figure 21-10. Machine bolts.

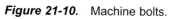

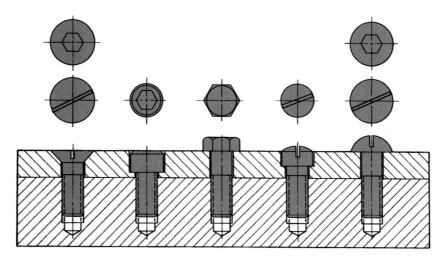

Figure 21-11. Types of cap screws.

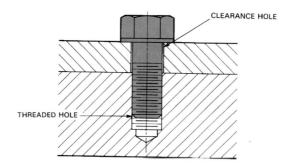

CLEARANCE HOLE

THREADED HOLE

Figure 21-12. How a cap screw works. What type of thread representation is used on this drawing?

Set Screws

Set screws prevent slippage of pulleys and gears on shafts, Figure 21-13. They are available in many different head and point styles. Set screws are made of heat-treated steel to make them stronger and less likely to fail.

Figure 21-13.   A typical set screw application.

Stud Bolts

A *stud bolt* is threaded on both ends. One end is threaded into a tapped (threaded) hole. The piece to be clamped is fitted over the stud bolt and the nut screwed on to clamp the two sections tightly together. Refer to Figure 21-14.

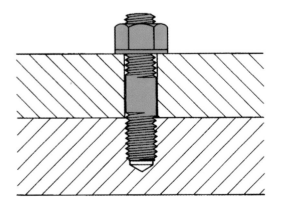

Figure 21-14.   Stud bolt. What type of thread representation is used on this drawing?

Nuts

Nuts are screwed down on bolts or screws to tighten or hold together the sections through which the bolt passes. A few of the many types of nuts available are shown in Figure 21-15.

Washers

Washers are used with nuts and bolts to distribute the clamping pressure over a larger area. They also prevent the fastener from marring the work surface when the nut is tightened. Many types are manufactured, Figure 21-16.

Nonthreaded Fasteners

Nonthreaded fasteners comprise a large group of holding devices.

Rivets

Permanent assemblies are made with *rivets*, Figure 21-17. Two or more sections of material are held together by these headed pins. Holes are drilled through the material to be riveted. The rivet shank passes through the hole. After aligning the sections, the plain end of the rivet is upset or headed by hammering to form a second head. The parts are drawn together by the heading operation.

Blind rivets are mechanical fasteners that have been developed for applications where the joint is accessible from one side only. They require special tools to put them in place, Figure 21-18. Blind rivet types are shown in Figure 21-19.

Cotter Pins

The *cotter pin* is fitted into a hole drilled crosswise in a shaft, Figure 21-20. The ends are bent over after assembly to prevent parts from slipping or turning off the shaft.

Keys, Keyways, and Keyseats

A *key* is a small piece of metal partially fitted into the shaft and partially into the hub to prevent rotation of a gear, wheel, or pulley on the shaft, Figure 21-21. The *keyway* is cut in the hub of the mating part. Different types of keys have been devised for special applications.

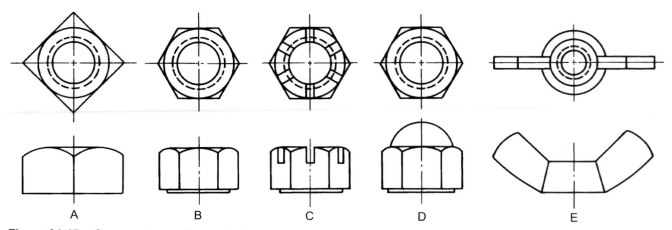

Figure 21-15. Common types of nuts. A—Square nut. Used with machine bolt with the same head shape. B—Semifinished nut has a machined bearing surface to provide a truer surface for a washer. C—Slotted nut. After the nut has been tightened, a cotter pin is fitted in one set of slots and through the hole in the shaft. The cotter pin prevents the bolt from turning loose. D—Acorn nut. Applied when the appearance is important or exposed sharp edges on the bolt must be avoided. E—Wing nut. Provides for rapid tightening without the use of a wrench.

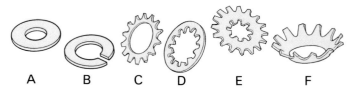

Figure 21-16. Washer types. A—Plain washer. B—Split lock washer. C—External lock washer. D—Internal lock washer. E—Internal-external washer; used when the mounting holes are oversize. F—Countersunk washer.

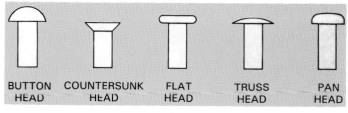

Figure 21-17. Rivet head styles.

Figure 21-18. Tool used to insert one type of blind rivet.

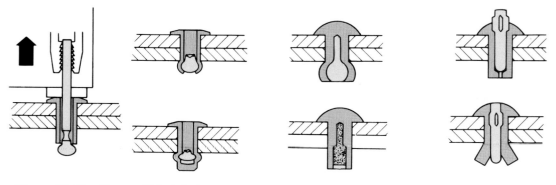

Figure 21-19. Types of blind rivets.

Figure 21-20. Cotter pin.

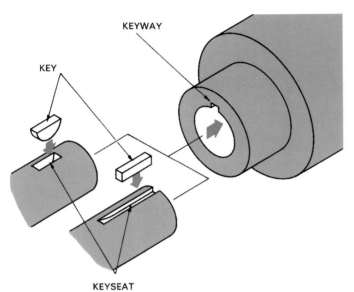

Figure 21-21. Two types of keys are shown. The half-round key is called a **Woodruff key.** The other is a **rectangular key.**

Fasteners for Wood

The most common fasteners used in wood are nails and screws.

Nails

Nails are an easy way to fasten wood pieces together, Figure 21-22. They are usually made of mild steel. For exterior work, nails made from mild steel are given a galvanized (zinc) coating so they will not rust. Nail size is given as "penny" and is abbreviated with the lowercase "d."

Wood Screws

Wood screws are manufactured from many kinds of metal. Screw size is indicated by the shank diameter and length. Head styles and how each is measured is shown in Figure 21-23. Wood screws are available in lengths from 1/4 in. to 6 in. (6.5 mm to 150.0 mm).

Figure 21-22. The building trades industry uses vast quantities of nails. More than a thousand different kinds are available.

How to Draw Bolts and Nuts

The dimensions given in Figure 21-24 are approximations but are acceptable for most drafting applications. Information needed to complete the drawing includes:

1. Bolt diameter.
2. Bolt length.
3. Type of head and/or nut.

To draw square and hexagonal headed bolts and nuts follow the procedure in Figure 21-25.

1. Draw centerlines and lines representing the diameter (D).

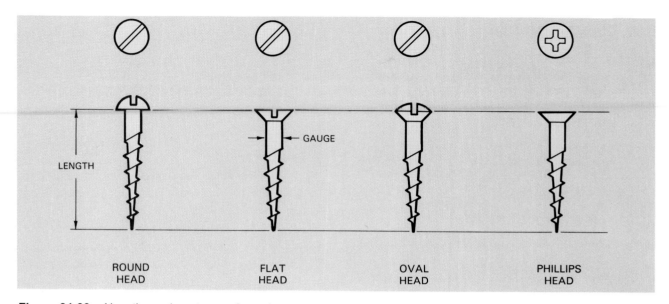

Figure 21-23. How the various types of wood screws are measured. Screw sizes and head styles do not change in metrics, but their sizes will be expressed in millimeters.

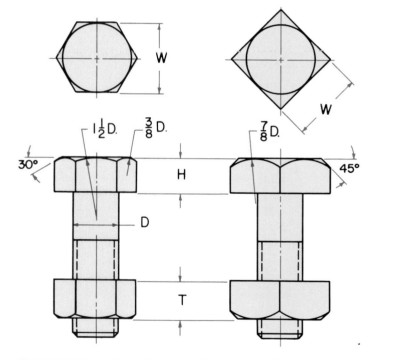

Figure 21-24. Information needed to draw a bolt and nut.

2. On centerline, draw a circle (diameter = 1 1/2D).

3. Using 30-60 degree triangles, circumscribe a hexagon (or with a 45 degree angle, a square) about the circle.

4. Develop the side view of the bolt.

5. Draw arcs in the bolt head and nut using radii given in Figure 21-24. (Templates for drawing

bolt heads and nuts are available and make the job faster and easier.)

6. Complete by drawing the chamfers on the bolt head and nut (not necessary if a template is used). Draw threads—either simplified or schematic—on the bolt. The dimensions for drawing schematic and simplified threads are shown in Figure 21-26.

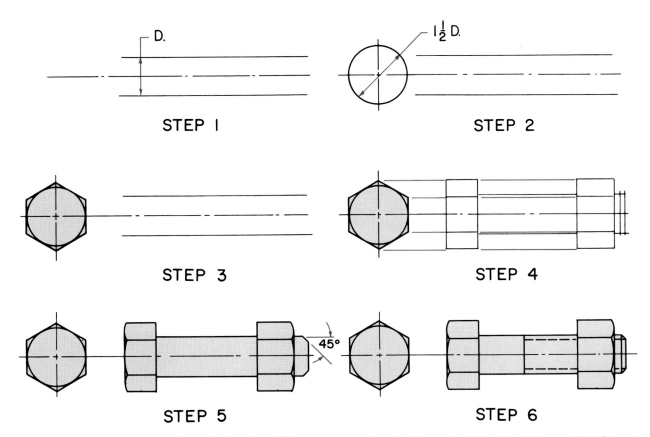

STEP I

STEP 2

STEP 3

STEP 4

45°

STEP 5

STEP 6

Figure 21-25. Steps in drawing a bolt and nut. The same procedure is followed whether the bolt or nut is to have a square or hexagonal head.

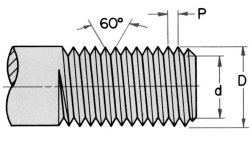

60° | P

D
d

DETAILED

P

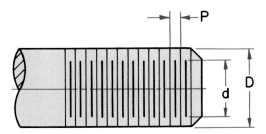

D
d

SCHEMATIC

D
d

SIMPLIFIED

$P = \text{PITCH} = \dfrac{1}{N}$

$N = \text{NUMBER OF THREADS PER INCH}$

$D = \text{MAJOR DIAMETER OF THREADS}$

$d = \text{MINOR DIAMETER OF THREADS}$

$d = D - \dfrac{1.300}{N}$

Figure 21-26. Dimensions used when drawing detailed, schematic, and simplified thread representations.

Drafting Vocabulary

Blind rivets	Keyway	Simplified representation
Cap screws	Machine bolts	Spline
Clearance hole	Machine screws	Stud bolt
Cotter pin	Nails	Tapped hole
Detailed representation	Nuts	Threaded hole
Fastener	Pan head	Tolerance
Fillister head	Pitch	Transmit motion
Galvanized	Profile	Unified Thread Series
Heat treated	Rectangular key	Washers
Hex head	Rivets	Wedging
Interchangeable	Schematic representation	Woodruff key
Key	Series	Wood screws
Keyseat	Set screws	

Test Your Knowledge—Unit 21

Please do not write in the book. Place your answers on another sheet of paper.

1. Identify 10 different fasteners. Underline the threaded fasteners.
2. What are fasteners?
3. List five applications that make use of screw threads.
4. List three uses of threaded fasteners.
5. What is the difference between a coarse thread and a fine thread on a given size and length bolt?
6. Make sketches of a detailed representation of a screw thread, a schematic representation and a simplified representation.
7. Explain the meaning of the following screw thread size: 1/2-20UNF-3LH.
8. What fasteners are most commonly used to join wood?

Match the correct lettered definition with the list of fastener terms.

9. Machine screw
10. Cap screw
11. Machine bolt
12. Set screw
13. Nut
14. Washer
15. Cotter pin
16. Rivets
17. Key
18. Keyseat

A. Threaded on both ends.
B. Fitted on a bolt.
C. Distributes clamping pressure of nut or bolt.
D. Prevents slippage of a pulley on a shaft.
E. Used for any general assembly work where screws less than 1/4 in. diameter are needed.
F. Used when close tolerance fasteners are not required.
G. Used when assembly requires a stronger, more precise and better appearing fastener.
H. Small piece of metal partially fitted into a shaft and pulley, gear, or wheel to prevent turning on a shaft.
I. Makes a permanent assembly.
J. Fits in a hole drilled near the end of a shaft.
K. Machined on a shaft for a key.

Outside Activities

1. Secure samples of machine screws, cap screws, machine bolts, set screws, stud bolts, nuts, washers, rivets, cotter pins, nails, and wood screws. Make a display for use in the drafting room. Label all samples.

2. Get a sample showing a key, keyseat and keyway in use. Describe how the components are assembled.

3. Secure an example of a set screw application. How does this example differ from the use of a key?

Problems

1. Draw 3/4-10UNC-2 × 4 in. long hexagonal and square head bolts and nuts on the same sheet. Allow 3 inches between the drawings. Use a simplified thread representation.

2. Draw 1-8UNC-2 × 3 in. long hexagonal and square head bolts and nuts on the same sheet. Allow 3 inches between the drawings. Use a schematic thread representation.

To produce this logic controller printed circuit board, detailed and accurate electronics drawings are required. Note the light blue lines on the circuit board, which connect the various components. (Wizdom Systems, Inc.)

Electrical and Electronics Drafting

After studying this unit and completing the assigned problems, you should be able to:

◆ Develop a basic understanding of electrical and electronics drafting.

◆ Explain why electrical and electronics drafting is diagrammatic and uses standard symbols to identify the various components.

◆ Describe several types of electrical and electronics diagrams.

Our world, as we know it today, could not exist without electricity and the electronic devices. You will realize how true this is by just listing the electrical and electronic devices you may depend upon in your home—TV set, radio, stereo, VCR, personal computer, Figure 22-1, washer, refrigerator, range, and the lighting to name but a few.

Figure 22-1. Typical personal computer.

There are many other electrical devices that may not be as familiar as those found at home or at school. Examples include the computers that are important to business and medicine, as well as the programmable devices which control the machines that make, inspect, assemble, and test so many of the products we use, Figure 22-2. Many hobbies are electronically oriented, Figure 22-3.

Figure 22-2. Drafters preparing drawings for this robotic welder (whether manually or with CAD) must have an extensive knowledge of electrical/electronics drafting techniques and standards. Electronic sensors on the robot "read" a code on the fixture holding the auto body components. The robot's computer adapts the welding sequence for the body design moving into position. This means that hatchbacks, station wagons, four-door sedans, and convertibles can be welded on the same assembly line. (Ford Motor Company)

Figure 22-3. Many hobbies are electronically oriented. Shown are radio-controlled model aircraft. The controls are activated electronically by radio signals. In flight, the model can duplicate the flight pattern of the real aircraft. These model aircraft have the power source completely enclosed in the fuselage; they have no propellers. (Frank Fanelli, *Flying Models*)

Electrical and Electronics Computer-Aided Design

Computer-aided design, Figure 22-4, has found extensive use in the design and layout of printed circuits, Figure 22-5, and integrated circuits (ICs), Figure 22-6, for electronic devices. CAD does the work in a fraction of the time required to do it by hand. Many complex integrated circuit designs would be obsolete before they could be put into production if the design and layout work were done manually.

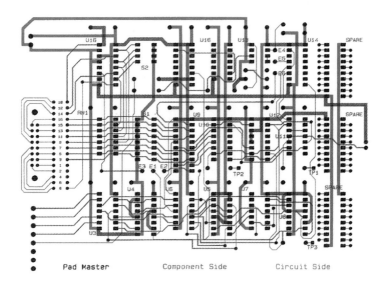

Pad Master   Component Side   Circuit Side

Figure 22-4. Computer-aided design (CAD) is used extensively in the design and layout of electronic circuits. CAD does the work in a fraction of the time needed to design manually and circuit modifications (changes) can be easily made. (Hewlett-Packard)

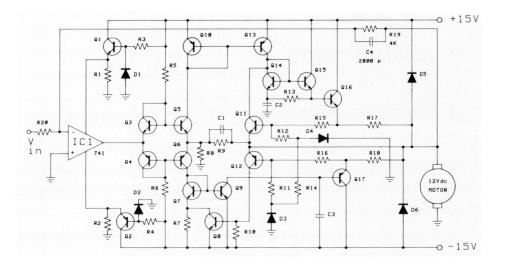

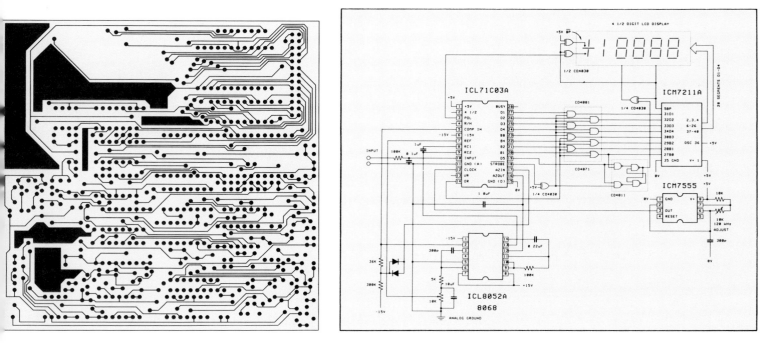

Figure 22-5. Samples of computer-designed circuits and circuit board. (ROBO Systems)

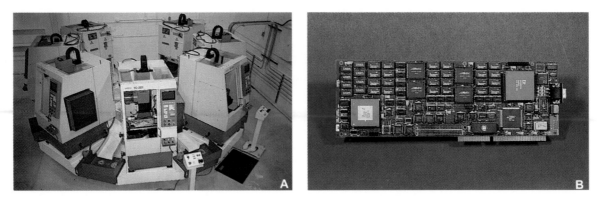

Figure 22-6. A—Computer-controlled machine tool. The machining units are arranged in a circle. Rather than have the operator go to each machine, the machines come to the operator. This increases productivity. (Kurt Manufacturing Company) B—Control unit similar to the one used on the CNC (Computer Numerical Control) machine tools shown in Figure 22-6A. It is so complex that it had to be designed by a computer.

As in other areas of computer-aided design, the designer must have a thorough knowledge of electrical and electronics drafting techniques and standards before they can become skilled in using CAD to its fullest potential.

Electrical and Electronics Drafting

Electrical and electronics drafting is done manually in much the same manner as conventional drafting. The same equipment is used.

Instead of using regular multiview drawings, a large portion of electrical and electronic drafting is diagrammatic in nature. That is, considerable use is made of symbols. The symbols represent the various components and wires that make up the electrical/electronic circuit, Figure 22-7. They are easier and quicker to draw than the actual part. Symbols are combined on a diagram that shows the function and relation of each component in the circuit.

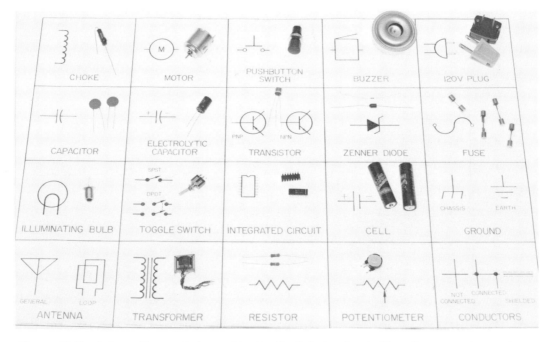

Figure 22-7. A few of the symbols used in electrical/electronics drafting. Components are shown with their symbols. These are but a few of the hundreds of symbols used.

Types of Diagrams Used in Electrical and Electronics Drafting

The drafter working in this area of drafting must be familiar with the different types of diagrams.

A *schematic diagram*, Figure 22-8, is a drawing with symbols and single lines to show electrical connections and functions of a specific circuit. The various components that make up the circuit are drawn without regard to their actual physical size, shape, or location.

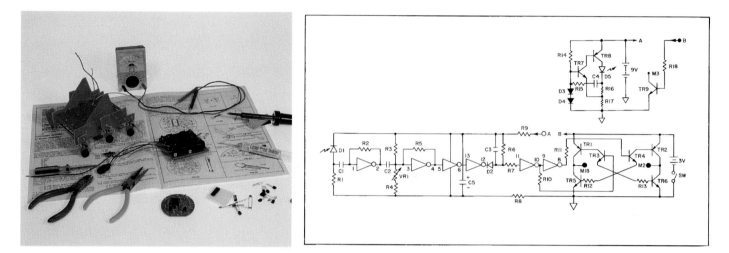

Figure 22-8.   All-terrain robot with schematic diagram. This miniature robot is similar to the spider-like robot being developed for lunar exploration. The six-legged robot shown here walks in a straight line until it "sees" an object in its path. The robot's state-of-the-art infrared beam detects the obstruction and alerts the motor control circuits, which alter the robot's course to avoid the obstacle. (Graymark International, Inc.)

Connection or wiring diagrams have several uses. First, they are commonly used to show the distribution of electricity on architectural drawings, Figure 22-9. The other type of diagram may be drawn to show the general physical arrangement of the ICs, transistors, diodes, resistors, switches, etc., that make up an electronic circuit, Figure 22-10.

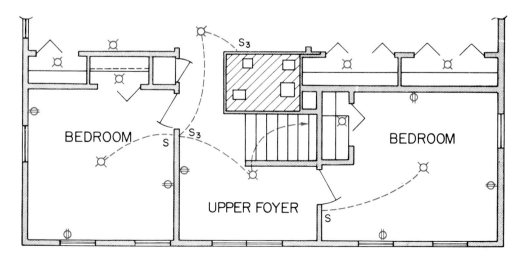

Figure 22-9.   A wiring diagram showing the location of switches, outlets, and lighting in a modern home.

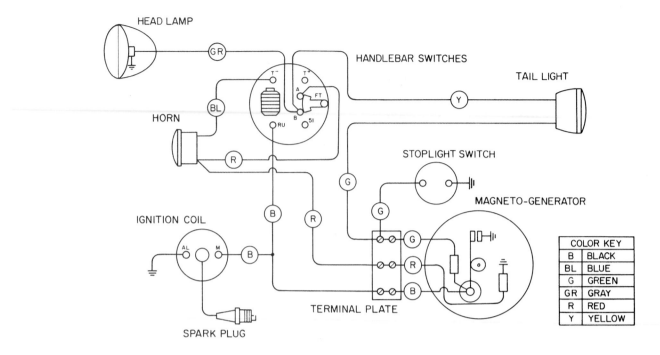

Figure 22-10. A wiring diagram for a motorcycle light. (Harley-Davidson)

Block diagrams, Figure 22-11, provide a simplified way to show the operation of an electronic device. This utilizes "blocks" (squares, rectangles, triangles, etc.) joined by a single line. It reads from left to right.

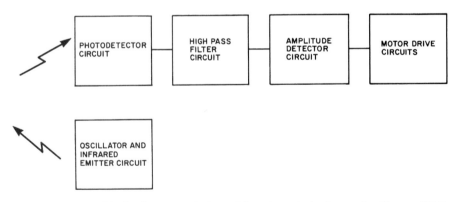

Figure 22-11. Block diagram of the all-terrain robot shown in Figure 22-8. (Graymark International, Inc.)

In ***pictorial diagrams***, the components are drawn in pictorial form and in their proper location, Figure 22-12. The pictorial diagram is used extensively by electronic kit manufacturers because it is so easy to understand.

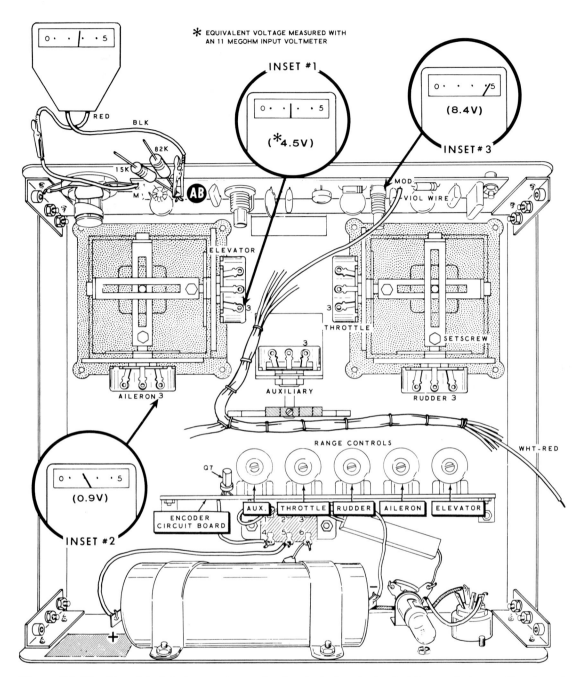

Figure 22-12. The pictorial diagram is used extensively by electronic kit manufacturers. You do not have to be an expert to construct most of these kits.

Drawing Electrical and Electronic Symbols

Symbols in circuit diagrams need not be drawn to any particular scale. However, they should be shaped correctly, large enough to be seen clearly, and in proportion, Figure 22-13. The easiest way to do this is to draw them with the aid of an electrical and electronic symbol template, Figure 22-14.

Symbols and lines are drawn the same weight as an object line. A darker line may be employed when a portion of the diagram must be emphasized.

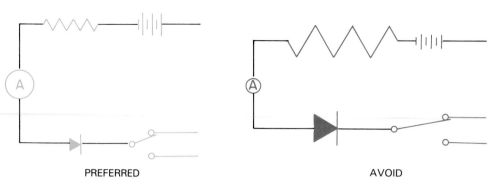

Figure 22-13. Draw symbols in proportion.

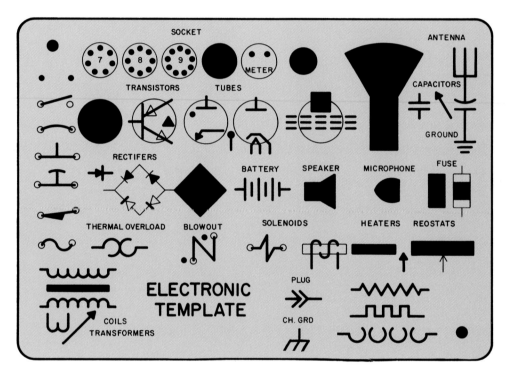

Figure 22-14. Typical electrical/electronics symbol template. This is only one template of a set. Templates are the easiest way to draw the symbols in proper proportion to each other.

Drafting Vocabulary

Block diagram
Connection or wiring
  diagram
Electrical

Electronic
Integrated circuit (IC)
Pictorial diagram
Printed circuit

Proportion
Schematic diagram
Symbols
Transistor

Test Your Knowledge—Unit 22

Please do not write in the book. Place your answers on another sheet of paper.

1. How does electrical and electronic drafting differ from conventional drafting?
2. Where has computer-aided design been used extensively by the electronics industry?

3. A schematic diagram is a drawing _____.
4. How does a block diagram differ from a wiring diagram?
5. What is a pictorial diagram?
6. Symbols and lines used in electrical/electronics drafting are drawn the same weight as a(n) _____ line.
7. Electrical diagrams used on house plans are called _____ diagrams.

Outside Activities

1. Make a schematic diagram of a desk lamp or light used on a drafting table.
2. Prepare a schematic diagram of a two-cell flashlight.
3. Make a complete wiring diagram of your bedroom.
4. Secure a small battery-powered toy. Examine how it operates. Make a suitable electronics diagram showing your findings.
5. Prepare a pictorial diagram of an inexpensive transistor radio.
6. Make a suitable diagram showing four batteries, switch, and lamp wired in series.
7. Make a suitable diagram showing four batteries, switch, and lamp wired in parallel.
8. Prepare the wiring diagram showing how to connect a receiver, amplifier, CD, tape deck, and two speakers.

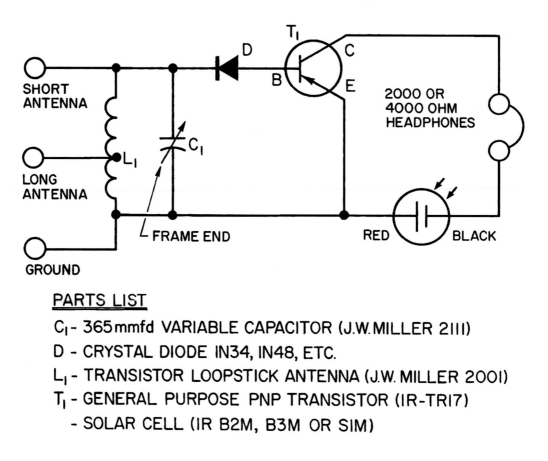

PARTS LIST

C_1 - 365 mmfd VARIABLE CAPACITOR (J.W. MILLER 2111)

D - CRYSTAL DIODE IN34, IN48, ETC.

L_1 - TRANSISTOR LOOPSTICK ANTENNA (J.W. MILLER 2001)

T_1 - GENERAL PURPOSE PNP TRANSISTOR (IR-TR17)

    - SOLAR CELL (IR B2M, B3M OR SIM)

Schematic diagram of a radio which is powered by a single solar cell. (International Rectifier)

Unit 23
Architectural Drafting

After studying this unit and completing the assigned problems, you should be able to:

Explain the importance of architectural plans.

Produce plans for a simple structure.

Read and use an architect's scale.

List the information usually required when building a home.

Identify common architectural abbreviations.

Apply architectural dimensioning rules.

Describe some of the advantages of CAD in preparing architectural plans.

Plans which provide tradesworkers with the information needed to construct buildings in which people live, work, and play are called *architectural drawings*. See Figure 23-1. Architectural drawings or plans are created, designed, and produced by architects.

Figure 23-1. Artist's sketch of a home to be built. Since this house was to be built on speculation (there was no buyer when construction started), this sketch was needed for advertising purposes.

Buying a home is probably one of the largest investments you will make in your lifetime. An understanding of the basic principles of architectural drafting will be a great help if you plan to design or construct a new home, remodel an existing home, or judge the soundness and value of a home offered for sale.

The ability to read and interpret architectural drawings is essential to those in the construction industry, such as carpenters, masons, plumbers, electricians, roofers, etc., Figure 23-2. It is also useful to workers in lumberyards or hardware and building supply stores.

Figure 23-2. The speculation house after it was constructed. The optional garage seen on the sketch was not built.

Building a Home

To make sure your home is built to your specific requirements, it is important that you have a good plan and a well-defined contract with your builder. Plans provide the vast amount of information needed to construct a modern home.

Building Codes

Plans should include all applicable aspects of the local and state building codes. **Building codes** are laws which provide for the health, safety, and general welfare of the people in the community. Building codes are based on standards developed by government and private agencies.

Building Permit

In most communities, the building contractor or owner must file a formal application for a building permit. Plans and specifications are submitted for the proposed structure. The plans are reviewed by building officials to determine whether they meet local building code requirements.

An inspection card is usually posted on the building site. Work is inspected by local building officials as construction progresses. The card is signed as each phase of construction is approved.

Plans

Since it is not possible to include all construction details on a single sheet, a set of typical house plans will include a plot plan, foundation and/or basement plan, floor plans, elevations (front, rear, and side views of the home), wall sections, built-in cabinet, and fireplace details.

Scale

The plans are generally drawn to a one-fourth inch scale (1/4" = 1'-0"). This means that 1/4 inch on the drawing equals 1 foot on the building being constructed.

A larger scale, 1" = 1'-0", is used when greater detail is needed on a structural part. Framing plans are often drawn to 1/8" = 1'-0" scale.

Plot Plan

The *plot plan* shows the location of the structure on the building site, Figure 23-3. Plot plans also show walks, driveways, and patios. Overall building and lot dimensions are included. *Contour lines* (lines indicating the slope of the site surface) are sometimes shown.

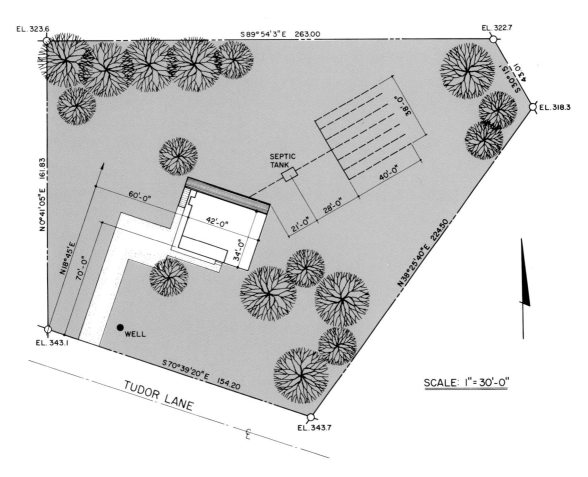

Figure 23-3. Plot plan for the house shown in Figure 23-2.

Elevations

Elevations are the front, rear, and side views of the house, Figure 23-4. They are made of lines which are visible when the building is viewed from various positions. Included on the elevations are the floor levels, grade lines, window and door heights, roof slope, and the types of materials to be used on the walls and roof. Foundation and footing lines below grade level are indicated with hidden lines.

Figure 23-4. Elevation drawing.

Floor Plans

The size and shape of the building, as well as the interior arrangement of the rooms, are shown on the *floor plans*, Figure 23-5. Additional information such as location and sizes of the interior partitions, doors, windows, stairs, and utility installations (plumbing, electrical, etc.) is also included.

Foundation and basement plans are frequently combined on a single sheet, Figure 23-6.

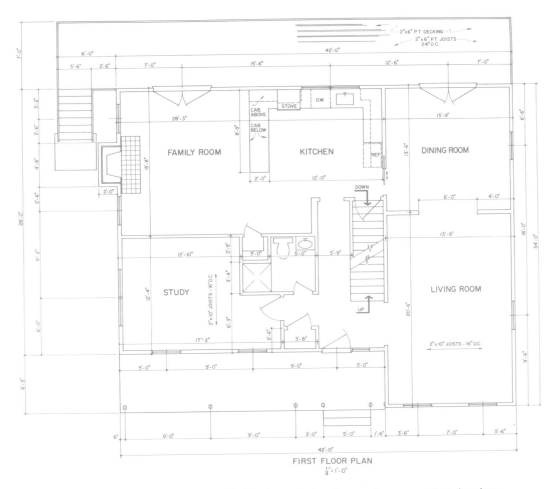

Figure 23-5. Floor plan. Electrical and climate control units are shown on other drawings.

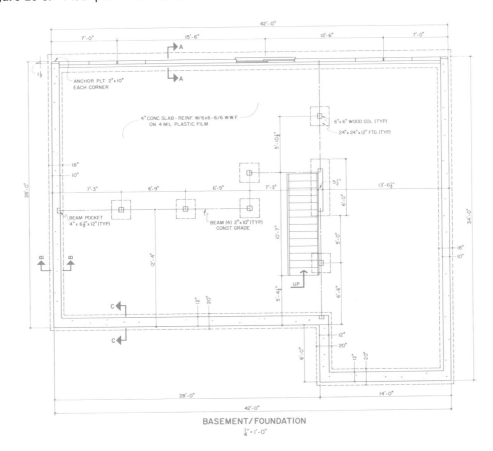

Figure 23-6. Foundation/basement plan.

Sections

Sectional views are used to give construction details by making an imaginary "cut" through the structure, Figure 23-7. Break lines are often incorporated into the section to reduce the drawing size and save time in drawing the plan.

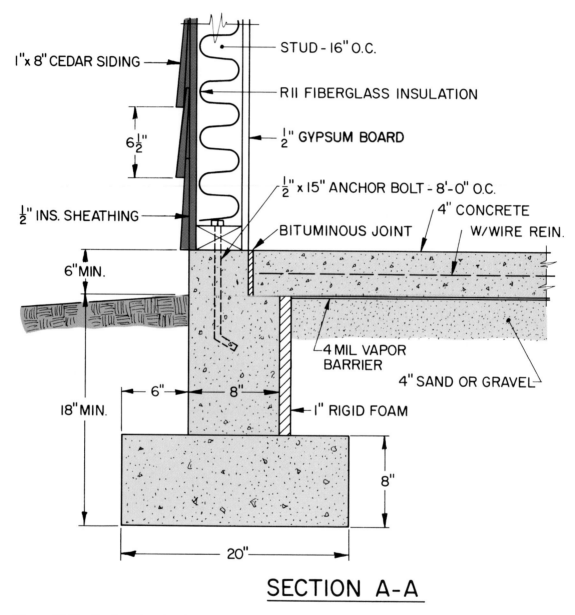

Figure 23-7. Section showing foundation details noted on Figure 23-6.

Along with the sizes of the framing materials, the types and kinds of materials to be used for sheathing, insulation, interior and exterior wall surfaces, and other miscellaneous information are also indicated.

Details

Information to assist the tradesworker in constructing such things as built-in cabinets, fireplaces, and other pertinent information are given on *detail sheets*, Figure 23-8.

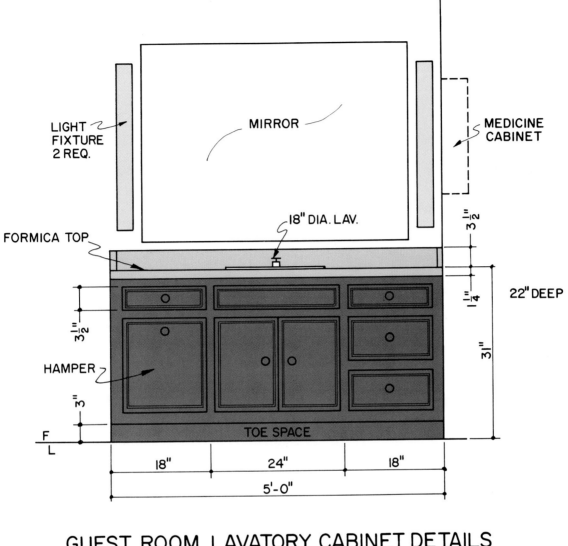

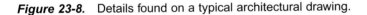

GUEST ROOM LAVATORY CABINET DETAILS
UNIT I - UNIT 2 IS REVERSED
SCALE: 1/2"= 1'-0"

Figure 23-8.   Details found on a typical architectural drawing.

Architectural Drafting Techniques

In general, most conventional drafting techniques will apply to architectural drafting.

Symbols, "Spec" Sheets

Extensive use of *symbols* will be noted in architectural drafting, Figure 23-9. Symbols are employed on plans because it is not practical to show items such as doors, windows, plumbing fixtures, etc., as they would actually appear in the structure.

Sheets of specifications usually accompany house plans. These *"spec" sheets* describe, in writing, things that cannot be easily indicated on the drawings such as quality of materials, how specific items are to be installed, etc.

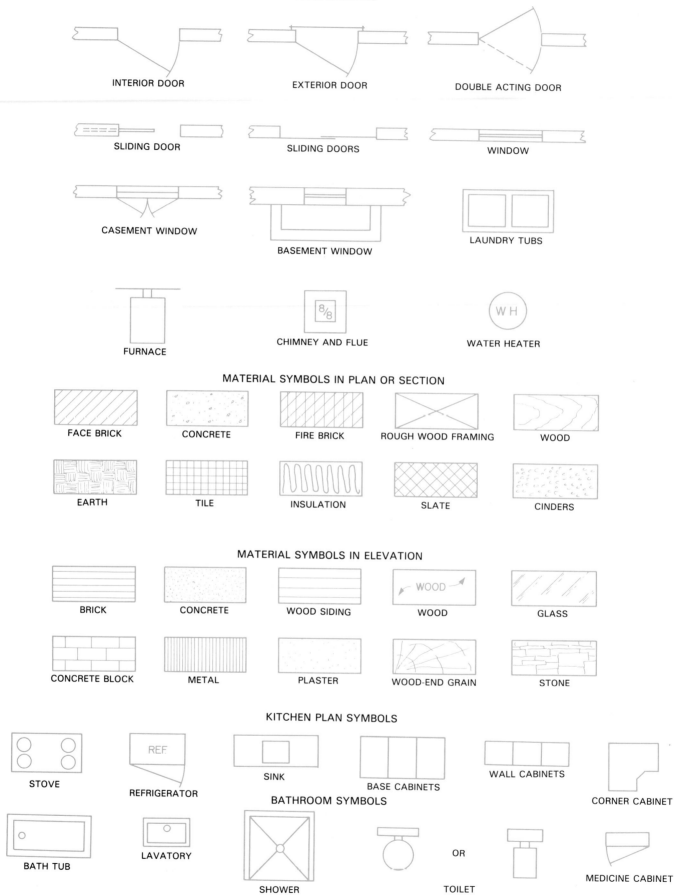

Figure 23-9. Symbols used to indicate materials and fixtures on drawings. (Continued)

ELECTRICAL SYMBOLS

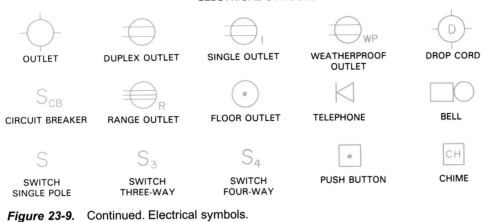

Figure 23-9.   Continued. Electrical symbols.

Scale

With few exceptions, architectural plans are drawn to scale. When drawn to scale, the views presented on the drawing are shown in an accurate proportion of the full-sized structure. A 1/4" = 1'-0" scale means that 1/4 inch on the drawing equals 1 foot in the structure to be built.

Scale drawings are relatively easy to make if an **architect's scale** is used. This scale is divided into various major units, each unit representing a distance of one foot. A 1/4" = 1'-0" unit of the scale is shown in Figure 23-10.

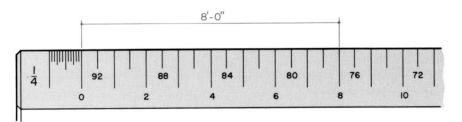

Figure 23-10.   Measuring feet on an architect's scale. The 1/4" = 1'-0" is being used. Count to the right of zero.

This unit and the other units on the scale are further subdivided into equal parts or multiples of twelve to represent inches, Figure 23-11.

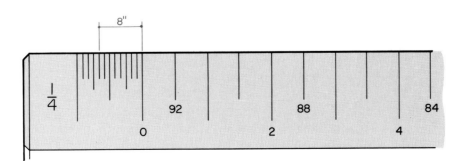

Figure 23-11.   Measuring inches on an architect's scale. Count to the left of zero.

When using the architect's scale to make a measurement, for example, 8'-8" (eight feet, eight inches), start at the 0 and go to the right to locate 8' (eight feet). Then, moving in the opposite direction from 0 locate the 8" (eight inches). The combined measurement would be 8'-8". See Figure 23-12.

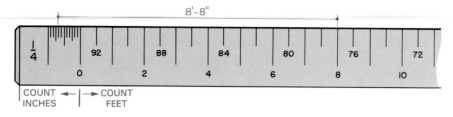

Figure 23-12. Measuring both feet and inches on an architect's scale.

Figure 23-13 will give you an opportunity to practice reading the architect's scale. List your answers on a sheet of notebook paper.

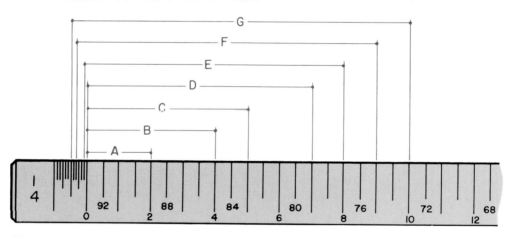

Figure 23-13. How many of these dimensions can you read correctly? Do not write in the book. Place your answers on a piece of notebook paper.

Title Block

A *title block*, Figure 23-14, for architectural drawings should include the following information:

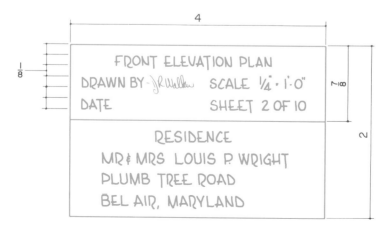

Figure 23-14. Plan title block.

1. Type of structure (house, garage, shed, etc.).
2. Where it is to be located.
3. Architect's (and/or drafter) name.
4. Date.
5. Drawing scale or scales used.
6. Sheet number and number of sheets making up the full set of drawings (Sheet 1 of 10, etc.).

The title block may be any convenient size; however, 2 in. by 4 in. is a suitable size for most home plans. It is usually located in the lower-right corner of the drawing sheet.

Lettering Architectural Plans

Most architectural lettering follows the Roman alphabet of letters and numbers, Figure 23-15. Conventional Single-Stroke Gothic lettering is sometimes used on architectural drawings. Many architects and drafters develop their own distinctive style of lettering. Whether you decide to use the Single-Stroke Gothic letter form or develop a style of your own, remember that your lettering must be legible.

Figure 23-15. One style of architectural lettering.

Dimensioning Architectural Plans

There is only a slight difference between the dimensioning techniques used on architectural drawings and on other forms of drafting. The dimensions are read from the bottom and right sides of the drawing sheet. Measurements over 12 in. are expressed in feet and inches. Foot and/or inch marks (' and ") are used on all dimensions.

Dimensions may be placed on all sides of the drawing and through the drawing itself. Dimension figures, usually 1/8 in. high, are written above the dimension line or in a break in the dimension line. The dimension line may be capped with arrowheads, small dark circles, or short 45 degrees slash marks, as shown in Figure 23-16.

Architectural Dimensioning Rules

1. Make sure that the dimensioning is complete. A builder should never have to assume or measure a distance on the drawing for a dimension.
2. *Always* letter dimensions full-size regardless of the drawing's scale.
3. Dimensions and notes should be at least 1/4 in. away from the view.
4. Align dimensions across the view whenever possible.
5. Place overall dimensions outside the view.
6. Overall dimensions are determined by adding detail dimensions. Do not scale the drawing.

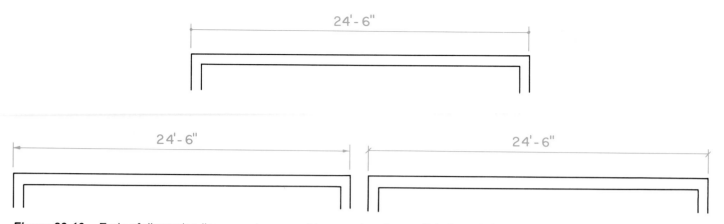

Figure 23-16. Ends of dimension lines may be capped by arrowheads, small darkened circles, or short 45 degree slash marks.

7. Dimension partitions center-to-center or from center to outside wall.
8. Room size may be indicated by stating the width and length of the room.
9. Windows, doors, beams, etc., are dimensioned to their centers.
10. Window and door sizes and types are given in a *schedule*, Figure 23-17.

WINDOW SCHEDULE				
MARK	REQ'D	OVERALL SASH SIZE WIDTH	HEIGHT	DESCRIPTION
W-1	4	5'-4" x	4'-0"	ALUMINUM SLIDING WINDOW WITH INSULATED SASH-FIBERGLASS SCREEN
W-2	3	5'-4" x	2'-0"	SAME AS ABOVE
W-3	I	4'-0" x	3'-0"	"
W-4	I	2'-7" x	5'-8"	1/4" PLATE GLASS
W-5	I	9'-6" x	2'-4"	FIXED, SAME AS ABOVE

Figure 23-17. Window schedule of the type found on a typical house plan.

Architectural Abbreviations

Time and space can be saved when lettering architectural drawings by using abbreviations. Figure 23-18 shows some commonly used abbreviations.

asphalt tile—AT   concrete-CONC   footing—FTG
beam—BM   drawing—DWG   grade line—GL
bedroom—BR   door—DR   lavatory—LAV
brick—BRK   elevation—EL   living room—LR
ceiling—CLG   exterior—EXT   plaster—PL
center line—CL or C̷   floor—FLR   room—RM

Figure 23-18. Common abbreviations used on architectural plans.

Obtaining Information

A great deal of research into catalogs and reference books must be made on the part of the student to find the many standard sizes that are essential in completing an accurate set of plans.

Additional information may be found in mail order catalogs, lumber and building supply company literature, from the Federal Housing Administration, and copies of your local building ordinances. Specialized textbooks in architecture, carpentry, plumbing, and other building areas are excellent sources of information when planning a home or other structure.

Planning a Home

When planning a home, you must determine how much money will be available for building purposes. Most people building homes must borrow a major portion of the construction money. Lending institutions use formulas based on the cash available for a down payment and the owner's income (salary, etc.) to establish how much money they will lend to build or purchase a home.

One way to ascertain the size of a home you plan to build is by the square foot method. A local contractor can give you the approximate cost per square foot of construction in your community.

Divide the cost of construction per square foot into the money available to build the house, and you will find the floor area of a home you can afford to build.

After determining the approximate size of the proposed house, you should try to develop a suitable floor plan. Say, for example, calculations show that you can afford a home with an area of 1200 square feet. Outlines of homes having 1200 square feet of floor space are shown in Figure 23-19. Room use and space available must be carefully considered and balanced.

Be sure to include closet and storage facilities. Using scale cutouts of furniture and appliances you plan to use in the various rooms will aid in your planning. The preliminary floor plan can be drawn on

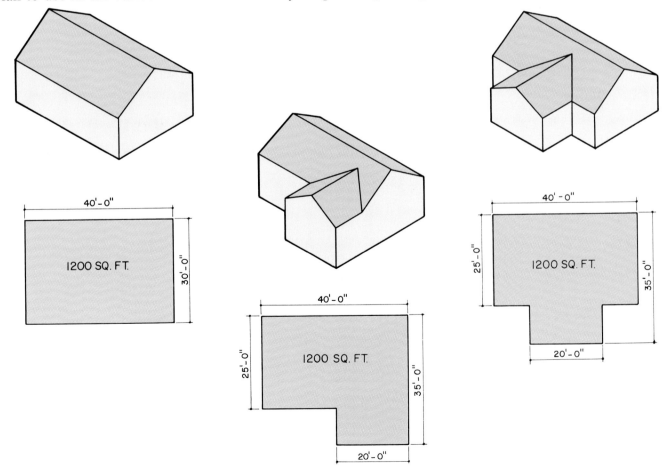

Figure 23-19. Examples of homes with floor areas of 1200 square feet.

Be sure to include closet and storage facilities. Using scale cutouts of furniture and appliances you plan to use in the various rooms will aid in your planning. The preliminary floor plan can be drawn on graph paper. Each square can equal one foot. Changes can be made easily on the grid.

The cutouts *must* be made to the same scale as the floor plan. Your drawing should also show all openings and the space doors will take up when opened.

By moving the cutouts, it is easy to find out whether the room is large enough for the intended use.

Sizes of some furniture and fixtures are given in Figure 23-20. Dimensions of items not shown may be found in manufacturers' catalogs, home magazines, and mail order catalogs.

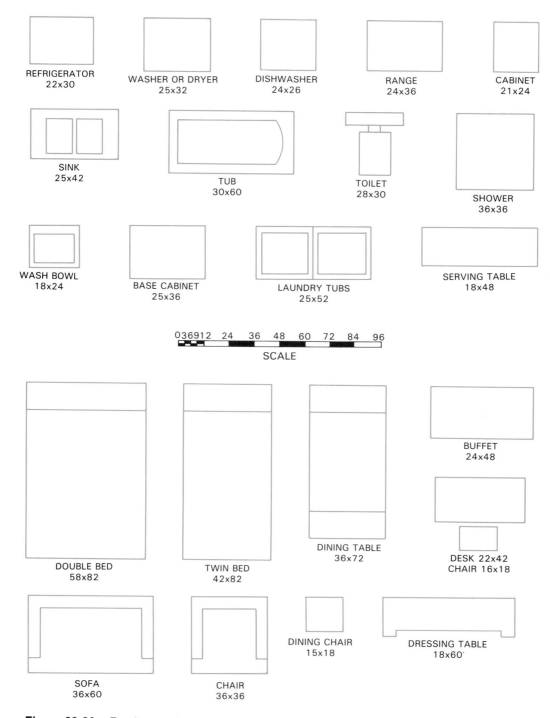

Figure 23-20. Furniture and appliance dimensions will aid in planning a home.

Cutout planning for a typical room is shown in Figure 23-21.

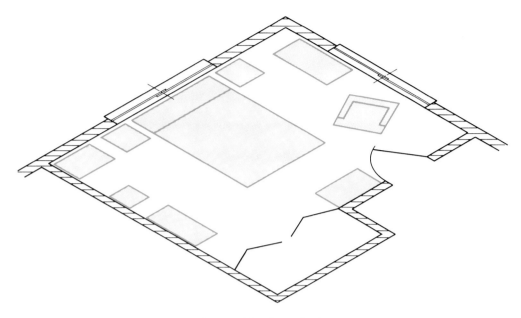

Figure 23-21. How cutouts may be used to plan a bedroom.

Metrication in Architecture

The building construction industry probably is one of the last business enterprises to "go metric" in the United States. Buildings for government contracts must now be drawn with metric dimensions.

Lumber producers have decided upon a "soft" conversion of existing lumber sizes. This will not require a basic redimensioning of lumber sizes as would the metric or "hard" conversion. A "soft" conversion simply means that inches have been changed to millimeters, pounds to kilograms, etc. A "hard" conversion with metric engineering standards is one in which all of the designing is done in preferred metric sizes.

Some plumbing fixtures made to metric standards are being imported. Problems have been encountered trying to join these fixtures to inch-based pipes and fittings.

In the United States, most buildings are designed to a 4-in. base module. See Figure 23-22. All lumber, blocks, bricks, panel stock (plywood, hardboard, etc.) and components such as windows and doors are specified in multiples of the 4-in. module. Windows, for example, usually are 2'-8" or 3'-0" wide.

A 100 mm by 100 mm module is being recommended as the basic metric module. All metric-based building materials and components will be based on multiples of this module.

A 100 mm by 100 mm module is almost, *but not quite*, the same size as the customary 4 in. by 4 in. module. The difference is small, but is enough to cause major problems if *all* materials used in constructing a building are not designed to the same basic module. See Figure 23-23.

Metric Architectural Drafting

There is little difference between metric and customary measuring systems in architectural drafting. Dimensions will be given in meters and millimeters instead of feet and inches.

Scales for use in metric-based architectural drafting are shown in Figure 23-24. All structures will be designed and constructed on the 100 mm module.

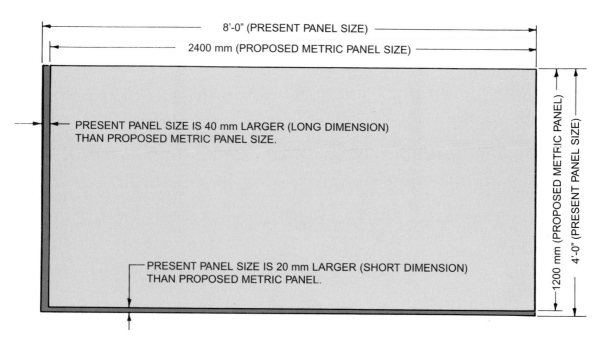

Figure 23-22. Most buildings and building materials now used are designed to multiples of the basic 4 in. module. A 100 mm x 100 mm module is the metric-based unit being proposed.

Figure 23-23. A comparison of the conventional size building panel and the proposed metric size panel. The proposed metric-size panel would be too small to use for a replacement of the conventional 4 foot x 8 foot panel.

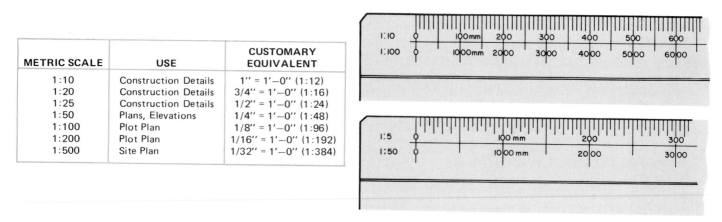

METRIC SCALE	USE	CUSTOMARY EQUIVALENT
1:10	Construction Details	1″ = 1′−0″ (1:12)
1:20	Construction Details	3/4″ = 1′−0″ (1:16)
1:25	Construction Details	1/2″ = 1′−0″ (1:24)
1:50	Plans, Elevations	1/4″ = 1′−0″ (1:48)
1:100	Plot Plan	1/8″ = 1′−0″ (1:96)
1:200	Plot Plan	1/16″ = 1′−0″ (1:192)
1:500	Site Plan	1/32″ = 1′−0″ (1:384)

Figure 23-24. A few of the metric scales recommended for use in metric-based architectural drafting.

Computer-Aided Design and Drafting

As with other areas of industry, computer-aided design (CAD) and computer-aided engineering (CAE) are changing the field of architecture, Figure 23-25. Talented people working with microcomputers are able to produce quality work, faster and more profitably than ever before. Time is greatly reduced between concept, design, working drawings, and construction scheduling of a building project.

Figure 23-25. As with other areas of the industry, computer-aided design (CAD) and computer-aided engineering (CAE) are changing the field of architecture. (Intergraph Corp.)

An important advantage of CAD in architecture allows two- and three-dimensional details to be developed and combined to create new structures and layouts, Figure 23-26. Ideas can be viewed in three-dimensional space and may be rotated and seen from any angle. Design changes can be viewed immediately.

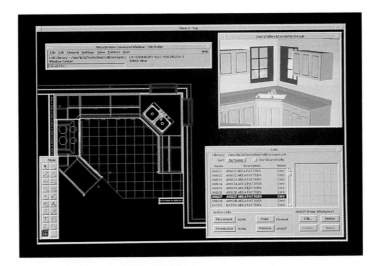

Figure 23-26. CAD can help plan room layout and show the home owner how it will look in 3-D. Shown is the layout of a portion of a kitchen. Changes can easily be made and shown in color. (Bentley Systems)

Architectural CAD programs are designed to reduce repetitive work to a minimum. See Figure 23-27. Other particulars such as the outline of a structure can be developed with the windows and doors located. Structural details, meeting engineering standards, and window and door details, will be inserted automatically. A bill of material and estimated cost of materials can be computed using the same information. New architectural CAD programs under development will offer architects advanced design and management capabilities.

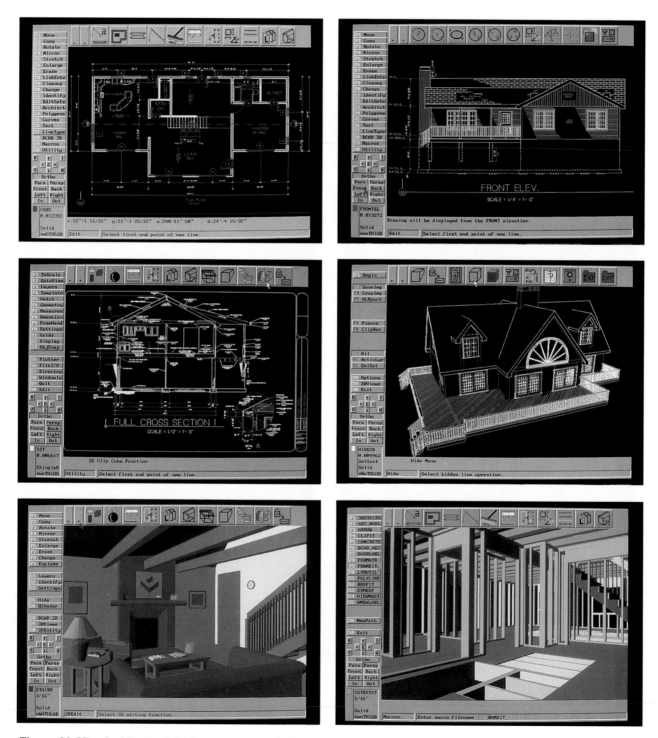

Figure 23-27. Architectural CAD programs are designed to reduce the number of repetitive tasks. Shown are a series of architectural drawings for the same house. Can you identify the drawings? (CADKEY, Inc.)

Drafting Vocabulary

Plot plan	Detail sheets	Ridge
Architect	Elevation	Scale
Architect's scale	Floor plan	Schedule
Architectural drawings	Foundation	Sectional view
Basement	Framing	Site
Building code	Grade line	Specification
Building permit	Inspection card	Standards
Built-in	Insulation	Structure
Construction	Module	Symbols
Contour lines	Ordinances	Title block
Contract	Partition	Wall sections

Test Your Knowledge—Unit 23

Please do not write in the book. Place your answers on another sheet of paper.

1. Architectural drawings are _____.
2. List five occupations to which the ability to read and interpret architectural drawings is essential.
3. What are building codes?
4. Architectural drawings are usually drawn to a scale of _____.
5. The location of the building(s) on the construction site is shown on the _____ _____. This drawing also shows the walks, driveways, and patios.
6. Drawings that show the various views of the structure (front, rear, and side views) are called _____.
7. The floor plan shows the _____.
   A. size and shape of the rooms
   B. location and sizes of the windows and doors
   C. location of utilities (plumbing, heating, and electrical fixtures)
   D. All of the above.
   E. None of the above.
8. Make a sketch showing three recommended methods for dimensioning architectural drawings. Refer to Figure 23-16.

Outside Activities

1. Make a scale drawing (1/4" = 1'-0") of the school drafting room. Use cutouts to determine an efficient layout for the furniture in the room.
2. Prepare a drawing of your bedroom showing the location of the furniture in the room. Use a scale of 1/2" = 1'-0".
3. Draw the floor plan of a two-car garage with a workshop at one end.
4. Design and draw a full set of plans for a two-room summer cottage or hunting cabin.
5. Draw the plans necessary to convert your basement or other space into a workshop, a recreation room, or a dark room.

6. Design and draw the plans needed to construct a storage or toolshed. See Figure 23-28.

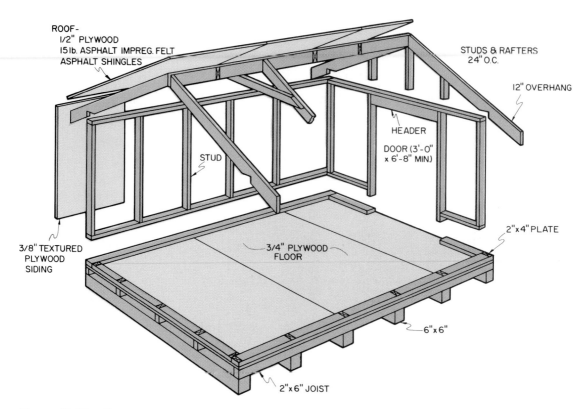

ROOF-
1/2" PLYWOOD
15 lb. ASPHALT IMPREG. FELT
ASPHALT SHINGLES

STUDS & RAFTERS
24" O.C.

12" OVERHANG

HEADER

DOOR (3'-0"
x 6'-8" MIN)

STUD

3/8" TEXTURED
PLYWOOD
SIDING

3/4" PLYWOOD
FLOOR

2"x4" PLATE

6"x 6"

2"x 6" JOIST

Figure 23-28. Storage or toolshed for Problem 6.

7. Design a small two-bedroom house that can be constructed at minimum cost. Construct a model of it.
8. Plan a structure to house a sports car agency.
9. Secure samples of computer-generated architectural drawings.

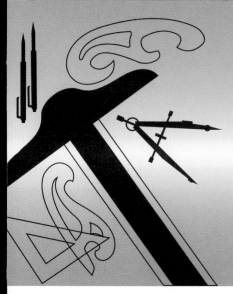

Unit 24
Computer-Aided
Drafting and Design

After studying this unit and completing the assigned problems, you should be able to:

◆ Explain how computer technology is revolutionizing drafting, design, and engineering.

◆ Describe the equipment used in computer-aided drafting and design.

◆ Define CAD terminology.

◆ Explain the difference between hardware and software.

Computer graphics is revolutionizing drafting and engineering, Figure 24-1. A *computer,* in its simplest form, is a device that performs mathematical calculations. The first computer was the human hand. It can solve many mathematical problems. The *abacus* was the earliest *mechanical* computer, Figure 24-2. No one knows when it was invented. The abacus is still widely used in many parts of the world.

A modern computer is electronic. It can perform high-speed computations, including mathematical calculations and logical decisions, under the direction of a program. A *program* is a series of instructions that "tells" the computer what needs to be done and how it is to be accomplished.

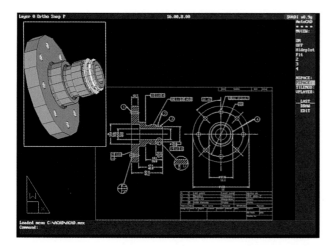

Figure 24-1. Computer graphics is revolutionizing drafting and engineering. With the proper CAD program, it is possible to generate a three-dimensional picture using the information developed on a two-dimensional drawing. (Autodesk, Inc.)

Figure 24-2. The abacus was the first mechanical computer. It is still used in many countries.

A computer system consists of a *central processing unit (CPU)* which processes all the instructions in the program and makes sure it is processed

345

properly, a *monitor* (display screen), a *keyboard, input devices, output devices,* and *auxiliary data storage units,* Figure 24-3.

Categories of Computers

Speed, memory, and size of programs that can be run are used to classify computers. There are three categories of computers—mainframe computers, minicomputers, and microcomputers.

Mainframe Computers

The *mainframe computer* is a large computer. It is capable of processing large amounts of data at high speeds with access to a very large database (electronic information). Some mainframe computers can process 200 million instructions a second. With a mainframe computer, drafters can work at terminals at locations remote from the computer, Figure 24-4. Terminals include an input device and a display screen. Each terminal is connected to the computer through a *network* of wires and cables, Figure 24-5. Networking permits different users to access the mainframe's database.

Minicomputers

Minicomputer falls somewhere between mainframe computers and microcomputers. They are smaller, less powerful, and have less storage capacity than a mainframe system. They are also less expensive.

While they are capable of being networked, not as many terminals can be used as compared to a mainframe. Minicomputers are powerful enough to run most complex CAD programs.

Microcomputers

Microcomputers are smaller and much less expensive than mainframe and minicomputers. Microcomputers are typically *stand-alone systems,* and allow only one user to run one program at a time. Each drafter has his or her own individual computer to perform work. Therefore, microcomputers are commonly referred to as *personal computers (PCs).* Most personal computers are networked to a larger computer to allow drafters, designers, and engineers to share drawing information.

Figure 24-3. Typical computer system. With the proper CAD program, two- and three-dimensional images of the object being designed can be generated. (Munro and Associates)

Figure 24-4.   Design/build team members use sophisticated computer workstations such as these to design and electrically pre-assemble the entire Boeing 777 aircraft. Design engineers create and manipulate full-color, 3-D solid images representing airplane sections. The Boeing 777 division currently has about 1700 workstations linked to the world's largest mainframe computer cluster dedicated to computer-aided design/computer-aided manufacturing (CAD/CAM). (The Boeing Company)

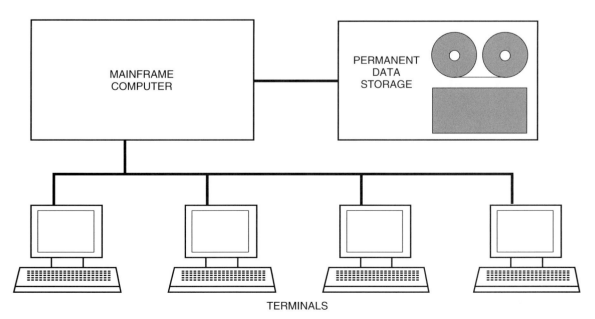

Figure 24-5.   With networking, each designer/engineer/drafter has access to common data storage to retrieve and save drawings. Each workstation can run on its own, or it can rely on the mainframe computer to process information.

A microcomputer consists of a monitor, keyboard, mouse, limited data storage, and a small silicon chip called a *microprocessor,* Figure 24-6. The microprocessor is considered to be the "brains" of the computer.

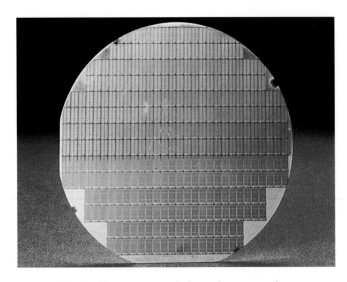

Figure 24-6. The computer is based upon a microprocessor similar to this tiny silicon chip. The 3-inch wafer shown produces 190 chips. (Motorola, Inc.)

Today, computers operate factories, diagnose medical problems, control fuel and air mixtures in automobile engines, and have thousands of other applications. They also aid in designing mechanical products, buildings and other structures, solving engineering problems, etc., through the use of computer-generated graphics, Figure 24-7. Since their products are usually the most sophisticated (complex) manufactured, the aerospace industry has been a pioneer in developing computer technology, Figure 24-8.

Computer Graphics

Computer technology has revolutionized (completely changed) all areas of engineering and drafting. It was first employed for aerospace design in the 1950s. Computer graphics is now a required tool of industrial technology, Figure 24-9.

Computer-aided design (CAD) is a computer graphics technology that places designs, drawings, graphs, and pictures (in color) on a monitor or dis-

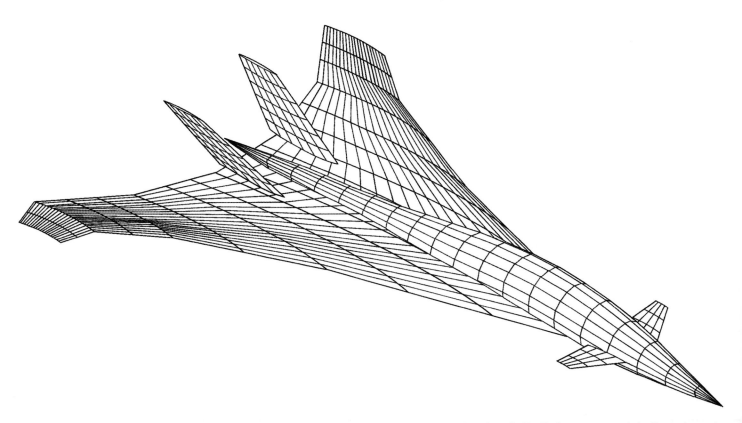

Figure 24-7. The start in the design process of developing a supersonic business jet aircraft. Preliminary research indicated an aircraft in this configuration. Much additional research and development (R&D) must be done before the first piece of metal can be cut to manufacture the aircraft.

Figure 24-8. During the design of the Boeing 777 (the first aircraft designed without first making conventional drawings) care was taken to assure that mechanics and technicians had easy access to all aircraft components. This computer-generated hard copy shows there is ample room to work on a portion of a hydraulic system in this area of the aircraft. (The Boeing Company)

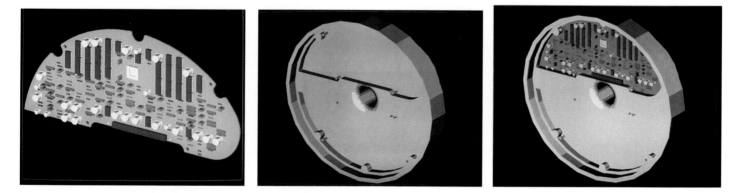

Figure 24-9. In the design of products with both electronic and mechanical components, like this aerospace assembly, the proper CAD program permits designs to be checked for fit as well as analyzed for thermal/structural characteristics. (Applicon)

play screen, Figure 24-10. A monitor is similar to a television screen but offers two-way "communication" between the engineer/designer/drafter and computer. The technology is often referred to as computer-aided design and drafting (CADD and/or CAD).

CAD does not require the engineer/designer/drafter to learn computer programming. These people are able to use computer graphics after a suitable training period. *However, the principles of drafting are common to traditional drafting and CAD.* A working knowledge of basic

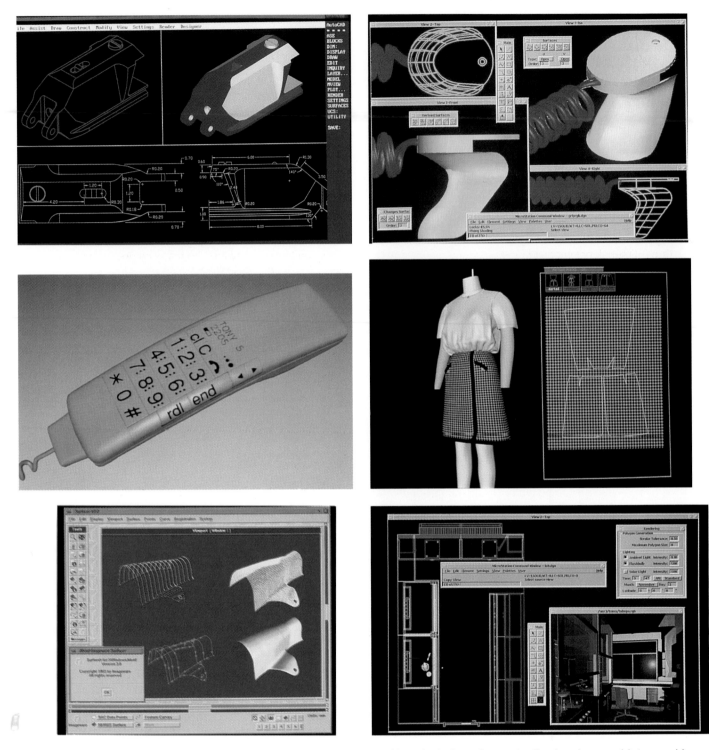

Figure 24-10. Samples of work generated by computer technology. (Autodesk, Inc.; Computer Design, Inc.; and Intergraph)

drafting standards, techniques, and procedures is absolutely necessary.

With computer graphics, two- and three-dimensional images of the object being designed can be generated, Figure 24-11. A wide range of colors can be employed in the images, Figure 24-

12. The computer can be instructed to rotate the object through various positions for study and evaluation, Figure 24-13.

Design changes can be made quickly and reevaluated. Depending upon the CAD system used, design changes can be made by touching a

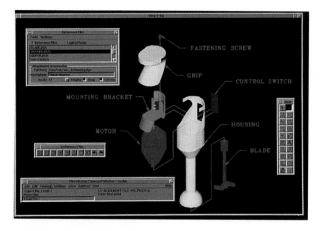

Figure 24-11. With computer graphics, two- and three-dimensional images of the object being designed can be generated. (Intergraph)

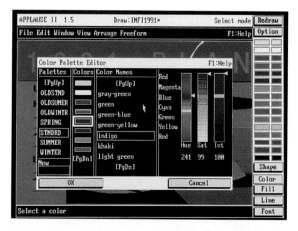

Figure 24-12. A wide range of colors can be employed in work designed on many computer systems. (Computer Design, Inc.)

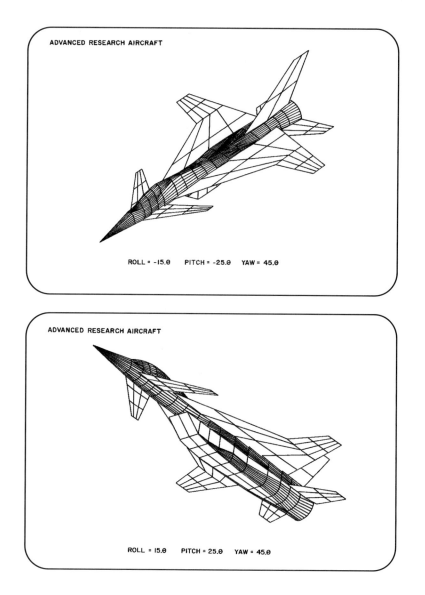

Figure 24-13. The computer can be instructed to rotate the object through various positions. The images have been redrawn from computer printouts to enhance the views for printing. (NASA)

light pen to the screen, Figure 24-14, or moving a cursor, Figure 24-15, to the area being altered (symbol, line, point, etc.) and selecting and keying in the necessary commands to make the required changes through the input device.

Figure 24-15. The cursor of this CAD system is controlled by moving the mouse (small input device to the right of the keyboard). (Intergraph)

SQUARE   UNDERLINE   POINTER   CROSSHAIR

Figure 24-16. A cursor indicates your point of interest on the monitor. Shown are examples of cursors used by various CAD systems.

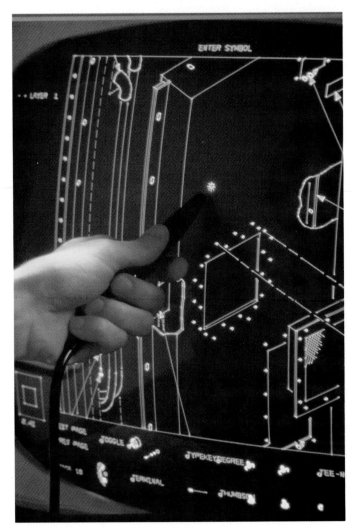

Figure 24-14. A light pen being used on a display screen.

A *light pen* is a stylus-shaped photosensitive pointing device. A *cursor* is a special character, Figure 24-16, that indicates your present point of interest on the display screen. By moving the pen or cursor to a position on the screen, the coordinates at that point are input as data to the computer memory. *Coordinates* are a series of points of measurement located along x, y, and z axes from a fixed origin, Figure 24-17. The z-axis is needed for three-dimensional graphics, Figure 24-18.

It is also possible to "zoom in" and enlarge a section on the proposed design for study and/or

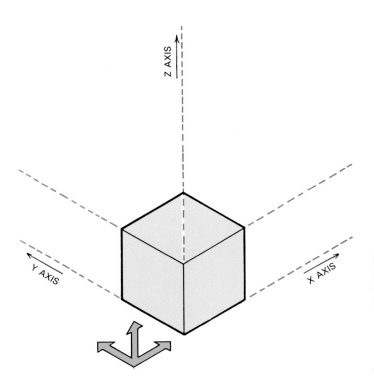

Figure 24-17. Coordinates are a series of points of measurement located along x, y, and z axes from a fixed point of origin.

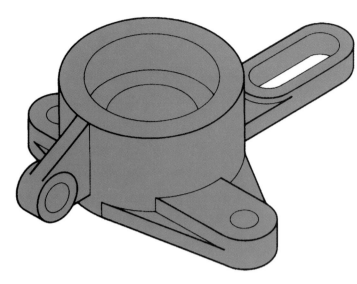

Figure 24-18. A third axis—the z axis—is needed to generate three-dimensional images like the one shown above. Some CAD systems are not powerful enough to generate true three-dimensional graphics. (ROBO Systems)

Figure 24-19. The digitizing tablet is an input device connected to a computer. It is especially useful for transferring conventional drawings into CAD files. The puck or stylus is used to "digitize" drawings. (CalComp)

modification. Other portions of a ***drawing database*** (organized information on a particular drawing) describing the part's design can be called up quickly on the screen.

A ***digitizing tablet,*** Figure 24-19, is frequently used to transfer existing drawings into CAD drawing files, Figure 24-20. A digitizing tablet is an input device connected to the computer. It is a flat, rectangular pad with hundreds of fine wires forming a grid just below the pad's plastic surface. A CAD drawing is created by moving a tracking device—usually a puck or a stylus—across the pad to which the drawing has been taped. Moving the tracking device across the pad causes the screen cursor to move. By selecting CAD commands and picking point positions on the existing drawing, a CAD drawing is created. The resulting CAD drawing can be edited and, if necessary, revised as would a drawing originally made using CAD.

When a design has been finalized (completed), CAD has the capability of turning design data into working drawings (called ***hard copy***) by means of ***high-speed plotters,*** Figure 24-21, or laser printers. The ***design data*** (mathematical shape and size description of the part) is retained in computer memory, or on a tape or disk, for later use in ***computer-aided manufacturing (CAM)*** systems that will cut the material and make the part, Figures 24-22 and 24-23.

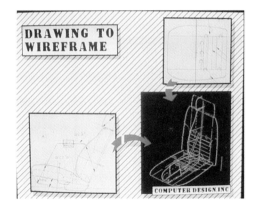

Figure 24-20. Conventional drawing that has been converted into computer graphics by means of a digitizing tablet. (Computer Design, Inc.)

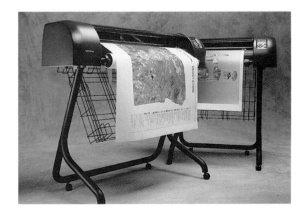

Figure 24-21. Two sizes of plotters. Plotters produce hard copy (working drawings) from CAD developed information. (CalComp)

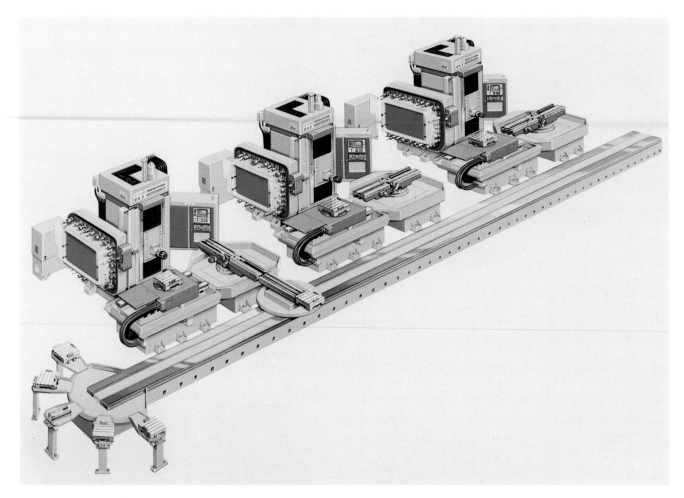

Figure 24-22. A group of machine tools that are part of a CAD/CAM system. Each machine has a microcomputer control system. Note how the work moves from station to station. (Kearney & Trecker Corp.)

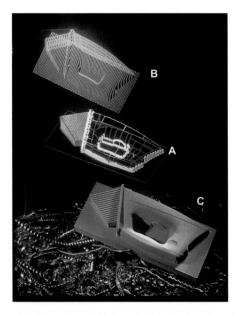

Figure 24-23. From CAD to CAM. A—Design data for the iron body is generated using a CAD system. B—Design data is then provided to a CAM program, which develops tool paths for the cutters and other tooling. C—Mold can be made without ever printing or plotting a single piece of paper. (Sescoi USA, Inc.)

Computer-aided manufacturing or CAM refers to automated manufacturing operations. Computers control the manufacturing processes. When computer-aided design and computer-aided manufacturing are used together it is referred to as *CAD/CAM.*

Computer graphics can also simulate how the part will "work" with other parts in an assembly, Figure 24-24. This permits the engineer/designer/drafter to determine whether possible conflicts exist (for example, two parts in the same place at the same time).

The most advanced form of computer graphics is known as *computer-aided imagery (CAI).* CAI simulates the object in full-color, three-dimensional form, and animates it so the parts that make up the design can be filmed as if they were in actual operation, Figure 24-25. While originally conceived for engineering purposes, CAI is used to create galactic special effects in motion pictures. Computer-aided imagery requires a supercomputer because each frame can require up to 72 million calculations.

Why Use Computer Graphics

The main function of an engineer/designer/drafter is to define (explain) the basic shape of a part, assembly, or product, Figure 24-26. This is

Figure 24-25. Computer-aided imagery (CAI) simulates how an Air Force tanker aircraft would behave under various weight and atmospheric conditions. There was no need to endanger the aircraft or lives of the crew by using an actual tanker to conduct the tests during marginal weather conditions. (Evans & Sutherland)

called *preliminary design.* The process involves many modifications (changes) and refinements (improvements) before a design is finalized.

Until the advent of CAD, engineers/designers/drafters had to imagine and then evaluate a three-dimensional object that was drawn in two-

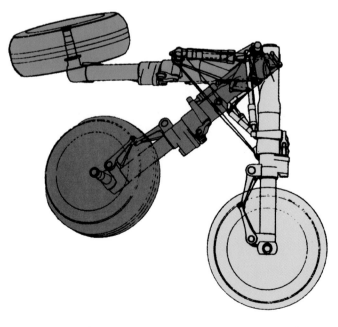

Figure 24-24. Computer graphics can simulate how CAD-developed parts will "work" together. This illustration shows a landing gear during retraction and extension. Note how it pivots as it is raised.

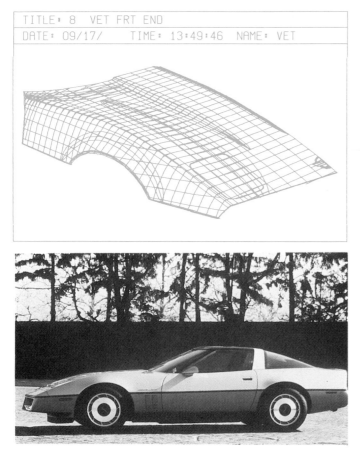

Figure 24-26. Preliminary design of the front end of a Corvette developed from a sketch. Final product shown below. (General Motors Advanced Concept Center)

How CAD Works

A normal sequence of CAD starts with the generation of a geometric model of the proposed design on the display screen, Figure 24-27. Generally, only the outline of the design is created of the geometric model. Details are added later.

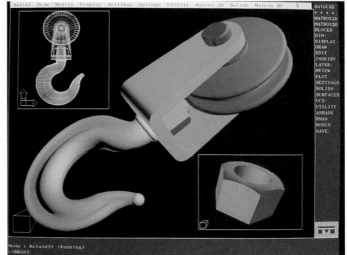

Figure 24-27. A normal sequence of CAD starts with the generation of a geometric model of the proposed design (upper left). Details are added when the design is approved. (Autodesk, Inc.)

A CAD program is activated through a *menu*, Figure 24-28. A menu is a program-generated list of options from which the drafter can select to execute desired procedures. A menu can appear as a list of commands on a digitizing tablet, or as a list of options on the monitor, Figure 24-29.

Digitizer menu selection is made using a light pen, stylus, mouse, or puck. Pointing at a command is the same as keying it on the keyboard. Pointing at a menu symbol selects it for placement on your drawing. Menu selection is made by moving the cursor to the appropriate block and picking with the input device, Figure 24-30.

Once graphic features are in the system, they can be moved, deleted, or mirrored (reversing a graphic image around one, or both axes), Figure 24-31. Once a feature is drawn using CAD, it can be manipulated in a variety of ways, making it unnecessary to draw the entity again.

dimensions on a flat sheet of paper. The only way a design could be verified in three-dimensions was to make a wood, clay, plaster, or plastic model. This is an expensive and time-consuming process.

The introduction of CAD provided the engineer/designer/drafter with a dynamic new tool. CAD technology permits more time for creative work. There is no need for the repetitive drafting required before CAD.

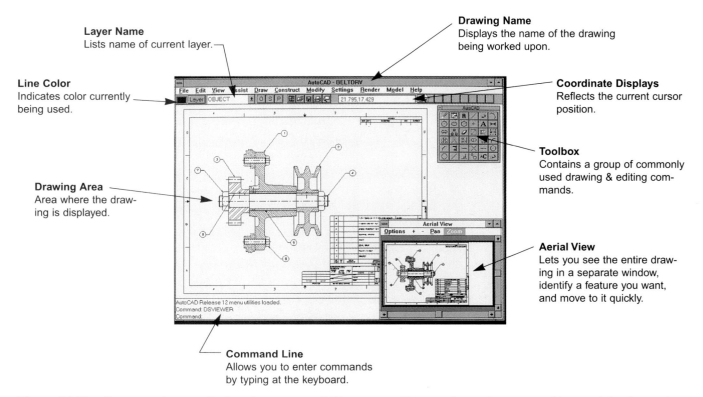

Layer Name
Lists name of current layer.

Line Color
Indicates color currently
being used.

Drawing Area
Area where the draw-
ing is displayed.

Drawing Name
Displays the name of the drawing
being worked upon.

Coordinate Displays
Reflects the current cursor
position.

Toolbox
Contains a group of commonly
used drawing & editing commands.

Aerial View
Lets you see the entire drawing in a separate window,
identify a feature you want,
and move to it quickly.

Command Line
Allows you to enter commands
by typing at the keyboard.

Figure 24-28.   Screen and menu display of a common CAD program. Commands can be accessed in a variety of ways to suit the user's needs. (Autodesk, Inc.)

Figure 24-29.   Commands in the CAD program can be accessed from the menu on screen using the mouse, or by using the function keyboard to the right of the computer. (Intergraph)

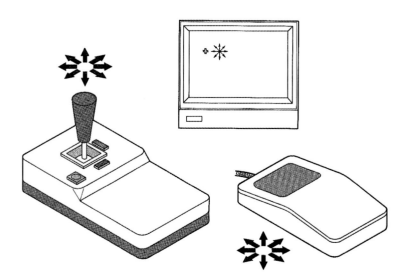

Figure 24-30.   Input devices, such as the joystick and mouse, are used to control cursor movement on the monitor. The selection button is pressed when a point placement is determined.

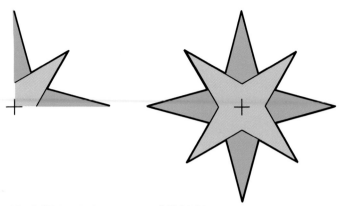

AS FIRST DEVELOPED   IMAGE MIRRORED ON TWO AXES

Figure 24-31. A symmetrical drawing can be mirrored around the x and y axes for perfect symmetry.

Extension lines, dimension lines, arrowheads, text, and dimensions are added by selecting the appropriate menu function and keying in the required material. The drafter can select text size, style, and orientation (location on the graphics), Figure 24-32.

How Computer Graphics are Generated

Numbers and letters (alphanumerics) are the language of computers, not pictures. Before a picture (drawing) can be developed or generated on the display screen, it must first be translated into computer "language."

There are two ways to convert a picture into numbers and letters—*raster* and *vector*, Figure 24-33. With *vector images,* each picture is broken into lines. Each line (vector) has two endpoints. It can be located on a *grid.* The grid is a drawing aid for determining distance. It may be shown on the screen as a network of uniformly spaced points (fine dots).

The technique is similar to drawing pictures by the graph method. See page 42 on enlarging or reducing by the graph method.

Horizontal measurement of the endpoint is its x-coordinate. The vertical measurement is its

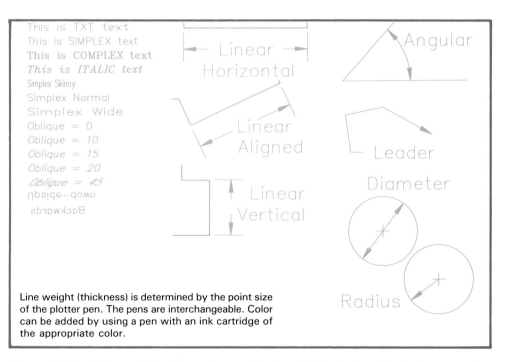

Line weight (thickness) is determined by the point size of the plotter pen. The pens are interchangeable. Color can be added by using a pen with an ink cartridge of the appropriate color.

Figure 24-32. When adding dimensions and text to a CAD drawing, the drafter can select text and numeral size, style, and orientation (position and/or location) on the graphics. The material shown above was prepared using the AutoCAD computer-aided design software. (Autodesk, Inc.)

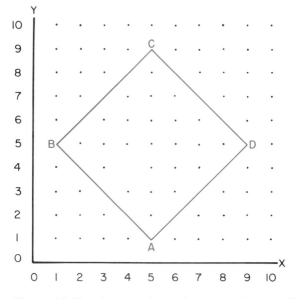

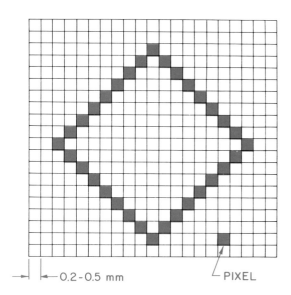

— 0.2–0.5 mm   └ PIXEL

Figure 24-33. A comparison of vector and raster displays.

y-coordinate. (Coordinates represent units of real measurement from a fixed point.) Most CAD programs permit grid size to be changed at will to determine sheet size of the hard copy (drawing printout). Plotter capacity limits actual printout size. Since alphanumerics are understood by the computer, a vector picture can be described by entering the x- and y-coordinates of the endpoints with the list of connections between the points.

Circles and arcs are generated by specifying three points or a center and a radius. Curves must be defined mathematically.

A *raster image* utilizes many tiny *picture elements,* called *pixels.* They are arranged in a fixed, precise manner. Each pixel is the same size and shape. They can be made to match a series of computer memory locations. Your TV is an example of a raster display.

Raster pictures are inserted by scanning, with either a TV camera or a specially designed *scanner.* Scanners are automatic digitizing devices. It analyzes the lines, circles, and other graphic elements of the drawing and converts it into computer data which are stored as a CAD drawing.

Drafting Vocabulary

Abacus	Database	Mirrored
Auxiliary data storage units	Design data	Monitor
Central processing unit	Digitizing tablet	Networking
Computer	Drawing database	Output device
Computer-aided design (CAD)	Grid	Personal computer (PC)
Computer-aided imagery (CAI)	Hard copy	Pixels
Computer-aided manufacturing (CAM)	Input device	Plotter
Coordinates	Keyboard	Preliminary design
Cursor	Light pen	Program
Data	Mainframe computer	Raster
	Menu	Raster image
	Microcomputer	Scanner
	Microprocessor	Stand-alone system
	Minicomputer	Vector
		Vector image

Test Your Knowledge—Unit 24

Please do not write in the book. Place your answers on another sheet of paper.

1. A computer, in its simplest form, is a device that performs _____ _____.
2. The oldest computer is the _____ .
3. A modern computer can perform high speed _____.
4. List the six general components that make up a CAD system.
5. A program is a series of instructions that "tells" a computer _____.
6. The term CAD means _____ _____ _____.
7. CAD is a computer technology that _____.
8. Preliminary design means _____.
   A. the main function of an engineer/designer/drafter
   B. to define the basic shape of a part, assembly, or product
   C. final design of a product
   D. All of the above.
   E. None of the above.
9. A CAD program is activated by a menu. A menu is a(n) _____.
10. A cursor is _____.
11. Computer-generated designs are converted into working drawings on a(n) _____ _____ _____.

Outside Activities

1. Secure samples of various types of drawings prepared using a computer. Prepare a bulletin board display around these drawings.
2. Visit an industrial drafting room that uses CAD. Interview an operator of the computer stations. How does the computer save time? How does the computer assist with repetitive work? How does the computer improve productivity? What did the computer installation cost the business? Report to the class what you observed and learned.
3. Visit a computer store that sells CAD equipment. Request literature on the equipment they sell. Prepare a bulletin board around the material.

Unit 25
Manufacturing Processes

After studying this unit and completing the assigned problems, you should be able to:

◆ Identify basic manufacturing processes.

◆ Explain how the basic manufacturing processes can be applied in product design.

◆ Explain and illustrate methods of cutting, shaping, forming, molding, and fabricating metals and plastics.

The purpose of most drawings is to describe a part or a product that is to be manufactured. Therefore, it is important that the drafter have an understanding of how the materials can be cut, shaped, formed, and fabricated. Metals, plastics, and other materials are available in a variety of shapes and sizes. See Figure 25-1. By understanding manufacturing processes, the drafter can better visualize the objects being drawn.

Machine Tools

The world of today could not exist without *machine tools*. They produce the accurate and uniform parts needed for the many products we use.

According to the National Machine Tool Builders' Association a *machine tool* is "a power-driven machine, not portable by hand, used to

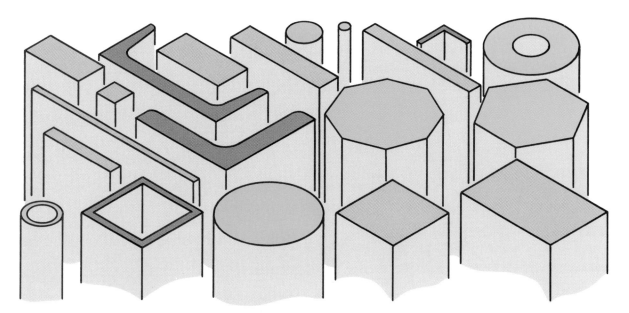

Figure 25-1. Metals, plastics, and other materials used in manufacturing are made in many sizes and shapes.

shape or form metal (or other materials) by cutting, impact, pressure, electrical techniques, or by a combination of these processes."

Machine tools are manufactured in a large range of styles and sizes. Only basic tools and manufacturing techniques will be covered in this unit.

Lathe

The *lathe* is one of the oldest and most important of the machine tools, Figures 25-2 and 25-3. It operates on the principle of the workpiece being rotated against the edge of a cutting tool, Figure 25-4. The cutting tool can be controlled to move lengthwise and across the face of the material being machined (turned).

Operations other than turning can also be performed on the lathe. It is possible to bore, Figure 25-5, ream (finishing a hole to exact size), and cut threads and tapers.

There are many variations of the basic lathe. One of the most complex is the *turning cell* shown in Figure 25-6. It is capable of automated machining without operator intervention (involvement). After the turning cell is set up and programmed, it changes parts and tools, monitors tool condition,

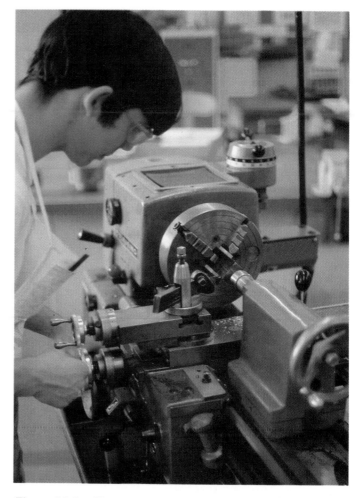

Figure 25-3. The lathe is a very important machine tool.

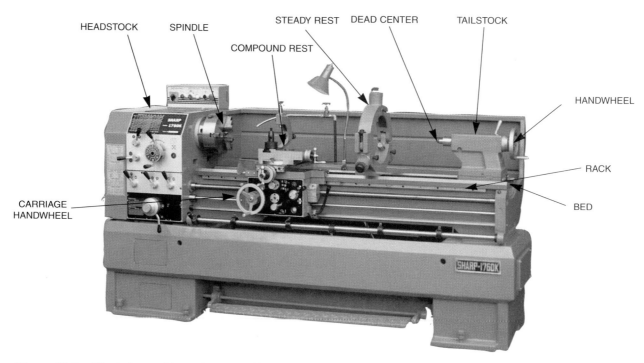

Figure 25-2. The lathe and its major parts. (Sharp Precision Machine Tools)

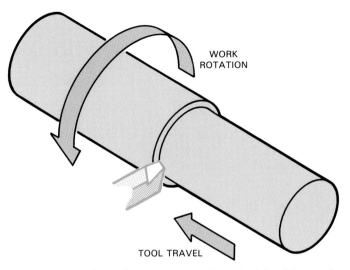

Figure 25-4. The lathe operates on the principle of the work-piece being rotated against the edge of a cutting tool.

Figure 25-5. Boring (internal machining) on a lathe. (Clausing Machine Tools)

Figure 25-6. Stand-alone turning cell. Tools and workpiece load/unload robotically. Machine monitors its own performance automatically under computer control. Lasers gage (measure) the workpiece to maintain part size. (Monarch Machine Tool Co.)

and determines when tools need to be changed. The finished machined parts are gaged (measured) automatically and the data is fed back into the machine's CNC controls, which makes necessary adjustments to maintain consistent machining accuracy. Material to be machined is commonly unloaded using a robot.

The **turret lathe** is used when a number of identical parts must be turned. It is a conventional lathe fitted with a six-sided tool holder called a **turret**. Different cutting tools fitted in the turret rotate into position for machining operations.

Other lathes range in size from the small lathe needed by the instrument and watchmaker, Figure 25-7, to the large lathes that machine the forming rolls for steel mills, Figure 25-8.

Drill Press

The **drill press** is probably the best known of the machine tools, Figure 25-9. A cutting tool called a **twist drill** is rotated against the workpiece with sufficient pressure to cut its way through the material, Figure 25-10. The spiral flutes on the twist drill do not pull the drill into the workpiece.

ACTUAL SIZE
OF PART

1/8 IN.

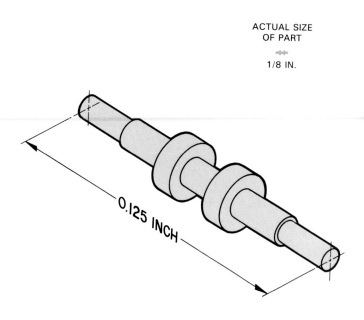

0.125 INCH

Figure 25-7. Part made on a lathe used by instrument and watchmakers.

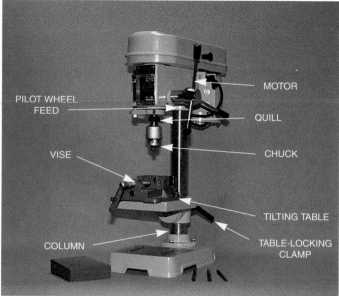

PILOT WHEEL
FEED

MOTOR

QUILL

VISE

CHUCK

TILTING TABLE

COLUMN

TABLE-LOCKING
CLAMP

Figure 25-9. Bench-model drill press with major parts identified.

Figure 25-8. Lathe for machining forming rolls for a steel mill.

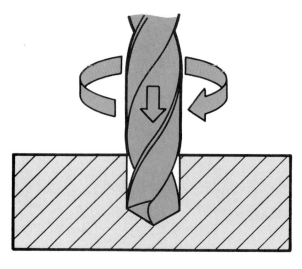

TOOL ROTATES
WHILE PRESSURE
IS APPLIED TO
FORCE DRILL INTO
MATERIAL

Figure 25-10. The operating principle of the drill press.

Pressure must be applied to the rotating drill to make it cut.

Other operations which can be performed on a drill press include: *reaming* (finishing a drilled hole to exact size), *countersinking* (cutting a chamfer on a hole so a flat head fastener can be inserted), and *tapping* (cutting internal threads in a drilled hole).

Milling Machine

A *milling machine* is a very versatile machine tool. It can be used to machine flat and irregularly shaped surfaces, drill, bore, and cut gears.

The *vertical milling machine* uses a cutter mounted *vertical* to the worktable, Figures 25-11 and 25-12. The *horizontal milling machine* uses a

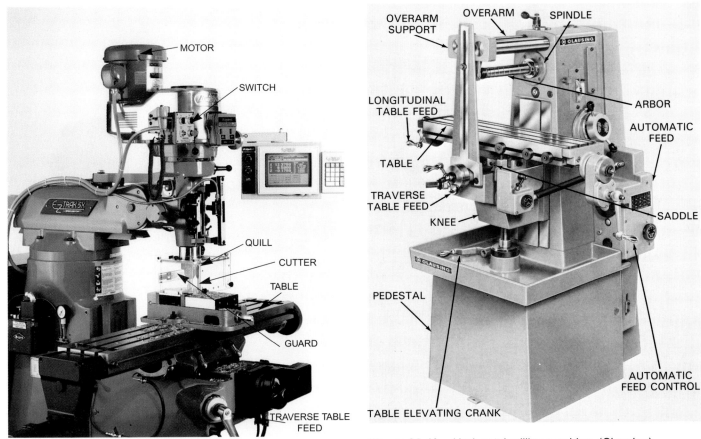

Figure 25-11. Vertical milling machine with retrofitted CNC controls. (Bridgeport Machines, Inc.)

Figure 25-13. Horizontal milling machine. (Clausing)

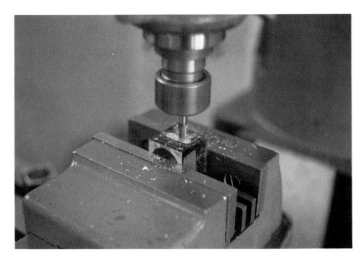

Figure 25-12. Cutter (called an end mill) is mounted in a vertical position. It cuts on both the face and periphery (side).

Figure 25-14. Cutter position on a horizontal milling machine.

cutter mounted *horizontal* to the worktable, Figures 25-13 and 25-14. Metal is removed by means of a rotating cutter that is fed into the moving work-

piece, Figure 25-15. There are many types of milling machines. They vary in size and by types of automatic controls.

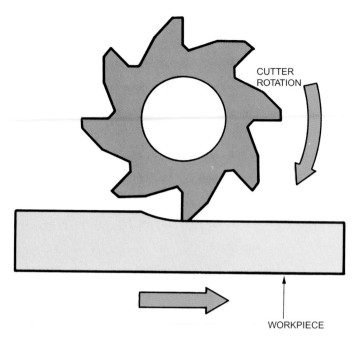

CUTTER
ROTATION

WORKPIECE

Figure 25-15. The operating principle of the horizontal milling machine.

Planer and Broach

Large flat surfaces are machined on a *planer*, Figure 25-16. It operates by moving the workpiece against a cutting tool or tools, Figure 25-17.

Broaching employs a multitooth cutting tool. Each tooth has a cutting edge that is a few thousandths of an inch higher than the one before and increases in size to the exact finished size required, Figure 25-18. The broach is pushed or pulled over the surface being machined. Many flat surfaces on automobile engines are broached. Broaching can also be utilized to do internal machining like cutting keyways, splines, and irregular-shaped openings, Figure 25-19.

Grinders

Grinding is an operation that removes material by rotating an abrasive wheel against the work-

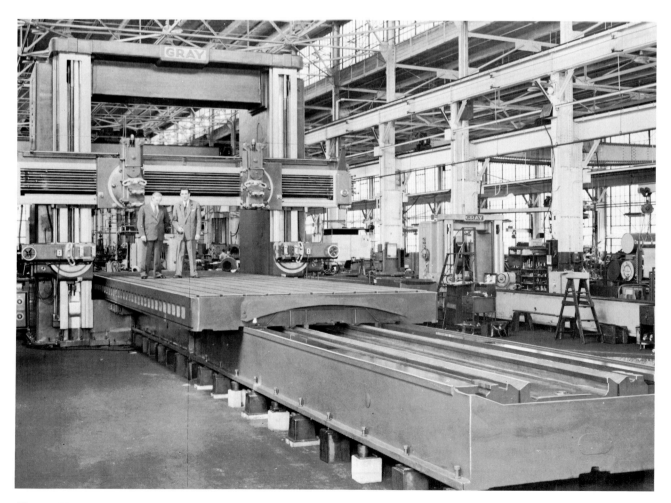

Figure 25-16. A 144 in. by 126 in. by 40 ft. (3.7 m by 3.2 m by 12.2 m) double-housing (two cutting heads) planer. The two people add perspective to the machine's size. (G.A. Gray Co.)

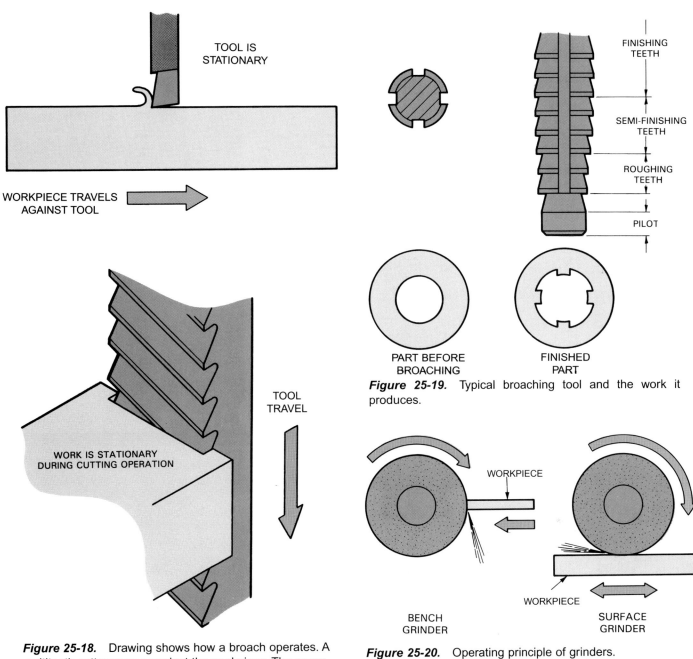

TOOL IS
STATIONARY

WORKPIECE TRAVELS
AGAINST TOOL

TOOL
TRAVEL

WORK IS STATIONARY
DURING CUTTING OPERATION

Figure 25-18.   Drawing shows how a broach operates. A multitooth cutter moves against the workpiece. The operation may be on a vertical or horizontal plane.

FINISHING
TEETH

SEMI-FINISHING
TEETH

ROUGHING
TEETH

PILOT

PART BEFORE
BROACHING

FINISHED
PART

Figure 25-19.   Typical broaching tool and the work it produces.

WORKPIECE

WORKPIECE

BENCH
GRINDER

SURFACE
GRINDER

Figure 25-20.   Operating principle of grinders.

piece, Figure 25-20. A **bench grinder** is the simplest and most widely used grinding machine, Figure 25-21.

Flat surfaces are ground to very close tolerances on a **surface grinder**, Figure 25-22. It is possible to surface grind hardened steel parts to tolerances of 1/100,000 (0.00001 in. or 0.0002 mm) with very fine (almost mirrorlike) surface finishes.

Round work can be ground to the same tolerances and surface finishes on a **cylindrical grinder**, Figure 25-23.

Figure 25-21.   Bench grinder.

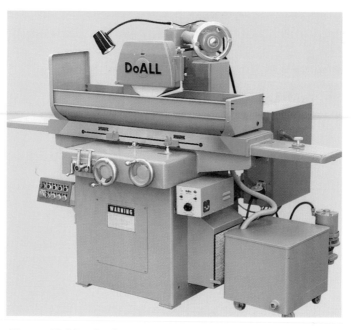

Figure 25-22.   Surface grinder. (The DoAll Co.)

Figure 25-23.   A cylindrical grinder with numerical control interface. Note the part being ground and the abrasive wheel doing the work. (K.O. Lee)

Other Machining Techniques

In addition to conventional machining, industry employs many unusual techniques to shape metal and other materials. There are electrical discharge machining (EDM), electrochemical machining (ECM), electron beam machining, laser machining, and ultrasonic machining to name but a few of the new techniques. They are usually used to machine materials that are difficult or impossible to machine by conventional methods. If interested, you might want to research these newer machining techniques.

Machine Tool Operation

Machine tools are controlled in a number of different manners.

1. *Manually.* A machinist feeds the cutter into the workpiece and manually guides it through the operations that will produce the part specified on the drawing, Figure 25-24.

2. *Numerical Control (NC).* A system of controlling a machine tool by means of numeric codes which direct commands to control devices (electric motors called *servos*) attached or built

Figure 25-24. The controls on this lathe are manually operated. (U.S. Army)

Figure 25-25. Numerically controlled (NC) machine tool. Instructions from magnetic tape or computer memory tell the servo motors when to start, in what direction to move, and how far they are to move the work being machined. Pictured is a plasma cutter, which produces parts with little waste. (Cybernation Cutting Systems, Inc.)

into the machine. Instructions from magnetic tape or directly from a computer, control electronic impulses that tell the servos when to start, in what direction to move, how far they are to move, cutter speed, and how deep the cutter is to cut, Figure 25-25.

3. *Computer Numerical Control (CNC).* CNC machine tools are designed around a microcomputer that is part of the machine's control unit (MCU). Using one of the many CNC languages, the machining instructions are entered directly into the system, Figure 25-26.

It is possible, using CAD, for an engineer/designer/drafter to design a part and see how it will fit and work with the other parts that make up the product. After the design has been confirmed, the computer can be programmed to analyze the geometry and calculate the tool paths required to make the part. The program is verified by machining a sample part from some inexpensive material like plastic or special wax.

The system that makes all of this possible is called CAD/CAM (computer-aided design/com-

Figure 25-26. A computer numerical controlled (CNC) bench-type verticle machining center is capable of all types of milling and drilling operations. (Light Machines Corp.)

puter-aided manufacturing). Some CAD/CAM operations employ robotics in their operation, Figure 25-27.

Shearing and Forming

Shearing is a process where the material (usually in sheet form) is cut to shape using actions similar to cutting paper with scissors.

Stamping is divided into two separate classifications: **cutting** and **forming**. The cutting operation is also known as **blanking**, Figure 25-28. It involves cutting flat sheets to the shape of the finished part. **Forming** is a process where flat metal is given three-dimensional form, Figure 25-29.

Casting Processes

Materials can also be given shape and form by reducing them to liquid in a molten state and pouring them into a mold of the desired shape. This process is called **casting**.

Sand Casting

In the **sand casting process** the mold that gives the molten metal shape is made of sand, Figure 25-30. This is one of the oldest metal forming techniques known.

A sand mold is made by packing sand in a box called a **flask**, around a **pattern**, Figure 25-31, of the shape to be cast, Figure 25-31. Since metal con-

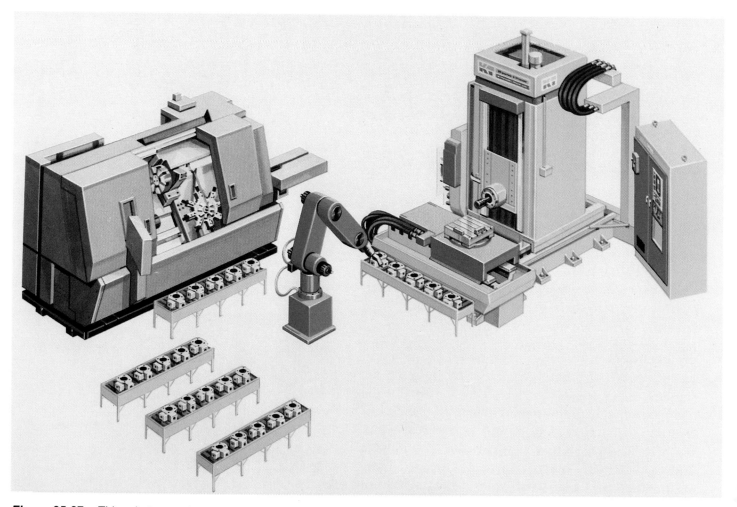

Figure 25-27. This robot, or automated cell, consists of several machine tools with robotic handling of work being machined. The cell becomes a fully automatic process through the application of robotics, power clamping of work, special tools, and other forms of automation. (Kearney & Trecker Corp.)

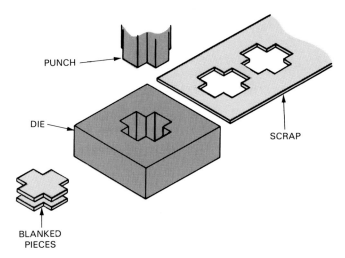

Figure 25-28. Blanking operation.

Figure 25-29. Forming gave flat metal sheet three-dimensional form for use in this panel. It also added strength and rigidity. (Lasedyne Div., Lumonics Corp.)

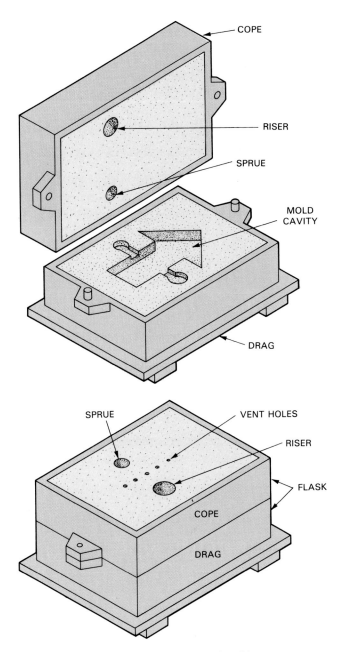

Figure 25-30. Parts of a typical sand mold.

tracts as it cools, patterns are made slightly oversize to allow for this shrinkage. After the pattern has been drawn (removed) from the sand, the mold halves, called the *cope* and the *drag*, are reassembled. A cavity or mold of the required shape remains in the sand.

Before assembling the mold, openings called *sprues, risers*, and *gates* are made in the mold. A *sprue* is an opening into which the molten metal is poured. *Risers* allow the hot gases to escape. *Gates* are trenches that run from the sprues and risers to the mold cavity. This permits molten metal to reach and fill the mold. Sand molds must be destroyed to remove the casting.

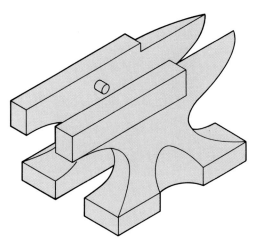

Figure 25-31. Two-piece pattern used to make cavity in a sand mold.

Permanent Mold Casting

Some molds used for casting metal are made from metal. These molds are called **permanent molds** because they do not have to be destroyed (like sand molds) to remove the casting. The process produces castings with a fine surface finish and a high degree of accuracy.

Auto pistons and fishing sinkers are familiar products made using the permanent mold process, Figure 25-32.

Figure 25-32. Familiar items made by the permanent mold process.

Die Casting

Die casting is a process where molten metal is forced into a **die** or **mold** under pressure, Figure 25-33. The pressure is maintained until the metal solidifies. The mold is opened and the casting is ejected, Figure 25-34. The mold or die is made of metal. Die castings are denser than sand castings. The castings are usually very accurate.

Plastics

Plastics are made into usable products by many different processes. Some plastics can be shaped directly into final form. Other plastics require several operations to transform them into usable products. Many of the processes for forming plastics are similar to those used to shape metal.

Compression Molding

In **compression molding** a measured amount of plastic is placed in a two-part mold. See Figure 25-35. The mold is heated and closed. Pressure is applied. Plastic, melted by the heat and pressure,

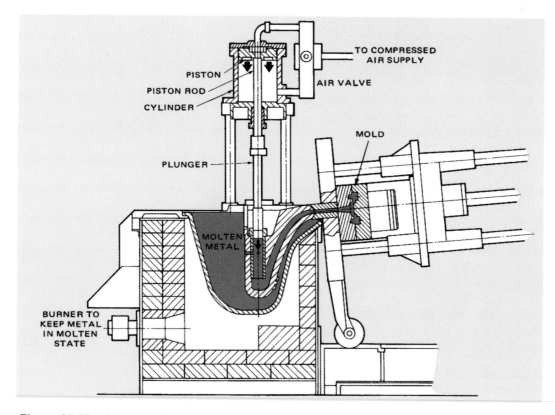

Figure 25-33. Diagram of a die casting machine. The molten metal is forced into the die cavity (mold) under pressure.

flows into all parts of the mold cavity. The mold is opened and the molded plastic is removed for trimming.

Transfer Molding

Transfer molding is similar to compression molding. It differs in that the plastic is heated in a separate section of the mold before it is forced into the mold cavity, Figure 25-36.

Injection Molding

Plastic model kits are one of the many plastic products made by *injection molding*, Figure 25-37. Plastic granules are placed in an injection molding machine and heated until they are soft enough to flow, Figure 25-38. The softened plastic is forced into the mold cavity, Figure 25-39. When cooled, the mold is opened, and the part is ejected.

Reinforced Plastics

Reinforced plastics are resins that have been strengthened with some type of fiberglass cloth, boron, and/or graphite, Figure 25-40. Resin-saturated fibers can be formed mechanically using heat and pressure. Body panels for the Chevrolet Corvette are made this way. They can also be laid-up by hand, Figure 25-41. Reinforced plastic products are usually formed in polished metal molds.

Figure 25-34. A die cast wheel emerging from the die. This wheel is more than a third lighter than a similar wheel produced by other casting techniques. (Kelsey-Hayes)

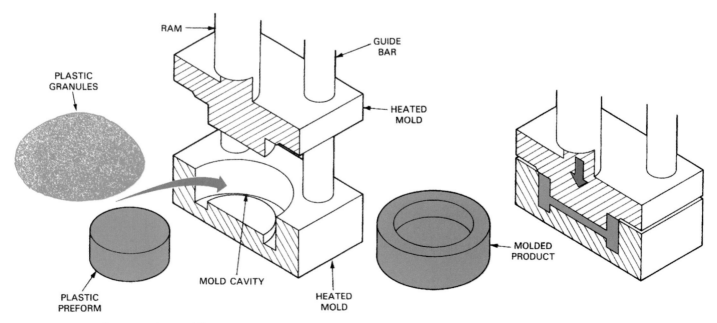

Figure 25-35. Compression molding process.

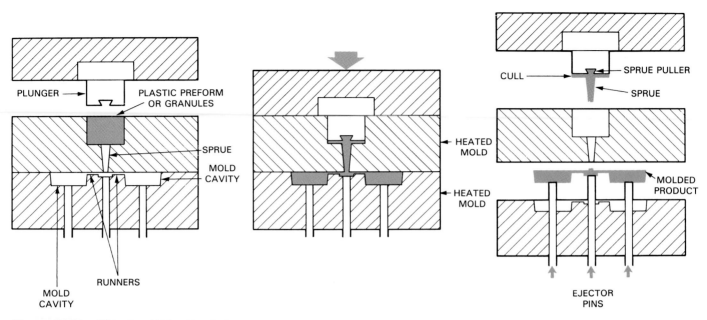

Figure 25-36. The transfer molding process.

Figure 25-37. Plastic model kits are typical products made by injection molding.

Figure 25-38. Injection molding machine. (Cincinnati Milacron)

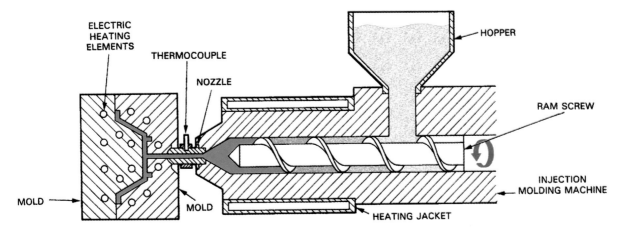

Figure 25-39. How injection molding is done.

Figure 25-40. Aircraft made from reinforced plastics (also known as *composites*). One engine has been mounted on the Avtek 400 prior to installing the elevators, flaps, and ailerons. The "Windecker Eagle," in the background, is the first and only all-composite aircraft to be awarded a Type certificate by the FAA. (Avtek Corp.)

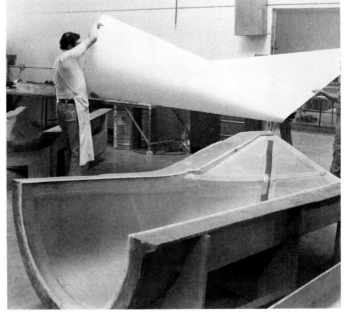

Figure 25-41. The aft section of fuselage and vertical stabilizer of the Avtek 400 being removed from the mold by two technicians. The plane's strong, lightweight airframe is made primarily of a honeycomb of "Nomex" aramid fiber sandwiched between skins of "Kevlar" aramid fiber. The plane's maximum weight is 5500 pounds, less than half that of most aluminum turboprop aircraft. (Avtek Corp.)

Stereolithography

A new plastics technology, called *stereolithography,* can produce complex design prototypes of castings and other components in hours instead of the days or weeks previously required. The three-dimensional hard plastic models can be viewed and studied to determine whether they are the best solution to the design problem. Since new models can be made quickly, design changes and modifications

Figure 25-42. Completed aircraft shown in preceding illustrations. Before being licensed for sale it must go through rigorous flight testing. (Avtek Corp.)

can be evaluated without the expense of making new patterns or molds.

The stereolithography process uses a computer-guided low-powered laser beam to harden a liquid photo-curable polymer plastic into the programmed shape, Figure 25-43. The process starts by creating the required design on a CAD system and orienting it in three-dimensional space, Figure 25-44. A support structure is added to hold the various elements of the design in place while the model is built up.

Thin vertical webs usually make up the support structure. The design is then sliced into cross sections of 0.005 in. to 0.020 in. (0.12 mm to 0.50 mm)

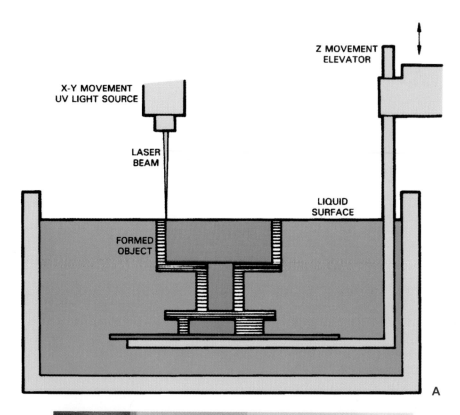

Figure 25-43. A—Basic diagram of how the stereolithography process works. B—This compact stereolithography unit provides a designer or engineer with a plastic prototype for evaluation. Note the three-dimensional image on the monitor. (Stratasys)

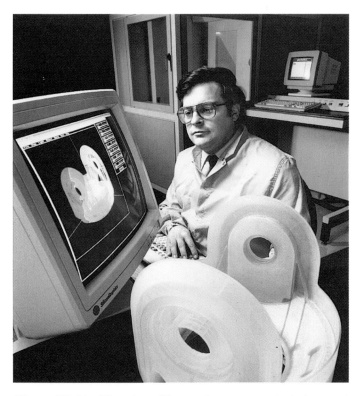

Figure 25-44. The stereolithography process starts by creating the required design on a CAD system. The CAD-designed part, in this case a yoke housing casting for a helicopter night-vision system, is shown on the computer screen. The prototype part made by the technique is shown in the foreground. (Hughes Electro-Optical & Data Systems Group)

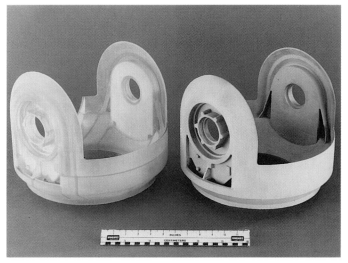

Figure 25-45. The casting design as made by the sterolithography process and the casting that was used in the night-vision system. The part design was proofed before the dies were made to produce it. (Hughes Electro-Optical & Data Systems Group)

A variation of the stereolithography technology has been developed that produces parts made of ceramic material, instead of plastic. Layers of a special ceramic powder are built up, using an inkjet-style printer to spray a quick-hardening binder to solidify each layer. The technology has been used to quickly produce shell molds for casting metals. The mold making process is called ***direct shell production casting (DSPC).***

thickness. The design data is downloaded into the stereolithography machine, which operates similarly to a CNC machine tool.

The machine's control unit guides a fine laser beam onto the surface of the vat containing the liquid photo-curable polymer. The liquid solidifies wherever it is struck by the laser beam. The model is constructed from the bottom up, on a platform located just below the surface of the liquid plastic. After each section is formed by the scanning laser, the platform drops a programmed distance. The process continues until the entire model is formed, Fig. 25-45.

The finished part requires ultraviolet curing. It is removed from the vat and excess plastic drained off. When completely cured, the model is removed from the platform and the support structure is clipped away. It can be finished by filing, sanding, and polishing, Fig. 25-46. Paint or dye can be applied.

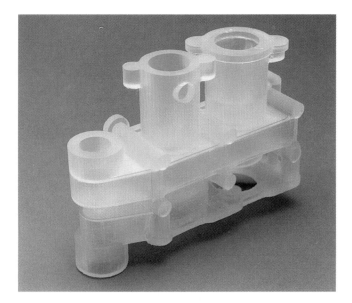

Figure 25-46. A complex part produced for evaluation by the stereolithography process. (Garrett Engine Division, Allied Signal Aerospace)

Drafting Vocabulary

Bench grinder
Blanking
Bore
Broach
Computer-aided
  design/computer-aided
  manufacturing
  (CAD/CAM)
Casting
Chamfer
Computer numerical
  control (CNC)
Compression molding
Cope
Countersinking
Cutting
Cylindrical grinding
Die
Die casting
Direct shell production
  casting (DSPC)
Drag
Drill
Drill press

Electrical discharge
  machining (EDM)
Electrochemical
  machining (ECM)
Electron beam machining
Electronic impulses
Flask
Forming
Gate
Graphite
Grinding
Horizontal milling
  machine
Injection molding
Keyway
Laser machining
Lathe
Machine tool
Milling machine
Mold
Mold cavity
Numerical control (NC)
Pattern
Permanent molds

Planer
Plastic
Ream
Reinforced plastics
Resin
Riser
Sand casting process
Servos
Shearing
Spline
Sprue
Stamping
Stereolithography
Surface grinder
Tapping
Transfer molding
Turning cell
Turret
Turret lathe
Twist drill
Ultrasonic machining
Vertical milling machine

Test Your Knowledge—Unit 25

Please do not write in the book. Place your answers on another sheet of paper.

1. Define "machine tool."
2. Make a sketch showing how a lathe operates.
3. Make a sketch showing how a drill press operates.
4. List three operations that can be performed on a drill press.
5. Make a sketch showing how a milling machine operates.
6. Make a sketch showing how a planer operates.
7. List three machine tools used to machine flat surfaces.
8. Make a sketch showing how a surface grinder operates.

Match each of the lettered descriptions with the appropriate numbered term.

9. Pattern
10. Sand casting
11. Manual machine operation
12. NC
13. CNC
14. CAD/CAM
15. Blanking
16. Forming
17. Permanent mold casting
18. Compression molding
19. Transfer molding
20. Injection molding

A. Sheet metal is given three-dimensional shape.

B. The mold is not destroyed when the casting is removed.

C. The machinist guides the machine through machining operation by hand.

D. Mold must be destroyed to remove casting.

E. Softened plastic is forced into the mold cavity by a revolving screw. Plastic model kits are made by this process.

F. A measured amount of plastic is placed into a two-part mold. The mold is heated and closed. Pressure is applied. The heat and pressure melts the plastic and forces it into all parts of the mold.

G. Computer-assisted numerical control. Machine movement is controlled by a computer that is part of the machine's control unit.

H. Used to make the mold cavity in a sand mold.

I. Cutting sheet metal to shape.

J. Computer-aided design/computer-aided manufacturing.

K. Machine operations controlled by electric motors called servos.

L. Plastic is heated in a separate section of the mold before it is forced into mold cavity.

Outside Activities

1. Prepare a bulletin board using clippings which illustrate basic machine tools.

2. Secure samples of work manufactured on various machine tools including the lathe, drill press, milling machine, and surface grinder. Make a display and call out the advantages of using the different machines.

3. Bring into class examples of products made that were cast in sand molds, die castings, and permanent molds. Describe the differences among the products.

4. Report to the class, using visual aids, the developments in Flexible Manufacturing Stations.

Quality control plays a major role in manufacturing processes. A—Inspecting body panels for fit on a test model vehicle before releasing the dies for production. (Ford Motor Company) B—Die casting being inspected to ensure it will meet design specifications. (American Foundrymen's Society)

Useful Information

ANSI DRAFTING STANDARDS

The American National Standards Institute (ANSI) revised the standards to be used on drawings. The new symbols and standards will replace the older abbreviations of certain English words. These new symbols will make drafting a more international language. Refer to the illustration on page 151 (Figure 9-20).

Drafters will use the symbols to dimension their drawings. In addition, drafters will use the symbols in the notes added to drawings. As older drawings are revised, drafters will add them to the revisions. CAD operators will be able to select the symbols from digitizer libraries or from menus on screens.

ANSI prefers unidirectional dimensioning over the aligned system of placing dimensions on a drawing. Fractional inch dimensions are only recommended for specialized fields which do not require decimal accuracy such as furniture making, sheet metal fabrication, and forging and foundry areas.

ANSI standards call for threads to be drawn using the simplified method.

The part's material should be described in the Bill of Materials. Thus, section lines should be constructed at a 45 degree angle regardless of the part's material.

Drawings done to ANSI standards should contain a note "Drawn in accordance to ANSI Y14.5M."

Major symbols and their meanings are shown in the chart. Changes to be noted include:
     The diameter symbol appears before each diameter value.
     The radius symbol appears before the radius value.
     New symbols are used to describe holes.

This is a very brief description of standards which industry used for geometric dimensioning and tolerancing. The new symbols will improve the international understanding of drafting. Additional information can be obtained from the American National Standards Institute, 1430 Broadway, New York, New York, 10018 or from the American Society of Mechanical Engineers, 345 East 47th Street, New York, New York, 10017.

Meaning		ANSI Standards	Old Method
Not to scale		12	12 or NTS
Diameter		⌀26	26 DIA
Radius		R13	13 R
Reference dimension	⏊	(14)	14 REF
Times/places		6X	6 PLACES
Counterbore	⏊	⌴⌀7	7 CBORE
Countersink		⋁⌀3	3 CSK
Deep	⏊	⫪14	14 DP
Square		☐ 14	14 SQ
		4X 1/2—12 UNC—3B	1/2—12 UNC—3B—4 HOLES
		⌀.4062 ⫪1.25	13/32 DRILL—1 1/4 DP
		⌴⌀.50 ⫪.062 —→	1/2 C BORE—1/16 DP —→

MEASUREMENT SYSTEMS

ENGLISH SYSTEM

MEASURES OF TIME
60 sec. = 1 min.
60 min. = 1 hr.
24 hr. = 1 day
365 dy. = 1 common yr.
366 dy. = 1 leap yr.

DRY MEASURES
2 pt.	= 1 qt.
8 qt.	= 1 pk.
4 pk.	= 1 bu.
2150.42 cu. in.	= 1 bu.

MEASURES OF LENGTH
12 in.	= 1 ft.
3 ft.	= 1 yd.
5 1/2 yd.	= 1 rod
320 rods	= 1 mile
5,280 ft.	= 1 mile
1,760 yd.	= 1 mile
6,080 ft.	= 1 knot

LIQUID MEASURES
16 fluid oz.	= 1 pt.
2 pt.	= 1 qt.
32 fl. oz.	= 1 qt.
4 qt.	= 1 gal.
31 1/2 gal.	= 1 bbl.
231 cu. in.	= 1 gal.
7 1/2 gal.	= 1 cu. ft.

MEASURES OF AREA
144 sq. in.	= 1 sq. ft.
9 sq. ft.	= 1 sq. yd.
30 1/4 sq. yd.	= 1 sq. rod
160 sq. rods	= 1 acre
640 acres	= 1 sq. mile

MEASURES OF WEIGHT (Avoirdupois)
7,000 grains (gr.)	= 1 lb.
16 oz.	= 1 lb.
100 lb.	= 1 cwt.
2,000 lb.	= 1 short ton
2,240 lb.	= 1 long ton

MEASURES OF VOLUME
1,728 cu. in.	= 1 cu. ft.
27 cu. ft.	= 1 cu. yd.
128 cu. ft.	= 1 cord

METRIC SYSTEM

The basic unit of the metric system is the meter (m). The meter is exactly 39.37 in. long. This is 3.37 in. longer than the English yard. Units that are multiples or fractional parts of the meter are designated as such by prefixes to the word "meter". For example:

1 millimeter (mm.) = 0.001 meter or 1/1000 meter
1 centimeter (cm.) = 0.01 meter or 1/100 meter
1 decimeter (dm.) = 0.1 meter or 1/10 meter
1 meter (m.)
1 decameter (dkm.) = 10 meters
1 hectometer (hm.) = 100 meters
1 kilometer (km.) = 1000 meters

These prefixes may be applied to any unit of length, weight, volume, etc. The meter is adopted as the basic unit of length, the gram for mass, and the liter for volume.

In the metric system, area is measured in square kilometers (sq. km. or km.^{2}), square centimeters (sq. cm. or cm.^{2}), etc. Volume is commonly measured in cubic centimeters, etc. One liter (l) is equal to 1,000 cubic centimeters.

The metric measurements in most common use are shown in the following tables:

MEASURES OF LENGTH
10 millimeters	= 1 centimeter
10 centimeters	= 1 decimeter
10 decimeters	= 1 meter
1000 meters	= 1 kilometer

MEASURES OF WEIGHT
100 milligrams	= 1 gram
1000 grams	= 1 kilogram
1000 kilograms	= 1 metric ton

MEASURES OF VOLUME
1000 cubic centimeters	= 1 liter
100 liters	= 1 hectoliter

METRIC PREFIXES, EXPONENTS, AND SYMBOLS

DECIMAL FORM	EXPONENT OR POWER	PREFIX	PRONUNCIATION	SYMBOL	MEANING
1 000 000 000 000 000 000	$= 10^{18}$	exa	ex'a	E	quintillion
1 000 000 000 000 000	$= 10^{15}$	peta	pet'a	P	quadrillion
1 000 000 000 000	$= 10^{12}$	tera	tĕr'á	T	trillion
1 000 000 000	$= 10^{9}$	giga	ji'gá	G	billion
1 000 000	$= 10^{6}$	mega	mĕg'á	M	million
1 000	$= 10^{3}$	kilo	kĭl'ō	k	thousand
100	$= 10^{2}$	hecto	hĕk'to	h	hundred
10	$= 10^{1}$	deka	dĕk'a	da	ten
1					base unit
0.1	$= 10^{-1}$	deci	dĕs'ĭ	d	tenth
0.01	$= 10^{-2}$	centi	sĕn'ti	c	hundredth
0.001	$= 10^{-3}$	milli	mĭl'ĭ	m	thousandths
0.000 001	$= 10^{-6}$	micro	mi'krō	μ	millionth
0.000 000 001	$= 10^{-9}$	nano	năn'ō	n	billionth
0.000 000 000 001	$= 10^{-12}$	pico	pēc'ō	p	trillionth
0.000 000 000 000 001	$= 10^{-15}$	femto	fĕm'tō	f	quadrillionth
0.000 000 000 000 000 001	$= 10^{-18}$	atto	ăt'tō	a	quintillionth

Most commonly used

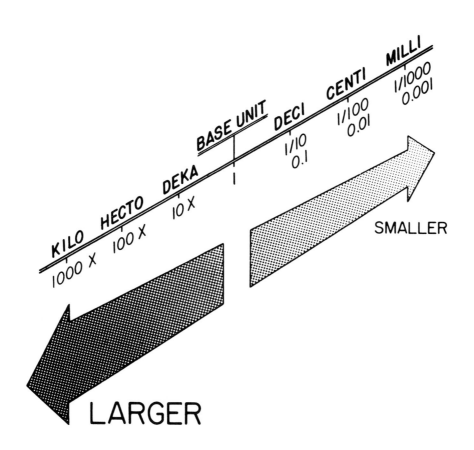

SI UNITS & CONVERSIONS

PROPERTY	UNIT	SYMBOL	EXACT CONVERSION			APPROXIMATE EQUIVALENCY
			FROM	TO	MULTIPLY BY	
length	meter	m	inch	mm	2.540×10	$25mm = 1$ in.
	centimeter	cm	inch	cm	2.540	$300mm = 1$ ft.
	millimeter	mm	foot	mm	3.048×10^{-4}	
mass	kilogram	kg	ounce	g	2.835×10	$2.8g - 1$ oz.
	gram	g	pound	kg	4.536×10^{-1}	$kg = 2.2$ lbs. $= 35$ oz.
	tonne (megagram)	t	ton (2000 lb)	kg	9.072×10^{2}	$1t = 2200$ lbs.
density	kilogram per cub. meter	kg/m^3	pounds per cu. ft.	kg/m^3	1.602×10	$16kg/M^3 = 1$ lb./ft^3
temperature	deg. Celsius	$^{\circ}C$	deg. Fahr.	$^{\circ}C$	$(^{\circ}F - 32) \times 5/9$	$0^{\circ}C = 32^{\circ}F$ $100^{\circ}C = 212^{\circ}F$
area	square meter	m^2	sq. inch	mm^2	6.452×10^{2}	$645mm^2 = 1$ in.^{2}
	square millimeter	mm^2	sq. ft.	m^2	9.290×10^{-2}	$1m^2 = 11$ ft.^{2}
volume	cubic meter	m^3	cu. in.	mm^3	1.639×10^{4}	$16400mm^3 = 1$ in.^{3}
	cubic centimeter	cm^3	cu. ft.	m^3	2.832×10^{-2}	$1m^3 = 35$ ft.^{3}
	cubic millimeter	mm^3	cu. yd.	m^3	7.645×10^{-1}	$1m^3 = 1.3$ yd.^{3}
force	newton	N	ounce (Force)	N	2.780×10^{-1}	$1N = 3.6$ oz.
	kilonewton	kN	pound (Force)	kN	4.448×10^{-3}	$4.4N = 1$ lb.
	meganewton	MN	Kip	MN	4.448	$1kN = 225$ lb.
stress	megapascal	MPa	pound/in^2 (psi)	MPa	6.895×10^{-3}	$1MPa = 145$ psi
			Kip/in^2 (ksi)	MPa	6.895	$7MPa = 1$ ksi
torque	newton-meters	$N.m$	in-ounce	N.m	7.062×10^{3}	$1N.m = 140$ in.oz.
			in.pound	N.m	1.130×10^{-1}	$1N.m = 9$ in.lb.
			ft pound	N.m	1.356	$1N.m = .75$ ft.lb.
						$1.4N.m = 1$ ft.lb.

FRACTIONAL INCHES
INTO DECIMALS AND MILLIMETERS

INCH	DECIMAL INCH	MILLIMETER	INCH	DECIMAL INCH	MILLIMETER
1/64	0.0156	0.3967	33/64	0.5162	13.0968
1/32	0.0312	0.7937	17/32	0.5312	13.4937
3/64	0.0468	1.1906	35/64	0.5468	13.8906
1/16	0.0625	1.5875	9/16	0.5625	14.2875
5/64	0.0781	1.9843	37/64	0.5781	14.6843
3/32	0.0937	2.3812	19/32	0.5937	15.0812
7/64	0.1093	2.7781	39/64	0.6093	15.4781
1/8	0.125	3.175	5/8	0.625	15.875
9/64	0.1406	3.5718	41/64	0.6406	16.2718
5/32	0.1562	3.9687	21/32	0.6562	16.6687
11/64	0.1718	4.3656	43/64	0.6718	17.0656
3/16	0.1875	4.7625	11/16	0.6875	17.4625
13/64	0.2031	5.1593	45/64	0.7031	17.8593
7/32	0.2187	5.5562	23/32	0.7187	18.2562
15/64	0.2343	5.9531	47/64	0.7343	18.6531
1/4	0.25	6.5	3/4	0.75	19.05
17/64	0.2656	6.7468	49/64	0.7656	19.4468
9/32	0.2812	7.1437	25/32	0.7812	19.8437
19/64	0.2968	7.5406	51/64	0.7968	20.2406
5/16	0.3125	7.9375	13/16	0.8125	20.6375
21/64	0.3281	8.3343	53/64	0.8281	21.0343
11/32	0.3437	8.7312	27/32	0.8437	21.4312
23/64	0.3593	9.1281	55/64	0.8593	21.8281
3/8	0.375	9.525	7/8	0.875	22.225
25/64	0.3906	9.9218	57/64	0.8906	22.6218
13/32	0.4062	10.3187	29/32	0.9062	23.0187
27/64	0.4218	10.7156	59/64	0.9218	23.4156
7/16	0.4375	11.1125	15/16	0.9375	23.8125
29/64	0.4531	11.5093	61/64	0.9531	24.2093
15/32	0.4687	11.9062	31/32	0.9687	24.6062
31/64	0.4843	12.3031	63/64	0.9843	25.0031
1/2	0.50	12.7	1	1.0000	25.4

PREFERRED METRIC SIZES FOR ENGINEERS

SIZES 1 mm to 10 mm

1st	2nd	3rd
1		
	1.1	
1.2		
		1.3
	1.4	
		1.5
1.6		
		1.7
	1.8	
		1.9
2		
		2.1
	2.2	
		2.4
2.5		
		2.6
	2.8	
3		
		3.2
	3.5	
		3.8
4		
		4.2
	4.5	
		4.8
5		
		5.2
	5.5	
		5.8
6		
		6.5
	7	
		7.5
8		
		8.5
	9	
		9.5
10		

SIZES 10 mm to 100 mm

1st	2nd	3rd
10		
	11	
12		
		13
	14	
		15
16		
		17
	18	
		19
20		
		21
	22	
		23
		24
25		
		26
	28	
30		
	32	
		34
	35	
		36
	38	
40		
	42	
		44
	45	
		46
	48	
50		
	52	
		54
	55	
		56
	58	
60		
		62
	65	
		68
	70	
		72
	75	
		78
80		
		82
	85	
		88
	90	
		92
	95	
		98
100		

SIZES 100 mm to 1000 mm

1st	2nd	3rd
100		
		105
110		
		115
120		
		125
	130	
		135
140		
		145
	150	
		155
160		
		165
	170	
		175
180		
		185
	190	
		195
200		
	210	
	220	
		230
	240	
250		
	260	
		270
280		
		290
300		
	320	
		340
350		
		360
	380	
400		
	420	
		440
450		
		460
	480	
500		
	520	
		540
550		
		560
	580	
600		
	620	
		640
	650	
		660
	680	
700		
	720	
		740
750		
		760
	780	
800		
		820
	850	
		880
900		
		920
	950	
		980
1000		

A COMPARISON OF INCH SERIES AND METRIC THREADS

INCH SERIES			METRIC			
Size	Dia. in.	TPI	Size	Dia. in.	Pitch (mm)	TPI (Approx.)
No. 0	.060	80	M1.4	.055	0.3	85
			M1.6	.063	0.35(a)	74
No. 1	.073	64				
		72	M2	.079	0.4(a)	64
No. 2	0.86	56				
		64	M2.5	.098	0.45(a)	56
No. 3	.099	48				
		56				
No. 4	.112	40				
		48	M3	.118	0.5(a)	51
No. 5	.125	40				
		44				
No. 6	.138	32	M3.5	.138	0.6(a)	42
		40	M4	.158	0.7(a)	36
No. 8	.164	32				
		36				
No. 10	.190	24				
		32	M5	.197	0.8(a)	32
			M6	.236	1.0	25
1/4	.250	20				
		28	M6.3	.248	1.0(a)	25
5/16	.312	18	M7	.276	1.0	25
		24	M8	.315	1.25(a)	20
					1.0	25
3/8	.375	16				
		24	M10	.394	1.5(a)	18
					1.25	20
7/16	.437	14				
		20	M12	.472	1.75(a)	14.5
					1.25	20
1/2	.500	13				
		20				
9/16	.562	12	M14	.551	2.0(a)	12.5
		18			1.5	17
5/8	.625	11				
		18	M16	.630	2.0(a)	12.5
					1.5	17
			M18	.709	2.5	10
					1.5	17
3/4	.750	10				
		16	M20	.787	2.5(a)	10
					1.5	17
			M22	.866	2.5	10
					1.5	17
7/8	.875	9				
		14	M24	.945	3.0(a)	8.5
					2.0	12.5
1	1.000	8				
		14	M27	1.063	3.0	8.5
					2.0	12.5

(a) Preferred series for use in the United States

A comparison of inch series and metric threads shows that even though many of the threads are similar in size and pitch they are not interchangeable. Extreme care must be observed to keep the two series separate.

CONVERSION TABLES

TO REDUCE	MULTIPLY BY	TO REDUCE	MULTIPLY BY
LENGTH			
miles to km	1.61	km to miles	0.62
miles to m	1609.35	m to miles	0.00062
yd. to m	0.9144	m to yd.	1.0936
in. to cm	2.54	cm to in.	0.3937
in. to mm	25.4	mm to in.	0.03937
VOLUME			
cu. in. to cm^3 or mL	16.387	cm^3 to cu. in.	0.061
cu. in. to L	0.0164	l to cu. in.	61.024
gal. to L	3.785	l to gal.	0.264
WEIGHT			
lb. to kg	0.4536	kg to lb.	2.2
oz. to gm	28.35	gm to oz.	0.0353
gr. to gm	0.0648	gm to gr.	15.432

METRIC SIZE DRAFTING SHEETS

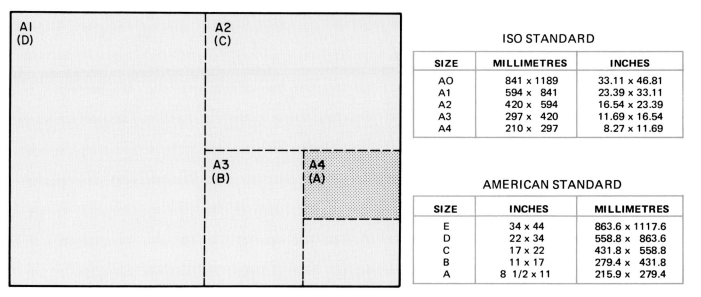

ISO STANDARD

SIZE	MILLIMETRES	INCHES
AO	841 x 1189	33.11 x 46.81
A1	594 x 841	23.39 x 33.11
A2	420 x 594	16.54 x 23.39
A3	297 x 420	11.69 x 16.54
A4	210 x 297	8.27 x 11.69

AMERICAN STANDARD

SIZE	INCHES	MILLIMETRES
E	34 x 44	863.6 x 1117.6
D	22 x 34	558.8 x 863.6
C	17 x 22	431.8 x 558.8
B	11 x 17	279.4 x 431.8
A	8 1/2 x 11	215.9 x 279.4

Left. Metric size drafting paper sheets are exactly proportional in size. Right. Charts show millimetre sizes and customary inch equivalent sizes of ISO papers.

ABBREVIATIONS

(SOME FORMER ABBREVIATIONS ARE NOW SYMBOLS)

Across flats	ACR FLT	Inside diameter	ID
Centers	CTR	Left hand	LH
Center line	CL	Material	MATL
Centimetre	cm	Metre	m
Chamfer	CHAM	Millimetre	mm
Counterbore	CBORE	Number	NO
Countersink	CSK	Outside diameter	OD
Countersunk head	CSK H	Pitch diameter	PD
Diameter (before dimension)	Ø	Radius	R
Diameter (in a note)	DIA	Right hand	RH
Drawing	DWG	Round	RD
Figure	FIG	Square (before dimension)	□
Hexagon	HEX	Square (in a note)	SQ
Hexagonal head	HEX HD	Thread	THD

FORMULAS

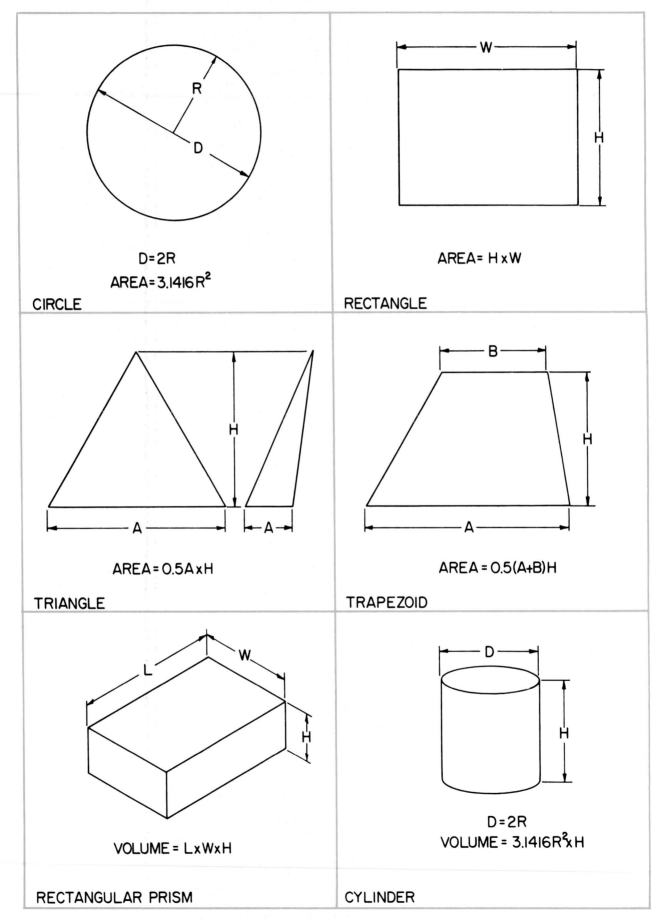

CIRCLE

D = 2R
AREA = 3.1416 R²

RECTANGLE

AREA = H x W

TRIANGLE

AREA = 0.5 A x H

TRAPEZOID

AREA = 0.5 (A+B) H

RECTANGULAR PRISM

VOLUME = L x W x H

CYLINDER

D = 2R
VOLUME = 3.1416 R² x H

FORMULAS

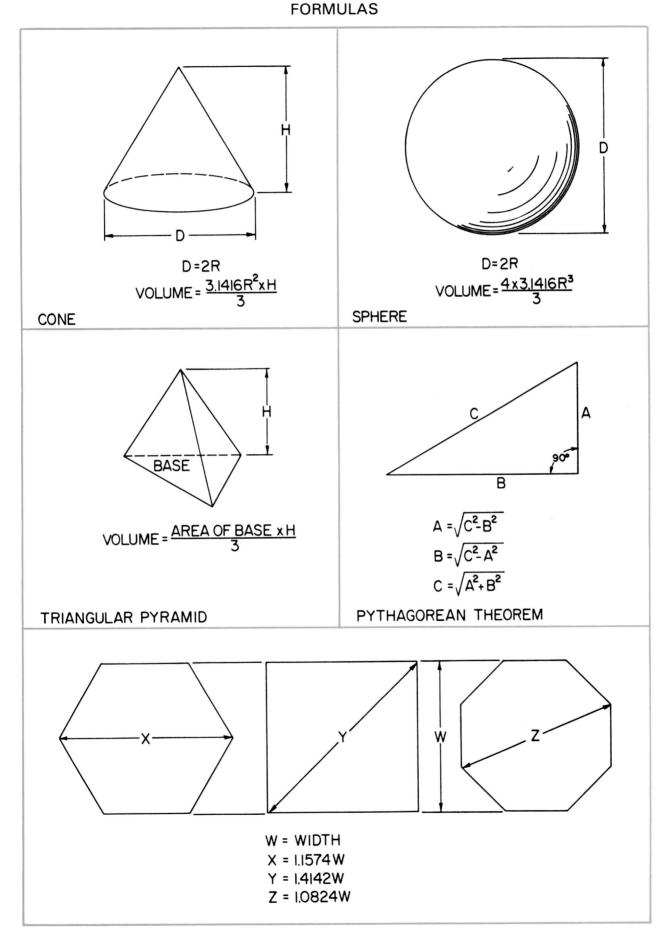

CONE

$D = 2R$

$$\text{VOLUME} = \frac{3.1416 R^2 \times H}{3}$$

SPHERE

$D = 2R$

$$\text{VOLUME} = \frac{4 \times 3.1416 R^3}{3}$$

TRIANGULAR PYRAMID

$$\text{VOLUME} = \frac{\text{AREA OF BASE} \times H}{3}$$

PYTHAGOREAN THEOREM

$$A = \sqrt{C^2 - B^2}$$
$$B = \sqrt{C^2 - A^2}$$
$$C = \sqrt{A^2 + B^2}$$

W = WIDTH
X = 1.1574 W
Y = 1.4142 W
Z = 1.0824 W

STANDARD ABBREVIATIONS
FOR USE ON DRAWINGS

A

Abrasive	ABRSV
Accessory	ACCESS
Accumulator	ACCUMR
Acetylene	ACET
Across Flats	ACR FLT
Actual	ACT
Actuator	ACTR
Addendum	ADD
Adhesive	ADH
Adjust	ADJ
Advance	ADV
Aeronautic	AERO
Alclad	CLAD
Alignment	ALIGN
Allowance	ALLOW
Alloy	ALY
Alteration	ALT
Alternate	ALT
Alternating Current	AC
Aluminum	AL
American National Standards Institute	ANSI
American Wire Gage	AWG
Ammeter	AMM
Amplifier	AMPL
Anneal	ANL
Anodize	ANOD
Antenna	ANT
Approved	APPD
Approximate	APPROX
Arrangement	ARR
As Required	AR
Assemble	ASSEM
Assembly	ASSY
Automatic	AUTO
Auxiliary	AUX
Average	AVG

B

Babbit	BAB
Base Line	BL
Battery	BAT
Bearing	BRG
Bend Radius	BR
Bevel	BEV
Bill of Material	B/M
Blueprint	BP or B/P
Bolt Circle	BC
Bracket	BRKT
Brass	BRS
Brazing	BRZG
Brinnell Hardness Number	BHN
Bronze	BRZ
Brown & Sharpe (Gage)	B&S
Burnish	BNH
Bushing	BUSH

C

Cabinet	CAB
Calculated	CACL
Cancelled	CANC
Capacitor	CAP
Capacity	CAP
Carburize	CARB
Case Harden	CH
Casting	CSTG
Cast Iron	CI
Cathode-Ray Tube	CRT
Center	CTR
Center to Center	C to C
Centigrade	C
Centimeter	CM
Centrifugal	CENT
Chamfer	CHAM
Circuit	CKT
Circular	CIR
Circumference	CIRC
Clearance	CL
Clockwise	CW
Closure	CLOS
Coated	CTD
Cold-Drawn Steel	CDS
Cold-Rolled Steel	CRS
Color Code	CC
Commercial	COMM
Concentric	CONC
Condition	COND
Conductor	CNDCT
Contour	CTR
Control	CONT
Copper	COP
Counterbore	CBORE
Counterclockwise	CCW
Counter-Drill	CDRILL
Countersink	CSK
Countersunk head	CSK H
Cubic	CU
Cylinder	CYL

D

Datum	DAT
Decimal	DEC
Decrease	DECR
Degree	DEG
Detail	DET
Detector	DET
Developed Length	DL
Developed Width	DW
Deviation	DEV
Diagonal	DIAG
Diagram	DIAG
Diameter	DIA
Diameter Bolt Circle	DBC
Diametral Pitch	DP
Dimension	DIM
Direct Current	DC
Disconnect	DISC
Double-Pole Double-Throw	DPDT
Double-Pole Single-Throw	DPST
Dowel	DWL
Draft	DFT
Drafting Room Manual	DRM
Drawing	DWG
Drawing Change Notice	DCN
Drill	DR
Drop Forge	DF
Duplicate	DUP

E

Each	EA
Eccentric	ECC
Effective	EFF
Electric	ELEC
Enclosure	ENCL
Engine	ENG
Engineer	ENGR
Engineering	ENGRG
Engineering Change Order	ECO
Engineering Order	EO
Equal	EQ
Equivalent	EQUIV
Estimate	EST

F

Fabricate	FAB
Figure	FIG
Fillet	FIL
Finish	FIN
Finish All Over	FAO
Fitting	FTG
Fixed	FXD
Fixture	FIX
Flange	FLG

STANDARD ABBREVIATIONS
(Continued)

Flat Head	FHD	J		Modification	MOD
Flat Pattern	F/P			Mold Line	ML
Flexible	FLEX	Joggle	JOG	Motor	MOT
Fluid	FL	Junction	JCT	Mounting	MTG
Forged Steel	FST			Multiple	MULT
Forging	FORG				
Furnish	FURN	K			
				N	
		Keyway	KWY		
G				Nickel Steel	NS
				Nomenclature	NOM
Gage	GA	L		Nominal	NOM
Gallon	GAL			Normalize	NORM
Galvanized	GALV	Laboratory	LAB	Not to Scale	NTS
Gasket	GSKT	Lacquer	LAQ	Number	NO.
Generator	GEN	Laminate	LAM		
Grind	GRD	Left Hand	LH		
Ground	GRD	Length	LG	O	
		Letter	LTR		
		Limited	LTD	Obsolete	OBS
H		Limit Switch	LS	Opposite	OPP
		Linear	LIN	Oscilloscope	SCOPE
Half-Hard	1/2H	Liquid	LIQ	Ounce	OZ
Handle	HDL	List of Material	L/M	Outside Diameter	OD
Harden	HDN	Long	LG	Over-All	OA
Head	HD	Low Carbon	LC		
Heat Treat	HT TR	Low Voltage	LV		
Hexagon	HEX	Lubricate	LUB	P	
Hexagonal head	HEX HD				
High Carbon Steel	HCS			Package	PKG
High Frequency	HF	M		Parting Line (Castings)	PL
High Speed	HS			Parts List	P/L
Horizontal	HOR	Machine(ing)	MACH	Pattern	PATT
Hot-Rolled Steel	HRS	Magnaflux	M	Piece	PC
Hour	HR	Magnesium	MAG	Pilot	PLT
Housing	HSG	Maintenance	MAINT	Pitch	P
Hydraulic	HYD	Major	MAJ	Pitch Circle	PC
		Malleable	MALL	Pitch Diameter	PD
		Malleable Iron	MI	Plan View	PV
I		Manual	MAN	Plastic	PLSTC
		Manufacturing (ed, er)	MFG	Plate	PL
Identification	IDENT	Mark	MK	Pneumatic	PNEU
Inch	IN	Master Switch	MS	Port	P
Inclined	INCL	Material	MATL	Positive	POS
Include, Including,	INCL	Maximum	MAX	Potentiometer	POT
Inclusive		Measure	MEAS	Pounds Per Square Inch	PSI
Increase	INCR	Mechanical	MECH	Pounds Per Square	PSIG
Independent	INDEP	Medium	MED	Inch Gage	
Indicator	IND	Meter	MTR	Power Amplifier	PA
Information	INFO	Middle	MID	Power Supply	PWR SPLY
Inside Diameter	ID	Military	MIL	Pressure	PRESS
Installation	INSTL	Millimeter	MM	Primary	PRI
International Standards	ISO	Minimum	MIN	Process, Procedure	PROC
Organization		Miscellaneous	MISC	Product, Production	PROD
Interrupt	INTER				

STANDARD ABBREVIATIONS
(Continued)

Q

Quality	QUAL
Quantity	QTY
Quarter-Hard	1/4H

R

Radar	RDR
Radio	RAD
Radio Frequency	RF
Radius	RAD or R
Ream	RM
Receptacle	RECP
Reference	REF
Regular	REG
Regulator	REG
Release	REL
Required	REQD
Resistor	RES
Revision	REV
Revolutions Per Minute	RPM
Right Hand	RH
Rivet	RIV
Rockwell Hardness	RH
Round	RD

S

Schedule	SCH
Schematic	SCHEM
Screw	SCR
Screw Threads	
American National Coarse	NC
American National Fine	NF
American National Extra Fine	NEF
American National 8 Pitch	8N
American Standard Taper Pipe	NTP
American Standard Straight Pipe	NPSC
American Standard Taper (Dryseal)	NPTF
American Standard Straight (Dryseal)	NPSF
Unified Screw Thread Coarse	UNC

Unified Screw Thread Fine	UNF
Unified Screw Thread Extra Fine	UNEF
Unified Screw Thread 8 Thread	8UN
Section	SECT
Sequence	SEQ
Serial	SER
Serrate	SERR
Sheathing	SHTHG
Sheet	SH
Silver Solder	SILS
Single-Pole Double-Throw	SPDT
Single-Pole Single-Throw	SPST
Society of Automotive Engineers	SAE
Solder	SLD
Solenoid	SOL
Speaker	SPKR
Special	SPL
Specification	SPEC
Spot Face	SF
Spring	SPG
Square	SQ
Stainless Steel	SST
Standard	STD
Steel	STL
Stock	STK
Support	SUP
Switch	SW
Symbol	SYM
Symmetrical	SYM
System	SYS

T

Tabulate	TAB
Tangent	TAN
Tapping	TAP
Technical Manual	TM
Teeth	T
Television	TV
Temper	TEM
Temperature	TEM
Tensile Strength	TS
Thick	THK
Thread	THD
Through	THRU

Tolerance	TOL
Tool Steel	TS
Torque	TOR
Total Indicator Reading	TIR
Transformer	XFMR
Transistor	XSTR
Transmitter	XMTR
Tungsten	TU
Typical	TYP

U

Ultra-High Frequency	UHF
Unit	U
Universal	UNIV
Unless Otherwise Specified	UOS

V

Vacuum	VAC
Vacuum Tube	VT
Variable	VAR
Vernier	VER
Vertical	VERT
Very High Frequency	VHF
Vibrate	VIB
Video	VD
Void	VD
Volt	V
Volume	VOL

W

Washer	WASH
Watt	W
Weatherproof	WP
Weight	WT
Wide, Width	W
Wire Wound	WW
Wood	WD
Wrought Iron	WI

Y

Yield Point (PSI)	YP
Yield Strength (PSI)	YS

NATIONAL COARSE AND NATIONAL FINE THREADS AND TAP DRILLS

SIZE	THREADS PER INCH	MAJOR DIA.	MINOR DIA.	PITCH DIA.	TAP DRILL 75 PERCENT THREAD	DECIMAL EQUIVALENT	CLEARANCE DRILL	DECIMAL EQUIVALENT
2	56	.0860	.0628	.0744	50	.0700	42	.0935
	64	.0860	.0657	.0759	50	.0700	42	.0935
3	48	.099	.0719	.0855	47	.0785	36	.1065
	56	.099	.0758	.0874	45	.0820	36	.1065
4	40	.112	.0795	.0958	43	.0890	31	.1200
	48	.112	.0849	.0985	42	.0935	31	.1200
6	32	.138	.0974	.1177	36	.1065	26	.1470
	40	.138	.1055	.1218	33	.1130	26	.1470
8	32	.164	.1234	.1437	29	.1360	17	.1730
	36	.164	.1279	.1460	29	.1360	17	.1730
10	24	.190	.1359	.1629	25	.1495	8	.1990
	32	.190	.1494	.1697	21	.1590	8	.1990
12	24	.216	.1619	.1889	16	.1770	1	.2280
	28	.216	.1696	.1928	14	.1820	2	.2210
1/4	20	.250	.1850	.2175	7	.2010	G	.2610
	28	.250	.2036	.2268	3	.2130	G	.2610
5/16	18	.3125	.2403	.2764	F	.2570	21/64	.3281
	24	.3125	.2584	.2854	I	.2720	21/64	.3281
3/8	16	.3750	.2938	.3344	5/16	.3125	25/64	.3906
	24	.3750	.3209	.3479	Q	.3320	25/64	.3906
7/16	14	.4375	.3447	.3911	U	.3680	15/32	.4687
	20	.4375	.3725	.4050	25/64	.3906	29/64	.4531
1/2	13	.5000	.4001	.4500	27/64	.4219	17/32	.5312
	20	.5000	.4350	.4675	29/64	.4531	33/64	.5156
9/16	12	.5625	.4542	.5084	31/64	.4844	19/32	.5937
	18	.5625	.4903	.5264	33/64	.5156	37/64	.5781
5/8	11	.6250	.5069	.5660	17/32	.5312	21/32	.6562
	18	.6250	.5528	.5889	37/64	.5781	41/64	.6406
3/4	10	.7500	.6201	.6850	21/32	.6562	25/32	.7812
	16	.7500	.6688	.7094	11/16	.6875	49/64	.7656
7/8	9	.8750	.7307	.8028	49/64	.7656	29/32	.9062
	14	.8750	.7822	.8286	13/16	.8125	57/64	.8906
1	8	1.0000	.8376	.9188	7/8	.8750	1-1/32	1.0312
	14	1.0000	.9072	.9536	15/16	.9375	1-1/64	1.0156
1-1/8	7	1.1250	.9394	1.0322	63/64	.9844	1-5/32	1.1562
	12	1.1250	1.0167	1.0709	1-3/64	1.0469	1-5/32	1.1562
1-1/4	7	1.2500	1.0644	1.1572	1-7/64	1.1094	1-9/32	1.2812
	12	1.2500	1.1417	1.1959	1-11/64	1.1719	1-9/32	1.2812
1-1/2	6	1.5000	1.2835	1.3917	1-11/32	1.3437	1-17/32	1.5312
	12	1.5000	1.3917	1.4459	1-27/64	1.4219	1-17/32	1.5312

SCREW THREAD ELEMENTS FOR UNIFIED AND NATIONAL FORM OF THREAD

THREADS PER INCH (n)	PITCH (p) $p = \frac{1}{n}$	SINGLE HEIGHT — SUBTRACT FROM BASIC MAJOR DIAMETER TO GET BASIC PITCH DIAMETER	DOUBLE HEIGHT — SUBTRACT FROM BASIC MAJOR DIAMETER TO GET BASIC MINOR DIAMETER	83 1/3 PERCENT DOUBLE HEIGHT — SUBTRACT FROM BASIC MAJOR DIAMETER TO GET MINOR DIAMETER OF RING GAGE	BASIC WIDTH OF CREST AND ROOT FLAT $\frac{p}{8}$	CONSTANT FOR BEST SIZE WIRE ALSO SINGLE HEIGHT OF 60 DEG. V–THREAD	DIAMETER OF BEST SIZE WIRE
3	.333333	.216506	.43301	.36084	.0417	.28868	.19245
3 1/4	.307692	.199852	.39970	.33309	.0385	.26647	.17765
3 1/2	.285714	.185577	.37115	.30929	.0357	.24744	.16496
4	.250000	.162379	.32476	.27063	.0312	.21651	.14434
4 1/2	.222222	.144337	.28867	.24056	.0278	.19245	.12830
5	.200000	.129903	.25981	.21650	.0250	.17321	.11547
5 1/2	.181818	.118093	.23619	.19682	.0227	.15746	.10497
6	.166666	.108253	.21651	.18042	.0208	.14434	.09623
7	.142857	.092788	.18558	.15465	.0179	.12372	.08248
8	.125000	.081189	.16238	.13531	.0156	.10825	.07217
9	.111111	.072168	.14434	.12028	.0139	.09623	.06415
10	.100000	.064952	.12990	.10825	.0125	.08660	.05774
11	.090909	.059046	.11809	.09841	.0114	.07873	.05249
11 1/2	.086956	.056480	.11296	.09413	.0109	.07531	.05020
12	.083333	.054127	.10826	.09021	.0104	.07217	.04811
13	.076923	.049963	.09993	.08327	.0096	.06662	.04441
14	.071428	.046394	.09279	.07732	.0089	.06186	.04124
16	.062500	.040595	.08119	.06766	.0078	.05413	.03608
18	.055555	.036086	.07217	.06014	.0069	.04811	.03200
20	.050000	.032475	.06495	.05412	.0062	.04330	.02887
22	.045454	.029523	.05905	.04920	.0057	.03936	.02624
24	.041666	.027063	.05413	.04510	.0052	.03608	.02406
27	.037037	.024056	.04811	.04009	.0046	.03208	.02138
28	.035714	.023197	.04639	.03866	.0045	.03093	.02062
30	.033333	.021651	.04330	.03608	.0042	.02887	.01925
32	.031250	.020297	.04059	.03383	.0039	.02706	.01804
36	.027777	.018042	.03608	.03007	.0035	.02406	.01604
40	.025000	.016237	.03247	.02706	.0031	.02165	.01443
44	.022727	.014761	.02952	.02460	.0028	.01968	.01312
48	.020833	.013531	.02706	.02255	.0026	.01804	.01203
50	.020000	.012990	.02598	.02165	.0025	.01732	.01155
56	.017857	.011598	.02320	.01933	.0022	.01546	.01031
60	.016666	.010825	.02165	.01804	.0021	.01443	.00962
64	.015625	.010148	.02030	.01691	.0020	.01353	.00902
72	.013888	.009021	.01804	.01503	.0017	.01203	.00802
80	.012500	.008118	.01624	.01353	.0016	.01083	.00722
90	.011111	.007217	.01443	.01202	.0014	.00962	.00642
96	.010417	.006766	.01353	.01127	.0013	.00902	.00601
100	.010000	.006495	.01299	.01082	.0012	.00866	.00577
120	.008333	.005413	.01083	.00902	.0010	.00722	.00481

Using the Best Size Wires, the measurement over three wires minus the Constant for Best Size Wire equals the Pitch Diameter.

MACHINE SCREW AND CAP SCREW HEADS

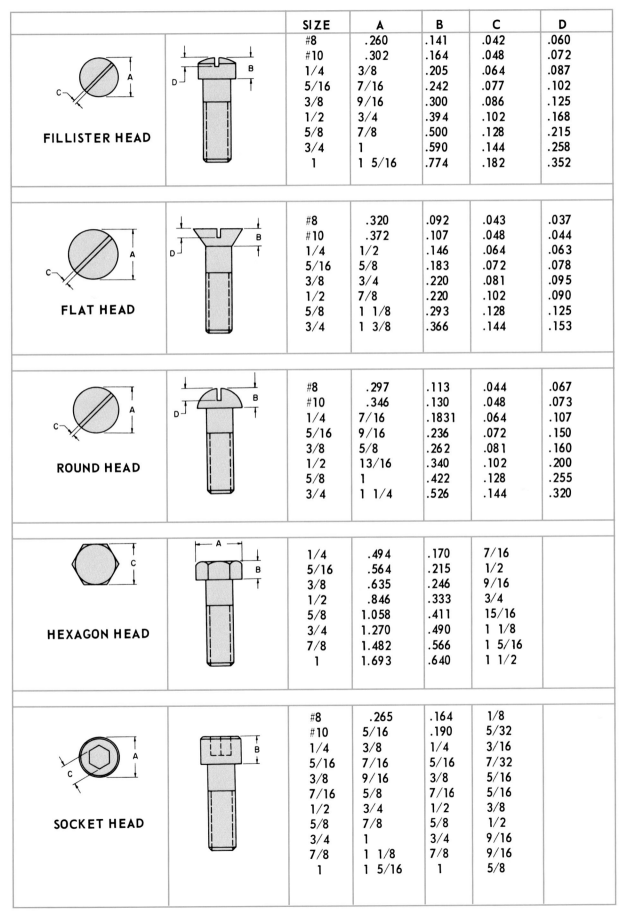

		SIZE	A	B	C	D
FILLISTER HEAD		#8	.260	.141	.042	.060
		#10	.302	.164	.048	.072
		1/4	3/8	.205	.064	.087
		5/16	7/16	.242	.077	.102
		3/8	9/16	.300	.086	.125
		1/2	3/4	.394	.102	.168
		5/8	7/8	.500	.128	.215
		3/4	1	.590	.144	.258
		1	1 5/16	.774	.182	.352
FLAT HEAD		#8	.320	.092	.043	.037
		#10	.372	.107	.048	.044
		1/4	1/2	.146	.064	.063
		5/16	5/8	.183	.072	.078
		3/8	3/4	.220	.081	.095
		1/2	7/8	.220	.102	.090
		5/8	1 1/8	.293	.128	.125
		3/4	1 3/8	.366	.144	.153
ROUND HEAD		#8	.297	.113	.044	.067
		#10	.346	.130	.048	.073
		1/4	7/16	.1831	.064	.107
		5/16	9/16	.236	.072	.150
		3/8	5/8	.262	.081	.160
		1/2	13/16	.340	.102	.200
		5/8	1	.422	.128	.255
		3/4	1 1/4	.526	.144	.320
HEXAGON HEAD		1/4	.494	.170	7/16	
		5/16	.564	.215	1/2	
		3/8	.635	.246	9/16	
		1/2	.846	.333	3/4	
		5/8	1.058	.411	15/16	
		3/4	1.270	.490	1 1/8	
		7/8	1.482	.566	1 5/16	
		1	1.693	.640	1 1/2	
SOCKET HEAD		#8	.265	.164	1/8	
		#10	5/16	.190	5/32	
		1/4	3/8	1/4	3/16	
		5/16	7/16	5/16	7/32	
		3/8	9/16	3/8	5/16	
		7/16	5/8	7/16	5/16	
		1/2	3/4	1/2	3/8	
		5/8	7/8	5/8	1/2	
		3/4	1	3/4	9/16	
		7/8	1 1/8	7/8	9/16	
		1	1 5/16	1	5/8	

TWIST DRILL DATA

METRIC DRILL SIZES (mm)[1]		Decimal Equivalent in Inches (Ref)
Preferred	Available	
	.40	.0157
	.42	.0165
	.45	.0177
	.48	.0189
.50		.0197
	.52	.0205
.55		.0217
	.58	.0228
.60		.0236
	.62	.0244
.65		.0256
	.68	.0268
.70		.0276
	.72	.0283
.75		.0295
	.78	.0307
.80		.0315
	.82	.0323
.85		.0335
	.88	.0346
.90		.0354
	.92	.0362
.95		.0374
	.98	.0386
1.00		.0394
	1.03	.0406
1.05		.0413
	1.08	.0425
1.10		.0433
	1.15	.0453
1.20		.0472
1.25		.0492
1.30		.0512
	1.35	.0531
1.40		.0551
	1.45	.0571
1.50		.0591
	1.55	.0610
1.60		.0630
	1.65	.0650

METRIC DRILL SIZES (mm)[1]		Decimal Equivalent in Inches (Ref)
Preferred	Available	
1.70		.0669
	1.75	.0689
1.80		.0709
	1.85	.0728
1.90		.0748
	1.95	.0768
2.00		.0787
	2.05	.0807
2.10		.0827
	2.15	.0846
2.20		.0866
	2.30	.0906
2.40		.0945
2.50		.0984
2.60		.1024
	2.70	.1063
2.80		.1102
	2.90	.1142
3.00		.1181
	3.10	.1220
3.20		.1260
	3.30	.1299
3.40		.1339
	3.50	.1378
3.60		.1417
	3.70	.1457
3.80		.1496
	3.90	.1535
4.00		.1575
	4.10	.1614
4.20		.1654
	4.40	.1732
4.50		.1772
	4.60	.1811
4.80		.1890
5.00		.1969
	5.20	.2047
5.30		.2087
	5.40	.2126
5.60		.2205
	5.80	.2283

[1] Metric drill sizes listed in the "Preferred" column are based on the R'40 series of preferred numbers shown in the ISO Standard R497. Those listed in the "Available" column are based on the R80 series from the same document.

TWIST DRILL DATA (Continued)

METRIC DRILL SIZES (mm)[1]		Decimal Equivalent in Inches (Ref)	METRIC DRILL SIZES (mm)[1]		Decimal Equivalent in Inches (Ref)
Preferred	Available		Preferred	Available	
6.00		.2362		19.50	.7677
	6.20	.2441	20.00		.7874
6.30		.2480		20.50	.8071
	6.50	.2559	21.00		.8268
6.70		.2638		21.50	.8465
	6.80[2]	.2677	22.00		.8661
	6.90	.2717		23.00	.9055
7.10		.2795	24.00		.9449
	7.30	.2874	25.00		.9843
7.50		.2953	26.00		1.0236
	7.80	.3071		27.00	1.0630
8.00		.3150	28.00		1.1024
	8.20	.3228		29.00	1.1417
8.50		.3346	30.00		1.1811
	8.80	.3465		31.00	1.2205
9.00		.3543	32.00		1.2598
	9.20	.3622		33.00	1.2992
9.50		.3740	34.00		1.3386
	9.80	.3858		35.00	1.3780
10.00		.3937	36.00		1.4173
	10.30	.4055		37.00	1.4567
10.50		.4134	38.00		1.4961
	10.80	.4252		39.00	1.5354
11.00		.4331	40.00		1.5748
	11.50	.4528		41.00	1.6142
12.00		.4724	42.00		1.6535
12.50		.4921		43.50	1.7126
13.00		.5118	45.00		1.7717
	13.50	.5315		46.50	1.8307
14.00		.5512	48.00		1.8898
	14.50	.5709	50.00		1.9685
15.00		.5906		51.50	2.0276
	15.50	.6102	53.00		2.0866
16.00		.6299		54.00	2.1260
	16.50	.6496	56.00		2.2047
17.00		.6693		58.00	2.2835
	17.50	.6890	60.00		2.3622
18.00		.7087			
	18.50	.7283			
19.00		.7480			

1 Metric drill sizes listed in the "Preferred" column are based on the R'40 series of preferred numbers shown in the ISO Standard R497. Those listed in the "Available" column are based on the R80 series from the same document.

2 Recommended only for use as a tap drill size.

LETTER SIZE DRILLS

A	0.234	J	0.277	S	0.348
B	0.238	K	0.281	T	0.358
C	0.242	L	0.290	U	0.368
D	0.246	M	0.295	V	0.377
E	0.250	N	0.302	W	0.386
F	0.257	O	0.316	X	0.397
G	0.261	P	0.323	Y	0.404
H	0.266	Q	0.332	Z	0.413
I	0.272	R	0.339		

WOOD SCREW TABLE

LENGTH	GAUGE STEEL SCREW	GAUGE BRASS SCREW	GAUGE NO.	DECIMAL	APPROX. FRACT.	DRILL SIZE A	DRILL SIZE B	DRILL SIZE C
1/4	0 to 4	0 to 4	0	.060	1/16	1/16		
3/8	0 to 8	0 to 6	1	.073	5/64	3/32		
1/2	1 to 10	1 to 8	2	.086	5/64	3/32	1/16	3/16
5/8	2 to 12	2 to 10	3	.099	3/32	1/8	1/16	1/4
3/4	2 to 14	2 to 12	4	.112	7/64	1/8	1/16	1/4
7/8	3 to 14	4 to 12	5	.125	1/8	1/8	3/32	1/4
1	3 to 16	4 to 14	6	.138	9/64	5/32	3/32	5/16
1 1/4	4 to 18	6 to 14	7	.151	5/32	5/32	1/8	5/16
1 1/2	4 to 20	6 to 14	8	.164	5/32	3/16	1/8	3/8
1 3/4	6 to 20	8 to 14	9	.177	11/64	3/16	1/8	3/8
2	6 to 20	8 to 18	10	.190	3/16	3/16	1/8	3/8
2 1/4	6 to 20	10 to 18	11	.203	13/64	7/32	5/32	7/16
2 1/2	8 to 20	10 to 18	12	.216	7/32	7/32	5/32	7/16
2 3/4	8 to 20	8 to 20	14	.242	15/64	1/4	3/16	1/2
3	8 to 24	12 to 18	16	.268	17/64	9/32	7/32	9/16
3 1/2	10 to 24	12 to 18	18	.294	19/64	5/16	1/4	5/8
4	12 to 24	12 to 24	20	.320	21/64	11/32	9/32	11/16
4 1/2	14 to 24	14 to 24	24	.372	3/8	3/8	5/16	3/4
5	14 to 24	14 to 24						

SCREWS ALSO AVAILABLE WITH PHILLIPS HEAD

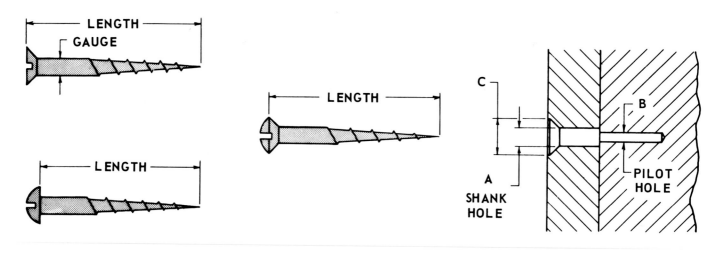

INCH SERIES			METRIC			
Size	Dia. (In.)	TPI	Size	Dia. (In.)	Pitch (mm)	TPI (Approx)
			M1.4	.055	.3 .2	85 127
#0	.060	80				
			M1.6	.063	.35 .2	74 127
#1	.073	64 72				
			M2	.079	.4 .25	64 101
#2	.086	56 64				
			M2.5	.098	.45 .35	56 74
#3	.099	48 56				
#4	.112	40 48				
			M3	.118	.5 .35	51 74
#5	.125	40 44				
#6	.138	32 40				
			M4	.157	.7 .5	36 51
#8	.164	32 36				
#10	.190	24 32				
			M5	.196	.8 .5	32 51
			M6	.236	1.0 .75	25 34
1/4	.250	20 28				
5/16	.312	18 24				
			M8	.315	1.25 1.0	20 25
3/8	.375	16 24				
			M10	.393	1.5 1.25	17 20
7/16	.437	14 20				
			M12	.472	1.75 1.25	14.5 20
1/2	.500	13 20				
			M14	.551	2 1.5	12.5 17
5/8	.625	11 18				
			M16	.630	2 1.5	12.5 17
			M18	.709	2.5 1.5	10 17
3/4	.750	10 16				
			M20	.787	2.5 1.5	10 17
			M22	.866	2.5 1.5	10 17
7/8	.875	9 14				
			M24	.945	3 2	8.5 12.5
1"	1.000	8 12				
			M27	1.063	3 2	8.5 12.5

CONVERSION CHART INCH/mm

Drill No. or Letter		Inch	mm
		.001	0.0254
		.002	0.0508
		.003	0.0762
		.004	0.1016
		.005	0.1270
		.006	0.1524
		.007	0.1778
		.008	0.2032
		.009	0.2286
		.010	0.2540
		.011	0.2794
		.012	0.3048
		.013	0.3302
80	.0135	.014	0.3556
79	.0145	.015	0.3810
	1/64	.0156	0.3969
78		.016	0.4064
		.017	0.4318
77		.018	0.4572
		.019	0.4826
76		.020	0.5080
75		.021	0.5334
		.022	0.5588
74	.0225	.023	0.5842
73		.024	0.6096
72		.025	0.6350
71		.026	0.6604
		.027	0.6858
70		.028	0.7112
		.029	0.7366
69	.0292	.030	0.7620
68		.031	0.7874
	1/32	.0312	0.7937
67		.032	0.8128
66		.033	0.8382
		.034	0.8636
65		.035	0.8890
64		.036	0.9144
63		.037	0.9398
62		.038	0.9652
61		.039	0.9906
		.0394	1.0000
60		.040	1.0160
59		.041	1.0414
58		.042	1.0668
57		.043	1.0922
		.044	1.1176
		.045	1.1430
56	.0465	.046	1.1684
	3/64	.0469	1.1906
		.047	1.1938
		.048	1.2192
		.049	1.2446
		.050	1.2700
		.051	1.2954
55		.052	1.3208
		.053	1.3462
		.054	1.3716
54		.055	1.3970
		.056	1.4224
		.057	1.4478
		.058	1.4732
		.059	1.4986
53	.0595	.060	1.5240
		.061	1.5494
		.062	1.5748
	1/16	.0625	1.5875
		.063	1.6002
52	.0635	.064	1.6256
		.065	1.6510
		.066	1.6764
51		.067	1.7018
		.068	1.7272
		.069	1.7526
50		.070	1.7780
		.071	1.8034
		.072	1.8288
49		.073	1.8542
		.074	1.8796
		.075	1.9050
48		.076	1.9304
		.077	1.9558
47	.0785	.078	1.9812
	5/64	.0781	1.9844
		.0787	2.0000
		.079	2.0066
		.080	2.0320
46		.081	2.0574
45		.082	2.0828
		.083	2.1082
		.084	2.1336
		.085	2.1590
44		.086	2.1844
		.087	2.2098
		.088	2.2352
43		.089	2.2606
		.090	2.2860
		.091	2.3114
		.092	2.3368
42	.0935	.093	2.3622
	3/32	.0937	2.3812
		.094	2.3876
		.095	2.4130
41		.096	2.4384
		.097	2.4638
40		.098	2.4892
		.099	2.5146
39	.0995	.100	2.5400

Drill No. or Letter		Inch	mm
		.101	2.5654
38	.1015	.102	2.5908
		.103	2.6162
37		.104	2.6416
		.105	2.6670
36	.1065	.106	2.6924
		.107	2.7178
		.108	2.7432
		.109	2.7686
	7/64	.1094	2.7781
35		.110	2.7490
34		.111	2.8194
		.112	2.8448
33		.113	2.8702
		.114	2.8956
		.115	2.9210
32		.116	2.9464
		.117	2.9718
		.118	2.9972
		.1181	3.0000
		.119	3.0226
31		.120	3.0480
		.121	3.0734
		.122	3.0988
		.123	3.1242
		.124	3.1496
	1/8	.125	3.1750
		.126	3.2004
		.127	3.2258
		.128	3.2512
30	.1285	.129	3.2766
		.130	3.3020
		.131	3.3274
		.132	3.3528
		.133	3.3782
		.134	3.4036
		.135	3.4290
29		.136	3.4544
		.137	3.4798
		.138	3.5052
		.139	3.5306
28	.1405	.140	3.5560
	9/64	.1406	3.5719
		.141	3.5814
		.142	3.6068
		.143	3.6322
27		.144	3.6576
		.145	3.6830
		.146	3.7084
26		.147	3.7338
		.148	3.7592
		.149	3.7846
25	.1495	.150	3.8100
		.151	3.8354
24		.152	3.8608
		.153	3.8862
23		.154	3.9116
		.155	3.9370
		.156	3.9624
	5/32	.1562	3.9687
22		.157	3.9878
		.1575	4.0000
		.158	4.0132
21		.159	4.0386
		.160	4.0640
20		.161	4.0894
		.162	4.1148
		.163	4.1402
		.164	4.1656
		.165	4.1910
19		.166	4.2164
		.167	4.2418
		.168	4.2672
		.169	4.2926
18	.1635	.170	4.3180
		.171	4.3434
	11/64	.1719	4.3656
		.172	4.3688
17		.173	4.3942
		.174	4.4196
		.175	4.4450
		.176	4.4704
16		.177	4.4958
		.178	4.5212
		.179	4.5466
15		.180	4.5720
		.181	4.5974
14		.182	4.6228
		.183	4.6482
		.184	4.6736
13		.185	4.6990
		.186	4.7244
		.187	4.7498
	3/16	.1875	4.7625
		.188	4.7752
12		.189	4.8006
		.190	4.8260
11		.191	4.8514
		.192	4.8768
		.193	4.9022
10	.1935	.194	4.9276
		.195	4.9530
9		.196	4.9784
		.1969	5.0000
		.197	5.0038
		.198	5.0292
		.199	5.0546
8		.200	5.0800

Drill No. or Letter		Inch	mm
7		.201	5.1054
		.202	5.1308
		.203	5.1562
	13/64	.2031	5.1594
6		.204	5.1816
5	.2055	.205	5.2070
		.206	5.2324
		.207	5.2578
		.208	5.2832
4		.209	5.3086
		.210	5.3340
		.211	5.3594
		.212	5.3848
3		.213	5.4102
		.214	5.4356
		.215	5.4610
		.216	5.4864
		.217	5.5118
		.218	5.5372
	7/32	.2187	5.5562
		.219	5.5626
		.220	5.5880
2		.221	5.6134
		.222	5.6388
		.223	5.6642
		.224	5.6896
		.225	5.7150
		.226	5.7404
		.227	5.7658
1		.228	5.7912
		.229	5.8166
		.230	5.8410
		.231	5.8674
		.232	5.8928
		.233	5.9182
A		.234	5.9436
	15/64	.2344	5.9531
		.235	5.9690
		.236	5.9944
		.2362	6.0000
		.237	6.0198
B		.238	6.0452
		.239	6.0706
		.240	6.0960
		.241	6.1214
C		.242	6.1468
		.243	6.1722
		.244	6.1976
		.245	6.2230
D		.246	6.2484
		.247	6.2738
		.248	6.2992
		.249	6.3246
E	1/4	.250	6.3500
		.251	6.3754
		.252	6.4008
		.253	6.4262
		.254	6.4516
		.255	6.4770
		.256	6.5024
F		.257	6.5278
		.258	6.5532
		.259	6.5786
		.260	6.6040
G		.261	6.6294
		.262	6.6548
		.263	6.6802
		.264	6.7056
		.265	6.7310
	17/64	.2656	6.7469
H		.266	6.7564
		.267	6.7818
		.268	6.8072
		.269	6.8326
		.270	6.8580
		.271	6.8834
I		.272	6.9088
		.273	6.9342
		.274	6.9596
		.275	6.9850
		.2756	7.0000
		.276	7.0104
J		.277	7.0358
		.278	7.0612
		.279	7.0866
		.280	7.1120
K		.281	7.1374
	9/32	.2812	7.1437
		.282	7.1628
		.283	7.1882
		.284	7.2136
		.285	7.2390
		.286	7.2644
		.287	7.2898
		.288	7.3152
		.289	7.3406
L		.290	7.3660
		.291	7.3914
		.292	7.4168
		.293	7.4422
		.294	7.4676
M		.295	7.4930
		.296	7.5184
	19/64	.2969	7.5406
		.297	7.5438
		.298	7.5692
		.299	7.5946
		.300	7.6200

Drill No. or Letter		Inch	mm
		.301	7.6454
N		.302	7.6708
		.303	7.6962
		.304	7.7216
		.305	7.7470
		.306	7.7724
		.307	7.7978
		.308	7.8232
		.309	7.8486
		.310	7.8740
		.311	7.8994
		.312	7.9248
	5/16	.3125	7.9375
		.313	7.9502
		.314	7.9756
		.3150	8.0000
		.315	8.0010
O		.316	8.0264
		.317	8.0518
		.318	8.0772
		.319	8.1026
		.320	8.1280
		.321	8.1534
		.322	8.1788
P		.323	8.2042
		.324	8.2296
		.325	8.2550
		.326	8.2804
		.327	8.3058
		.328	8.3312
	21/64	.3281	8.3344
		.329	8.3566
		.330	8.3820
		.331	8.4074
Q		.332	8.4328
		.333	8.4582
		.334	8.4836
		.335	8.5090
		.336	8.5344
		.337	8.5598
		.338	8.5852
R		.339	8.6106
		.340	8.6360
		.341	8.6614
		.342	8.6868
		.343	8.7122
	11/32	.3437	8.7312
		.344	8.7376
		.345	8.7630
		.346	8.7884
		.347	8.8138
S		.348	8.8392
		.349	8.8646
		.350	8.8900
		.351	8.9154
		.352	8.9408
		.353	8.9662
		.354	8.9916
		.3543	9.0000
		.355	9.0170
		.356	9.0424
		.357	9.0678
T		.358	9.0932
		.359	9.1186
	23/64	.3594	9.1281
		.360	9.1440
		.361	9.1694
		.362	9.1948
		.363	9.2202
		.364	9.2456
		.365	9.2710
		.366	9.2964
		.367	9.3218
U		.368	9.3472
		.369	9.3726
		.370	9.3980
		.371	9.4234
		.372	9.4488
		.373	9.4742
		.374	9.4996
	3/8	.375	9.5250
		.376	9.5504
V		.377	9.5758
		.378	9.6012
		.379	9.6266
		.380	9.6520
		.381	9.6774
		.382	9.7028
		.383	9.7282
		.384	9.7536
		.385	9.7790
W		.386	9.8044
		.387	9.8298
		.388	9.8552
		.389	9.8806
		.390	9.9060
	25/64	.3906	9.9219
		.391	9.9314
		.392	9.9568
		.393	9.9822
		.3937	10.0000
		.394	10.0076
		.395	10.0330
		.396	10.0584
X		.397	10.0838
		.398	10.1092
		.399	10.1346
		.400	10.1600

Drill No. or Letter		Inch	mm
		.401	10.1854
		.402	10.2108
		.403	10.2362
Y		.404	10.2616
		.405	10.2870
		.406	10.3124
	13/32	.4062	10.3187
		.407	10.3378
		.408	10.3632
		.409	10.3886
		.410	10.4140
		.411	10.4394
		.412	10.4648
Z		.413	10.4902
		.414	10.5156
		.415	10.5410
		.416	10.5664
		.417	10.5918
		.418	10.6172
		.419	10.6426
		.420	10.6680
		.421	10.6934
	27/64	.4219	10.7156
		.422	10.7188
		.423	10.7442
		.424	10.7696
		.425	10.7950
		.426	10.8204
		.427	10.8458
		.428	10.8712
		.429	10.8966
		.430	10.9220
		.431	10.9474
		.432	10.9728
		.433	10.9982
		.4331	11.0000
		.434	11.0236
		.435	11.0490
		.436	11.0744
		.437	11.0998
	7/16	.4375	11.1125
		.438	11.1252
		.439	11.1506
		.440	11.1760
		.441	11.2014
		.442	11.2268
		.443	11.2522
		.444	11.2776
		.445	11.3030
		.446	11.3284
		.447	11.3538
		.448	11.3792
		.449	11.4046
		.450	11.4300
		.451	11.4554
		.452	11.4808
		.453	11.5062
	29/64	.4531	11.5094
		.454	11.5316
		.455	11.5570
		.456	11.5824
		.457	11.6078
		.458	11.6332
		.459	11.6586
		.460	11.6840
		.461	11.7094
		.462	11.7348
		.463	11.7602
		.464	11.7856
		.465	11.8110
		.466	11.8364
		.467	11.8618
		.468	11.8872
	15/32	.4687	11.9062
		.469	11.9126
		.470	11.9380
		.471	11.9634
		.472	11.9888
		.4724	12.0000
		.473	12.0142
		.474	12.0396
		.475	12.0650
		.476	12.0904
		.477	12.1158
		.478	12.1412
		.479	12.1666
		.480	12.1920
		.481	12.2174
		.482	12.2428
		.483	12.2682
		.484	12.2936
	31/64	.4844	12.3031
		.485	12.3190
		.486	12.3444
		.487	12.3698
		.488	12.3952
		.489	12.4206
		.490	12.4460
		.491	12.4714
		.492	12.4968
		.493	12.5222
		.494	12.5476
		.495	12.5730
		.496	12.5984
		.497	12.6238
		.498	12.6492
		.499	12.6746
	1/2	.500	12.7000

Glossary

Acute angle: An angle less than 90°.

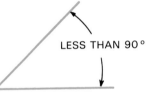

LESS THAN 90°

ACUTE ANGLE

Air brush: A device used to spray paint by means of compressed air.

Alloy: A mixture of two or more metals fused or melted together to form a new metal.

Allowance: The limits permitted for satisfactory performance of machined parts.

Alphabet of lines: Collection of the different lines which vary in line weight (thickness) and line type. Also see *Line conventions.*

Alphanumeric: Pertaining to a set of characters that contains both letters and numbers.

Angle: The figure formed by two lines coming together to a point.

Annealing: The process of heating metal to a given temperature (the exact temperature and period the temperature is held depends upon the composition of the metal being annealed) and cooling it slowly to remove stresses and induce softness.

ANSI: American National Standards Institute. A nongovernmental organization that proposes, modifies, approves, and publishes drafting and manufacturing standards for voluntary use in the United States.

Arc: Portion of a circle.

ARC

Artificial intelligence (AI): Computer techniques that mimic certain functions typically associated with human intelligence. Also refers to a computer or machine capable of improving their operation as a result of repeated experience.

Asphalt roofing: A roofing material made by saturating felt with asphalt.

Assembly: A unit fitted together from manufactured parts.

Automatic dimensioning: A form of dimensioning where a CAD program automatically measures distances, and assigns arrows and dimensions to the display and/or hard copy.

Auxiliary View: A view showing the true length and true width of an angular surface.

Axis: The center line of a view or of a geometric figure. (Plural—axes)

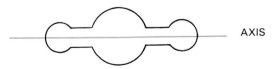

AXIS

Beam: A horizontal structural support member. Examples include joists, rafters, purlins, and girders.

Beam compass: Compass used to draw large circles and arcs.

Bevel: A surface formed or trimmed so as to not be perpendicular to adjoining surfaces.

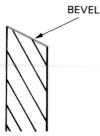

BEVEL

Bisect: To divide into two parts of equal size or length.

Bit: The smallest unit of information for a computer. It is represented by a zero (0) or a one (1) in a computer by the absence (0) or presence (1) of an electric current.

Blowhole: A hole produced in a casting when gasses are entrapped during the pouring operation.

Blueprint: A reproduction of a drawing that has a blue background with white lines. Also, a general term used to describe any reproduced drawing.

Border line: The heaviest line used in drafting, acts as a "frame" for the drawing and establishes a space between the drawing and the edge of the paper.

BORDER LINE (VERY THICK)

Brazing: Joining metals by the fusion of nonferrous alloys that have melting temperatures above 800° F but lower than the metals being joined.

Burnisher: A tool used to transfer rub-on (dry transfer) materials to a drawing.

Bushing: A bearing for a revolving shaft. A hardened steel tube used on jigs to guide drills and reamers.

Byte: A group of 8 bits treated as a unit and often used to represent one alphanumeric character.

CAD: Computer-aided drafting. Sometimes listed as CADD, (Computer-aided design and drafting).

Cartography: The science/art of map-making.

Casting: An object made by pouring molten metal into a mold.

Case harden: A process of surface hardening iron base alloys so that the surface layer or case is made substantially harder than the interior or core of the metal.

HARDENED
AREA OF
METAL

CASEHARDENING

Centerline: A fine line composed of long and short dashes with spaces in between used to indicate the center of a symmetrical object.

CENTER LINE (THIN)

Chamfer: An angled cut to remove the "corner" of two perpendicular surfaces.

CHAMFER

Chip: Small piece of material, usually silicon, in which tiny amounts of other elements (with desired electronic properties) are deposited to form integrated circuits (IC). Also called a microchip.

Circumference: The perimeter of a circle.

Circumscribe: To draw a line around.

Clearance: The distance by which one part clears another part.

Clockwise: From left to right in a circular motion. The direction clock hands move.

CLOCKWISE

Computer: An electronic device capable of solving problems, accepting and processing data according to previous instructions of a program.

Concave surface: A curved depression in the surface of an object.

CONCAVE SURFACE

Concentric: Having a common center.

CONCENTRIC CIRCLES

Conical: Having a shape with a circular base and sides converging to a single point.

CONICAL

Construction line: Very light lines used as guides for lettering or positioning an object on a drawing.

CONSTRUCTION AND GUIDE LINES (VERY THIN)

Contour: The outline of an object.

Contour lines: Lines on a plot plan to show elevation. Each contour line shows all points where the ground has a single elevation.

Conventional: Not original; customary, or traditional.

Convex surface: A rounded surface on an object.

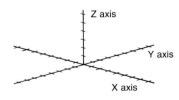

CONVEX SURFACE

Coordinates: The positions or locations of points on the X, Y, and Z planes.

Z axis
Y axis
X axis

Core: A body of sand or other material that is formed to a desired shape and placed in a mold to produce a cavity or opening in a casting.

Counterbore: To enlarge a hole to a given depth.

COUNTERBORED HOLE

Counterclockwise: From right to left in a circular motion.

COUNTERCLOCKWISE

Countersink: To chamfer a hole to receive a flat or fillister head fastener.

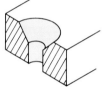

COUNTERSUNK HOLE

Central processing unit (CPU): The part of the computer that processes input information, manipulates data, and retrieves information from memory storage. (The "brain" of the computer.)

Crosshair: Crossed vertical and horizontal lines on a CAD graphic display that represents the cursor or current location. Also, the vertical and horizontal lines in an input device like a digitizing puck.

Cursor: A special character on the video display that indicates the next position at which a character will be entered or deleted.

Cutting-plane line: Heavy dashed line used to show sectional views. Arrows at the end of the line illustrate the direction of the sectional view.

CUTTING PLANE LINE (THICK)

Cylinder: A geometric figure with a uniform circular cross-section through its entire length.

CYLINDER

Diagonal: A line connecting the opposite corners of a figure.

DIAGONAL

Diameter: The distance from one side of a circle to the other, running directly through the circle's center.

DIAMETER

Die: The tool used to cut external threads. A tool used to shape materials.

Die casting: A method of casting metal under pressure by injecting it into metal dies of a die casting machine.

Digitizer tablet: A CAD input device on which a puck is maneuvered to select commands and position the on-screen crosshairs.

Dimension line: A light line normally runs between two extension lines, has arrows or ticks at its ends, and has a dimension at its center.

— 8 —

DIMENSION LINE (THIN)

Draft: The clearance on a pattern or mold that allows easy withdrawal of the pattern from the mold.

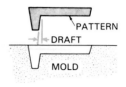

PATTERN
DRAFT
MOLD

Drilling: Cutting round holes by use of a cutting tool called a drill.

Drill rod: Accurately ground and polished tool steel rods.

Drive fit: Using force or pressure to fit two pieces together. One of several classes of fits.

Eccentric: Not on a common center.

ECCENTRIC CIRCLES

Ellipse: A closed curve in the form of a symmetrical oval.

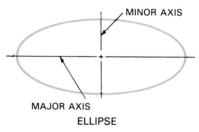

MINOR AXIS

MAJOR AXIS
ELLIPSE

Engineering drawing: The graphic language of the engineer, also known as a technical drawing or drafting.

Equilateral: A figure having equal length sides.

Expansion fit: The reverse of shrink fit. The piece to be fitted is placed in liquid nitrogen or dry ice until it shrinks enough to fit into the mating piece. Interference develops between the fitted pieces as the cooled piece expands to normal size.

Extension line: This light line is used to attach dimension lines to specific points on an object.

EXTENSION LINE (THIN)

Ferrous metal: A metal that contains iron as its major ingredient.

Fillet: A curved surface connecting two interior surfaces that form an angle. Also see *Round.*

FILLET

Fixture: A device for holding metal while it is being machined.

Forge: To form material using heat and pressure.

Flash: A thin fin of metal formed at the parting line of a forging or casting where a small portion of metal is forced out between the edges of the die.

Flask: A frame of wood or metal consisting of a cope (the top portion) and a drag (the bottom portion) used to hold sand that forms the mold used in the foundry.

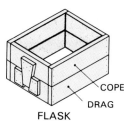
COPE
DRAG
FLASK

Floor plan: The drawing that shows the exact shape, dimensions, and arrangements of the rooms of a building.

Force fit: The interference between two mating parts sufficient to require force to press the pieces together. The joined pieces are considered permanently assembled.

Free fit: Used when tolerances are liberal. Clearance is sufficient to permit a shaft to turn freely without binding or overheating when properly lubricated.

Frustum: The figure formed by removing the upper portion of a cone or pyramid, leaving a top parallel to its base.

FRUSTUM

Gate: The opening that guides the molten metal into the cavity that forms the mold. See *Sprue.*

Gears: Toothed wheels that transmit rotary motion without slippage.

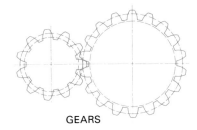
GEARS

Graphic: Written or drawn.

Grid: A CAD drawing aid formed by a network of uniformly spaced points or lines on the screen.

Hardcopy: Any printed or plotter output.

Hardening: The heating and quenching of certain iron-base alloys for the purpose of producing a hardness superior to that of the untreated metal.

Hardware: The components that make up a computer system.

Heat treatment: The careful application of a combination of heating and cooling cycles to a metal or alloy to bring about certain desirable conditions such as hardness and toughness.

Hexagon: A six-sided figure with each side forming a 60° angle.

HEXAGON

Hidden line: A dashed medium-weight line used to identify features that cannot be seen in a given view.

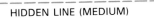
HIDDEN LINE (MEDIUM)

ID: Abbreviation for inside diameter.

Inclined: Making an angle with another line or plane.

INCLINE

Inscribe: To draw one figure within another figure.

STAR INSCRIBED IN PENTAGON

Inspection: The measuring and checking of finished parts to determine whether they have been made to specifications.

Interchangeable: Refers to a part that has been made to specific dimensions and tolerances and is capable of being fitted in a mechanism in place of a similarly made part.

Investment casting: A process that involves making a wax, plastic, or a frozen mercury pattern, surrounding it with a wet refractory material, melting or burning the pattern out after the investment material has dried and set, and finally pouring molten metal into the cavity.

Isometric drawing: A three-dimensional pictorial drawing in which the horizontal axes form a 30° angle with a true horizontal line.

ISOMETRIC

Jig: A device that holds the work in position and guides the cutting tool.

Key: A small piece of metal imbedded partially in the shaft and partially in the hub to prevent rotation of a gear or pulley on the shaft.

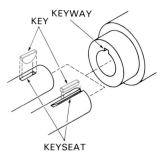

Keyway: The slot or recess in the shaft that holds the key.

Lay out: To locate and scribe points for machining and forming operations.

Lettering: Text on a drawing. Different lettering styles include Gothic, Freehand, and Roman.

Line conventions: Symbols that furnish a means of representing or describing some part of an object. It is expressed by a combination of line weight and appearance. Also see specific line types— *Construction line, Border line, Visible line, Dimension line, Extension line, Hidden line, Centerline, Cutting-plane line, Section line,* and *Phantom line.*

Machine tool: The name given to that class of machines which taken as a group, can reproduce themselves.

Major diameter: The largest diameter of a thread measured perpendicular to the axis.

Manual drafting: Drafting that is performed without the aid of a computer. Also called "board drafting."

Mesh: To engage gears to a working contact.

Mill: To remove metal with a rotating cutter on a milling machine.

Minor diameter: The smallest diameter on a screw thread measured across the root of the thread and perpendicular to the axis. Also known as the "root diameter."

Mold: The material that forms the cavity into which molten metal is poured.

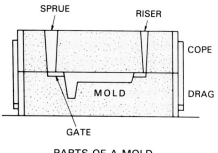

PARTS OF A MOLD

Mouse: Computer input device with a trackball that allows the user to move the on-screen cursor quickly and easily.

Oblique drawing: A pictorial drawing in which the front view of an object is shown as true size and true shape.

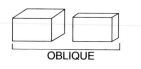

OBLIQUE

Multiview drawing: A drawing that uses more than one view to depict a single object.

Obtuse angle: An angle more than 90°.

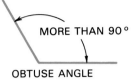
OBTUSE ANGLE

Octagon: An eight-sided geometric figure with each side forming a 45° angle.

OCTAGON

OD: Abbreviation for outside diameter.

Orthographic projection: Method of showing a three-dimensional object in two dimensions by displaying various views.

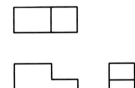
ORTHOGRAPHIC PROJECTION

Pentagon: A five-sided geometric figure with each side forming a 72° angle.

PENTAGON

Perimeter: The boundary of a geometric figure.

Permanent mold: Mold ordinarily made of metal that is used for the repeated production of similar castings.

Perpendicular: A line at right angles to a given line.

Perspective drawing: A pictorial drawing in which lines moving "away" from the viewer converge.

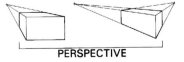
PERSPECTIVE

Phantom line: A light line composed of long dashes separated by two short dashes used to indicate alternate positions for moving parts.

PHANTOM LINE (THIN)

Photo drawing: A drawing prepared using a photograph on which dimensions, notes, and specifications have been added.

Pictorial drawing: A drawing of an object as it appears to the eye.

Pitch: The distance from a point on one thread to a corresponding point on the next thread.

Plan: A drawing showing the top view of something relatively large, such as a building or plot of land. Specific plan drawings include foundation plans, structural plans, and floor plans.

Plat: A drawing of a piece of property showing the exact boundary locations.

Plate: Another name given to a drawing.

Plotter: An output device for creating hardcopy of graphic images by controlling pens on a drawing media.

Printout: Hardcopy produced by a printer, plotter, or automated drafting machine.

Print: A reproduced copy of an original drawing. Also called "blueprint" or "blueline."

Profile: The outline of an object.

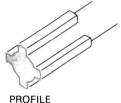
PROFILE

Project: To extend from.

Prototype: A full-size, operating model of an item to be produced.

Puck: An input device used with a digitizing tablet to input commands and maneuver the on-screen cursor.

Quadrant: One-fourth of a plane figure, such as a circle or ellipse. Quadrants are separated by the figure's axes.

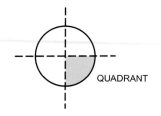

QUADRANT

Rack: A flat strip with teeth designed to mesh with teeth on a gear. Used to change rotary motion to a reciprocating motion.

Radius: The distance from any point on a circle to the circle's center. (Plural—radii.)

RADIUS

Ream: To finish a drilled hole to an exact size.

Rectangle: A geometric figure with opposite sides equal in length and each corner forming a 90° angle.

RECTANGLE

Right angle: A 90° angle. The angle that is formed by lines that are perpendicular to one another.

90°

RIGHT ANGLE

Riser: An opening in a mold that permits the gases to escape. The gases are formed when molten metal is poured into the mold.

Round: A curved surface connecting two exterior surfaces that form an angle. Also see *Fillet*.

Rough layout: A rough pencil plan that arranges lines and symbols so that they have a pleasing relation to one another.

Scale: A measuring device used to convert full-size dimensions to the drawing scale. Different types of scales include an architect's scale, engineer's scale, and mechanical engineer's scale.

Schedule: In architectural drawings, a table listing required materials and items. Common schedules include door schedules, window schedules, and lighting schedules.

Seam: The line formed where two edges are joined together.

Section line: Light lines used when drawing inside features of an object. Different patterns correspond to different materials.

SECTION LINE (THIN)

Sectional view: A secondary drawing showing one aspect or direction of the primary drawing.

Segment: Any part of a divided line.

Sketch: To draw without the aid of drafting instruments.

Software: The means of communicating with a computer. The program.

Spline: A series of grooves, cut lengthwise, around a shaft or hole.

Spotface: To machine a circular spot on the surface of a casting to furnish a bearing surface for the head of a bolt or a nut.

SPOTFACED HOLE

Sprue: The opening in a mold that leads to the gate, which in turn leads to the cavity into which molten metal is poured.

Square: To machine or cut at right angles. A geometric figure with four equal length sides and four right (90°) angles.

Stylus: A hand-held input device used with a digitizing tablet.

Symbol: A figure or character used in place of a word or group of words.

Tap: The tool used to cut internal threads.

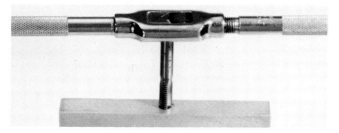

Taper: A piece that increases or decreases in size at a uniform rate to assume a wedge or conical shape.

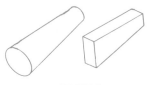

TAPERS

Template: A pattern or guide.

Thread: The act of cutting a screw thread.

Tracing: A drawing made on translucent paper, normally used as an original for making copies.

Train: A series of meshed gears.

Triangle: A three-sided geometric figure.

EQUILATERAL TRIANGLE   ISOSCELES TRIANGLE   IRREGULAR TRIANGLE

Truncate: To cut off a geometric solid at an angle to its base.

TRUNCATED PRISM

Unified Coarse (UNC): Coarse series of American Standard for Unified Screw Threads.

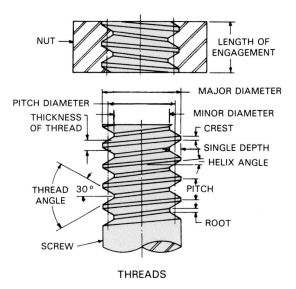

THREADS

Unified Fine (UNF): Fine series of American Standard for Unified Screw Threads.

UNC: Abbreviation for coarse series of American Standard for Unified Screw Threads.

UNF: Abbreviation for fine series of American Standard for Unified Screw Threads.

Unified threads: A series of screw threads that have been adopted by the United States, Canada, and Great Britain to attain interchangeability of certain screw threads.

Vanishing point: The point to which lines converge in a perspective drawing.

Vertical: At right angles to a horizontal line or plane.

Visible line: Heavy line used to outline visible edges of an object.

VISIBLE LINE (THICK)

Working drawing: A drawing that gives the craftworker the necessary information to make and assemble a product.

ACKNOWLEDGEMENTS

While it would be a most pleasant task, it would be impossible for one person to develop the material included in this text by visiting the various industries represented, and observing, studying, and taking photos firsthand.

My sincere thanks to those in the industry who helped in gathering the necessary material, information, and photographs. Their cooperation is most appreciated.

I would also like to thank the teachers using the text who were kind enough to offer suggestions for improving *Exploring Drafting*.

John R. Walker

Index

The following Index is useful for finding topics in the body of the textbook and also provides a method of quickly locating definitions of technical terms. Page numbers shown in the **bold** typeface indicate the locations of definitions of technical terms.